Transit-time effects in unipolar solid-state devices

Dr. D. DASCĂLU
Department of Electronics and Telecommunications
Polytechnical Institute of Bucharest

TRANSIT-TIME EFFECTS IN UNIPOLAR SOLID-STATE DEVICES

EDITURA ACADEMIEI
BUCUREȘTI 1974
ROMÂNIA

ABACUS PRESS
TUNBRIDGE WELLS, KENT
ENGLAND

First published in 1974 from the author's original
English language manuscript
by
EDITURA ACADEMIEI ROMÂNE
Bucureşti, str. Gutenberg 3 bis

and
ABACUS PRESS, TUNBRIDGE WELLS, KENT

ISBN 0 85626 007 X

PRINTED IN ROMANIA

Preface

The delay introduced by the finite time required for charge transport is known as the transit-time effect. This phenomenon is peculiar to charge-controlled electronic devices. The non-stationary behaviour of *unipolar* (one-carrier) vacuum and solid-state devices is dominated by the transit-time effect.

The dynamics of electrons in high-frequency vacuum devices has been subjected to extensive theoretical research. The mathematical analysis of electron dynamics is complicated by the fact that the electrons move in a non-stationary and non-uniform electric field, each electron having a different 'history', which depends upon its emission moment. The space-charge set up by the charge carriers modifies the field in which these carriers move. In fact, we are encountering the more general problem of interaction between charges and fields.

A recent monograph*) summarized the earlier work on electron dynamics in diode regions. The last chapter of this monograph was devoted to the transit-time effects in semiconductor regions. Since then, a number of unipolar solid-state devices experienced a significant development and the importance of studying the transit-time effects in solid-state devices increased considerably.

A variety of phenomena are connected to the interaction between non-stationary fields and a flow of carriers moving in a solid, and the electrical behaviour of charge-controlled solid-state devices can be extremely diverse. The transit-time effects determine an upper limit in frequency for certain devices, whose normal operation is quasi-stationary (a typical example is the transistor). Other devices operate in a non-stationary mode (negative-resistance transit-time devices and transferred-electron (TE) devices) and the transit-time effects are essential.

This book is devoted to transit-time phenomena in *unipolar* solid-state devices. The field-effect (FE) transistors, the TE devices, the space-charge-limited current (SCLC) diodes and triodes, the barrier-injection transit-time (BARITT) devices are discussed. The avalanche transit-time diodes are bipolar devices and, although they may contain separate regions where one type of carrier experiences a transit-time delay, they are not considered explicitly in this book.

The emphasis is placed upon the mathematical approach and upon the understanding of the phenomenology of carrier dynamics. The book is, accordingly,

---

*) Ch. K. Birdsall and W. B. Bridges, 'Electrodynamics of diode regions', Academic Press, New York—London, 1966.

organized by having in view typical modes of operation rather than classes of devices. The effect of the bulk space charge and the injection mechanism upon the steady-state, the alternating current and transient small-signal regime, some noise properties, large-signal behaviour of negative-resistance oscillators and amplifiers, the turn-on transient of SCLC and FE devices, the propagation of high-field domains and accumulation layers in TE devices are studied *).

The author puts forward a theory of the steady-state and alternating-current behaviour of the one-dimensional solid-state regions. This theory represents a generalization of a number of papers by the present author, as well as of a number of studies made by other authors. Most of the ideas developed in these papers and studies originate from Shockley's outstanding work on high-frequency negative resistance semiconductor devices. The theory presented here benefits from one basic feature of Llewellyn's theory on small-signal a.c. behaviour of planar vacuum regions, in the sense that the mathematical analysis separates the bulk effects, which are described by the basic differential equations, by electrode or interface effects, described by boundary conditions to the above equations. However, the mathematical approach in solving these equations differs from that known as the 'Llewellyn method'. The method to deal with small-signal problems is simpler and more efficient than Llewellyn's. This new method was almost exclusively used during the last few years, together with various large signal approaches. The aim of this book is to facilitate the spread and use of these analytical methods in similar problems, as well as to contribute to their development. Although more complex problems will be necessarily solved by using a computer, the analytical methods will still retain advantages. A combination between the analytical treatment of the basic equations (leading for example to the selection of a minimum number of parameters) and numerical methods should be, probably, ideal.

Due to the actual development of the very-high-frequency solid-state devices involving charge transport, the study of transit-time effects will surely continue, and the present book will represent only an introduction to this interesting topic of theoretical and experimental research.

D. DASCALU

---

*) A more complete description of TE device behaviour may be found elsewhere (see for example J. E. Carroll : 'Hot electron microwave generators', E. Arnold (Publishers), London, 1970).

Acknowledgements

My gratitude goes to Professor Mihai Drăgănescu, of the Polytechnical Institute of Bucharest, for suggesting the injection currents in solids as a field of study, for the initial guidance and the permission to use before publication his work on high-frequency behaviour of SCLC diodes, as well as for his continuous encouragement during the last ten years. His personality deeply influenced my scientific activity and also the present study.

I wish to thank Dr. G. T. Wright of the University of Birmingham (England) for clarifying discussions on transit-time devices, during my stay with the Solid-State Devices Group, of the Electronic and Electrical Engineering Department, in that University.

I am grateful to a number of authors and journals for their permission to reproduce from published papers, as acknowledged in the text.

The author owes thanks to his earlier students and his present co-workers N. Marin and Gh. Brezeanu for fruitful co-operation on transit-time devices. Some results of their analytical and computer calculations are included in this book, particularly in Chapters 6, 10 and 11. The author is indebted to his earlier colleague, Dr. C. Bulucea (now Scientific Director of CCPCE, Băneasa, Romania), for suggesting the idea of writing this book. The author is grateful to his wife and his children for their patience and understanding during the time this book was prepared for publication.

Contents

Part II

Linear Theory: Small-signal Operation, Stability and Noise

Chapter 4. Transport Coefficients and General Linear Analysis

Chapter 5. Small-signal Behaviour of Space-charge-limited Current Devices

Chapter 6. Small-signal Negative Resistance of Injection-controlled Devices

List of Symbols

Symbol	*Explanation*	*First occurrence in (defined by)*
A	area of injecting contact in a device with plane-parallel electrodes	
$\mathbf{B}$	magnetic induction	Equation (2.1)
C	dynamic small-signal capacitance	
C_d	domain capacitance	Equation (13.73)
C_{gc}	gate-to-channel capacitance	Equation (8.43)
C_i	insulator capacitance (per unit area)	Equation (8.1)
C_{ij}	a.c. transport coefficients	Equations (4.19), (4.20)
C_0	geometrical capacitance (usual, per unit of electrode area)	Equation (P.1.3)
C_{st}	stationary capacitance (per unit area)	Equation (5.26)
D	displacement vector	Sections 2.1 and 8.1 only
$D\ (D_n, D_p)$	diffusion constant (for electrons, holes)	Equations (2.12), (2.13)
d	domain thickness	Equation (13.32)
E	electric field intensity	
E_a	anode field	Equation (3.38)
E_b	bias field	Equation (13.40)
E_c	cathode field	Equation (3.17)
E_D	maximum field in a high-field domain	Equation (13.26)
E_d		Equation (3.87)
E_L	anode field	Equation (12.15)
E_l		Equation (5.17)
E_M	threshold field, or field intensity determining the peak velocity in negative-mobility materials	Section 1.4 and Figure 13.4.
E_m	field intensity corresponding to the valley velocity	Figure 13.4
E_N	neutral field or external field	Equations (3.51) and (13.5)
E_R	critical field for Schottky effect	Equation (3.22)
e	electronic charge	

F_0	short-circuit noise factor	Equation (9.12)
F_∞	open-circuit noise factor	Equation (9.14)
f	frequency	
f_T	transit-time frequency	Equation (1.8)
f_p, f_v, f_d	'f' functions	Equation (7.11)
G	a.c. conductance	Equation (2.45)
$G_{p\infty}$	v.h.f. parallel conductance	Equation (6.51)
g_d	drain (small-signal) conductance	Equation (8.76)
g_{d0}	drain conductance in the linear region (small drain currents)	Equation (8.42)
g_m	transconductance	Equation (8.65)
g_{ms}	transconductance in saturation	Equation (8.38)
$\mathbf{H}$	magnetic field	Equation (2.2)
h	insulator thickness	Equation (8.1)
h	Planck constant	Equation (3.12)
I_c	channel (particle) current	Equation (8.8)
I_D	drain current	Figure 8.1.
I_{Dsat}	saturation drain current	Equation (8.15)
i	$\sqrt{-1}$ (imaginary unit)	
J	current density	
$\bar{J}$	normalized current density	Equation (3.77) (except for ch. 12)
J_R	Richardson current density	Equation (3.11)
J_{sat}	saturation current density of a metal-semiconductor contact	Equation (3.21)
j	particle current	Equation (2.2)
$j_c(E)$	control characteristic	Section 3.6.
$j_n(E)$	neutral characteristic	Equation (3.42)
K_{ij}	a.c. transport coefficients	Equations (4.14) and (4.15)
K_{21x}, K_{22x}		Equation (4.64)
k	Boltzmann constant	
L	device or region length (drift length)	Figure 1.1, etc.
L_{DE}	extrinsic Debye length	Equation (1.4)
m	carrier mass	
$(-N)$	fixed space-charge density	Equation (2.37)
N_A	acceptor density	Equation (2.16)
N_D	donor density	Equation (1.3)
n	electron density	Equation (2.12)
p	hole density	Equation (2.13)
p_t	trapped hole density	Equation (2.17)
Q	electric charge or quality factor	
Q_D	domain charge (per unit area)	Equation (13.61)
Q_{mob}	mobile electric charge (per unit area of injecting contact)	Equation (1.2)
q	carrier charge	

$R(\omega)$	real part of impedance	
R_0		Equation (5.47)
R_0		Equation (11.31)
R_d		Equation (6.10)
R_i	incremental resistance (usually, for a device with injecting area equal to unity)	(P.1.3)
S_i	spectral intensity	Equation (9.1)
s	Laplace variable	Equation (4.37)
T	absolute temperature	
T_L	carrier transit-time	Equation (1.2)
T_x	carrier transit-time until the plane x	Equation (3.65)
t	time	
V	applied voltage bias	Figure 1.1.
$V(x, t)$	channel potential with respect to source	Equation (8.9)
$\bar{V}$	normalized applied voltage	Equation (3.78)
$\bar{V}'$	normalized 'effective' voltage	Equation (3.105)
V_D	drain-to-source voltage	Figure 8.1.
$V_{D\,sat}$	saturation drain voltage	Equation (8.14)
V_d	domain excess voltage	Equation (13.9)
V_G	gate-to-source voltage	Figure 8.1.
V_{FB}	flat-band voltage	Figure 1.3.
V_{PT}	punch-through voltage	Figure 1.1.
V_{RT}	reach-through voltage	Figure 1.3.
V_T	threshold voltage	Equation (8.7)
v	carrier velocity	
v_D	domain velocity	
v_d		Equation (13.15)
v_l	limit (saturation) velocity	Equation (5.17) etc.
v_M	peak velocity	Equation (1.6)
v_m	valley velocity	Equation (3.101)
v_N	external velocity	Equation (13.21)
W	electron energy	Figure 1.3.
W	source width	Equation (8.8)
W_D	donor level	Equation (3.114)
W_c	bottom of the conduction band	Equation (3.114)
W_F	Fermi level	Equation (3.13)
W_t	trapping level	Equation (3.13)
W_v	top of the valence band	Equation (3.14)
x	distance from the injecting plane (one-dimensional model)	
x'	coordinate moving with the domain	Equation (13.15)
x_D	coordinate of the peak field	Equation (13.52)
x'_D	position of the peak field	Equation (13.30)
Y	small-signal admittance (usually, per unit of electrode area)	
Z	small-signal impedance (for a unit area emission electrode)	

$\overline{Z}$	normalized impedance	Equation (5.69)
Γ		Equation (6.7)
Δ	injection level	Equation (3.72)
Δf	frequency band	Equation (9.35)
ΔJ_0	detected current	Chapter 11.
δ	trapping factor	Equation (3.128)
ε	semiconductor permittivity	Equation (1.4)
ε_i	insulator permittivity	Equation (8.1)
ζ	space-charge factor	Equation (3.75)
η_c	conversion efficiency	Equation (14.11)
θ	transit angle (ωT_L)	Equation (4.83)
θ_d		Equation (3.93)
θ_r		Equation (3.69)
θ_x	transit angle (ωT_x)	Equation (7.19)
μ	d.c. mobility (v/E)	Section 1.4.
μ	magnetic permeability	Equation (2.6) only
μ_d	differential (a.c.) mobility	Equation (1.7)
ν		Equation (5.66)
ν_c	collision frequency	Equation (2.22)
ρ_s	surface charge density	Equation (8.4)
Σ	density of surface traps	Equation (3.13)
σ	surface charge density	
σ_t	density of trapped charge	Equation (3.13)
$\sigma_1(0)$	a.c. cathode conductivity	Equations (3.54) and (4.16)
τ_c	cathode relaxation time	Equation (6.8)
τ_d	differential relaxation time	Equations (1.7) and (3.94)
τ_f	lifetime of free carriers	Equation (3.126)
τ_t	lifetime of trapped carriers	Equation (3.125)
τ_r	dielectric relaxation time	Equation (1.5)
Φ	electric potential	
φ_0		Equation (3.99)
φ_L		Equation (3.100)
ω	angular frequency of the applied signal	Equation (2.40)
ω_r	relaxation frequency	Equation (3.66)

Boldface symbols stand for vector quantities.

Boldface underlined symbols denote phasors.

Subscript 0 (zero) indicates a zero-order of magnitude (of direct current, steady-state) quantity.

Subscript 1 denotes a first-order (or small-signal) quantity.

D.c. means direct current, a.c. – alternating current, h.f. – high frequency, etc.

Part I

Introduction: Basic Devices and Phenomena

1.

Unipolar Solid-state Devices

1.1 Introduction

The present book deals, primarily, with a number of solid-state devices which are relatively new or are still under development. They belong to a large class of electronic devices, namely the unipolar (solid-state) devices. The concept of unipolar solid-state device was established as a result of the pioneering studies on carrier injection in *insulators* made by Mott and Gurney [1], Rose and Lampert [2]—[5] and then by Wright and his co-workers [6]—[10], as well as of the several outstanding papers written by Shockley [11]—[14] on unipolar *semiconductor* devices.

Among the unipolar devices which are used in electronics today the most important are: the field-effect transistor (FET) (including the MOS transistor), the transferred-electron devices, the Schottky diode. A number of two-terminal unipolar devices are active: i.e. they exhibit a negative resistance and can be used for alternating power generation and signal amplification. This negative resistance behaviour is very often restricted to certain (high-)frequency ranges and cannot be explained in a quasi-stationary manner. The inertia of carrier transport (the so-called transit-time effect) plays a major rôle in these high-frequency phenomena. The greatest part of this book is devoted to transit-time effects in negative-resistance devices. The transient response and the h.f. detection properties of certain unipolar diodes, as well as the very high frequency (v.h.f.) amplification properties of the FET (a simplified model) are also studied.

In this chapter we present certain devices which are typical for the problems studied in the book. This presentation should be considered merely as an introduction to the more elaborate treatment given later. This introductory discussion allows us to formulate an analytical model which is gradually introduced in the following chapters.

1.2 Punch-through Diode and the Space-charge-limited Current in Semiconductors

The punch-through diode was first suggested by Shockley and Prim [13], and is a familiar p^+np^+ structure biased as a floating base transistor (Figure 1.1a). The external voltage applied between 'emitter' (positively biased) and 'collector' drops across the barrier regions of the emitter-base and base-collector junctions, respectively.

The base-collector junction is reversely biased and its barrier region (which, for a $p^+\nu p^+$ structure, is located almost entirely in the lowly doped central region) extends through the ν base as the applied voltage increases. The current flowing in the external circuit (which is exactly the residual collector current in the grounded emitter configuration and consists mainly of holes) is very small and increases slowly with increasing voltage. If the applied voltage is sufficiently high, the neutral-base width will decrease to zero. This phenomenon, called punch-through, does actually occur only if the base resistivity is sufficiently high. Conversely, the avalanche multiplication will occur before punch-through.

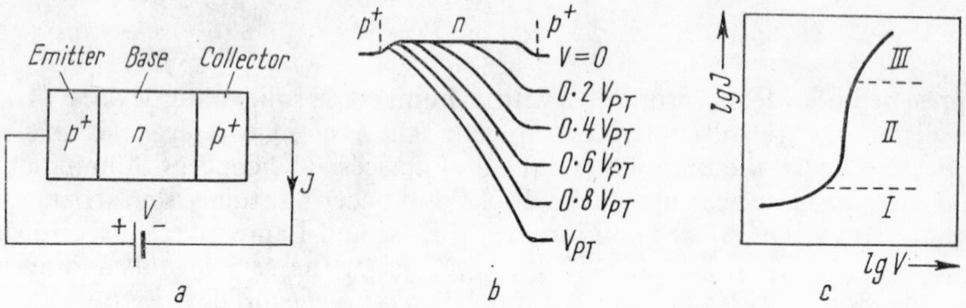

Figure 1.1. The punch-through diode:
(a) a p^+np^+ structure biased as a floating base transistor, (b) the potential distribution inside the p^+np^+ structure for an applied bias below or equal to the punch-through voltage V_{PT}, (c) the shape of the $J - V$ characteristic indicating the following operating regions: (I) the operation like a floating base transistor, (II) the device 'punch-through', (III) the space-charge-limited current (SCLC) regime (J is the current flowing through a device having the junction area equal to unity).

The potential diagram shown in Figure 1.1b indicates that at punch-through the holes flowing from emitter to collector have to surpass a potential barrier and then they are injected *directly* into the collector barrier region, where they are accelerated by a high electric field towards the p^+ collector. If the voltage is slightly increased above the punch-through value V_{PT} (Figure 1.1b), the height of this potential barrier decreases and the hole current increases rapidly. This rapid increase is shown quantitatively by the log J — log V characteristic plotted in Figure 1.1c. At larger applied voltages this characteristic follows a $J \propto V^2$ law [13]. This behaviour is peculiar to space-charge-limited (SCL) currents in solids, as we shall show below qualitatively.

The SCL emission is characterized by the existence of a potential minimum (for electron emission) or maximum (for hole emission) in front of the emitting electrode. Only a part of the current which can be supplied by this emitting electrode, does really flow in the external circuit, because only a part of the 'emitted' carriers can traverse (against the electric field) the region between the emitting surface and the surface (sometimes called virtual cathode) with a zero electric field. Here, there is a difference between SCL emission in vacuum and in solids, respectively. Whereas the electrons in a vacuum device surpass this potential barrier due to their finite emission velocities, in solids, the emitted carriers go beyond the virtual cathode by the process of diffusion.

From Figure 1.1b we conclude, intuitively, that as the applied voltage increases well above the punch-through value V_{PT}, both the height of the potential maximum and the distance between this maximum and the emitting (left-hand-side) junction should become negligibly small. This conclusion is not rigourously valid but it will be accepted until a more detailed discussion of SCL injection in the next chapters. Therefore we state that if the applied voltage is sufficiently high, the electric field at the emitting junction ($x = 0$) will become zero

$$E(0) = 0 \tag{1.1}$$

(the virtual cathode of the punch-through diode will coincide with the emitting junction). Because the electric field is high, the current will be carried predominantly *by drift*, except in the immediate vicinity of the emitter ($x = 0$). We stress the fact that condition (1.1) should be satisfied continuously as the voltage increases: this requires an *infinite source* of charge carriers at $x = 0$. This condition can be satisfied approximately because the doping of the p^+ is very high, much higher than in the central ν region.

The proportionality between the current density J and the square of the applied voltage can be understood as follows. We start from the equation

$$J = Q_{\text{mob}}/T_L \tag{1.2}$$

(where Q_{mob} is the total mobile charge contained in the drift region of length L, and unity cross-section and T_L the transit time, or the time taken by carriers to travel across this same region). This equation is quite general for semiconductor regions bounded by plane-parallel electrodes and can be obtained by integrating $ep = J/v$ (see List of Symbols) from $x = 0$ to $x = L$ (Figure 1.1a). The mobile charge is approximately equal to the total charge because if the applied voltage is high the injected mobile charge predominates over the fixed charge of the ionized impurities. Then Q_{mob} is proportional to the 'anode' field (or the field at the collecting electrode) because the 'cathode' ($x = 0$) field is zero (condition 1.1). Hence $Q_{\text{mob}} \propto V$. On the other hand, T_L is inversely proportional to the average (drift) velocity and, if the carrier mobility is constant, we will have $T_L \propto V^{-1}$. Therefore, J should be proportional to V^2 for large V. At still higher fields the mobility becomes field dependent (decreases) and the parabolic dependence $J \propto V^2$ breaks down.

We note the fact that the magnitude of the SCL current is determined (limited, at a given applied voltage) by the bulk space charge and not by the injecting contact (which is, however, assumed to be ideal, or ohmic, being an 'infinite' reservoir of charge carriers).

Figure 1.2 shows an experimental $J - V$ characteristic for a silicon $p^+\mu p^+$ structure, which is in good agreement [15] with the simple theory of the punch-through diode [13]. The deviation noticed at higher bias fields is due to the field-dependence of the carrier mobility.

Shockley [14], [16], and then Yoshimura [17] calculated a *negative resistance* for the $p^+\nu p^+$ structure biased in the 'punch-through' region. The dynamic resistance becomes negative in a frequency range in which the signal period is comparable to the carrier transit-time. It was not until recently that experimental indication about the existence of such a h.f. negative resistance was obtained [18]--[20].

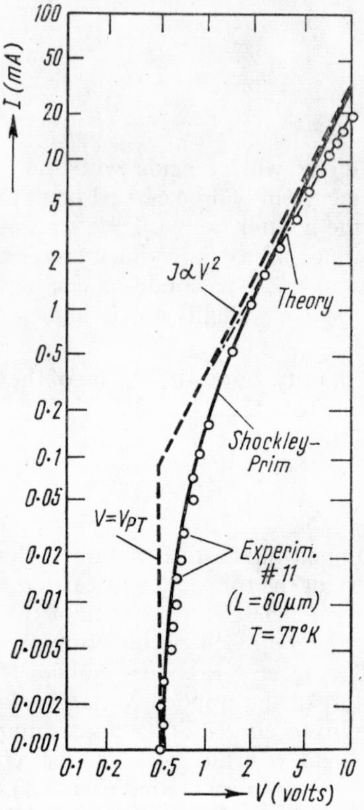

Figure 1.2. Comparison between the experimental I-V characteristic of a $p^+\nu p^+$ silicon structure, measured at liquid nitrogen temperature [15], and the theoretical characteristic derived by Shockley and Prim [13] by assuming $\mu =$ const (solid line), as well as the theory which takes into account the field dependence of hole mobility which occurs at higher voltage bias.

1.3 Injection-controlled Current in Metal-semiconductor-metal Structures

The metal-semiconductor-metal (MSM) structure considered here is a uniformly doped semiconductor slice with metal contacts on the opposite sides of the slice. The corresponding energy-band diagram at thermal equilibrium is shown in Figure 1.3a. The semiconductor is n-type (moderate doping) and both metal semiconductor contacts are rectifying. This MSM structure consists in two Schottky barriers connected back to back [21]. If a potential difference is applied between the metal contacts, one contact is forward biased (say the right hand contact) and the second one is reverse biased. As the applied voltage increases, the total length occupied by the two depletion regions increases. By definition, when the applied voltage reaches the reach-through voltage, V_{RT}, the two depletion regions touch each other. This situation is illustrated in Figure 1.3b.

We assume now that the voltage increases above V_{RT}, and reaches the value V_{FB} (the flat-band voltage) defined by the condition that the electric field becomes zero and the energy band becomes flat at the second contact (Figure 1.3c). If $V >$

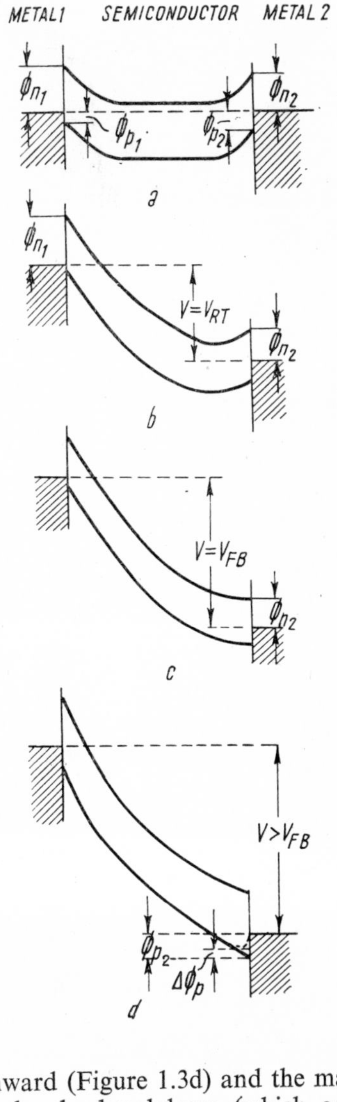

Figure 1.3. Energy band diagram of a metal-semiconductor-metal structure: (a) at thermal equilibrium (Φ_{n1} and Φ_{n2} are the electron barrier heights, etc.); (b) when the bias voltage (the metal 2 is positively biased) is equal to the reach-through voltage V_{RT} (the two depletion regions of the metal-semiconductor junctions just touch each other); (c) the flat-band condition at the interface between the semiconductor and the metal; (d) the applied voltage is larger than the flat-band voltage V_{FB}; $\Delta\Phi_p$ is the image-force lowering of Φ_{p2}.

$> V_{FB}$, the energy band is bent further downward (Figure 1.3d) and the maximum applied voltage V_B is determined by the avalanche breakdown (which occurs at the reverse biased contact, where the electric field is maximum).

The $J - V$ characteristic shows a very rapid increase of the current as the applied voltage reaches V_{RT}. It can be shown that, for a situation like that depicted in Figure 1.4, the hole current (emitted by the forward biased contact in Figure 1.3) becomes much larger than the electron current at $V > V_{RT}$. The very steep increase of the hole current at $V > V_{RT}$ is due to its exponential dependence upon

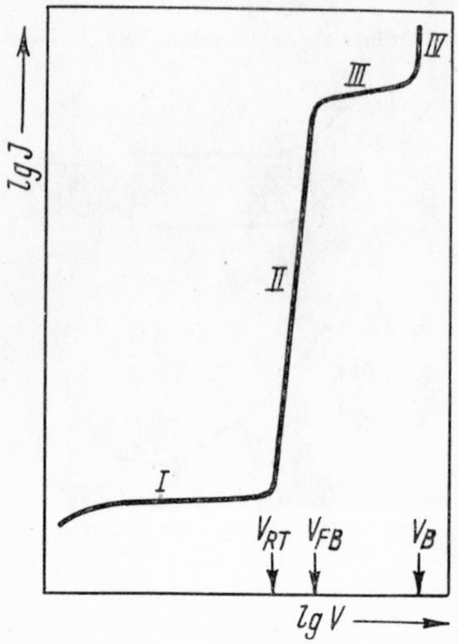

Figure 1.4. A typical J-V characteristic of the metal-semiconductor structure from Figure 1.3 a, showing the following regions: Region (I) ($V < V_{RT}$) — the structure behaves essentially as two Schottky barriers connected back to back, and a very small current flow; Region (II) ($V_{RT} < V < V_{FB}$) — the entire semiconductor is depleted of majority carriers (electrons), holes are injected by thermionic emission over the potential barrier existing in the vicinity of the contact on the right; Region (III) ($V_{FB} < V < V_B$) — the potential barrier disappeared and the total current is equal to the saturation current of contact 2, which increases slowly due to the Schottky effect (the image-force lowering of the barrier, see Fig. 1.3.d); Region (IV) ($V > V_B$) — the current rises steeply due to avalanche breakdown.

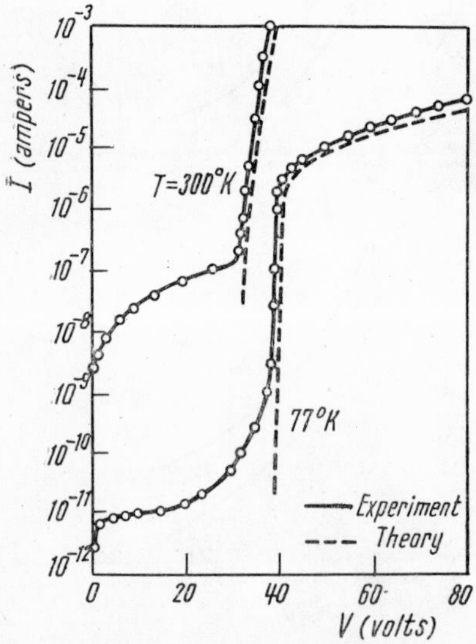

Figure 1.5. Comparison between the theory of the MSM structure (dashed curves) and the experimental $J - V$ characteristics (solid curves) measured on PtSi — Si — PtSi structures made from n-type silicon with doping of $4 \cdot 10^{14}$ cm^{-3} and thickness of 12 μm. The departures between theory and experiment in the low current region are mainly due to surface leakage currents [21].

the height of the potential barrier. However, in contrast to the SCL emission case, here the metal-semiconductor cannot be assumed an infinite reservoir of charge carriers. At $V = V_{FB}$ the current through the device (which is practically equal to the hole current) becomes equal to the hole saturation current of the emitting contact. Above V_{FB} the current increases slowly with V (until the avalanche multiplication, $V < V_B$). This increase is due to the field dependence of the barrier height *at* the emitting contact (for example, due to image-force lowering of the barrier, sometimes called the Schottky effect).

Figure 1.5 shows the measured characteristic of an experimental device [21] and gives a comparison with theory. It may be shown that both the semiconductor doping and semiconductor length, L, are properly chosen such as to avoid both carrier multiplication and SCL emission (see also Chapter 6). Coleman and Sze [22] reported microwave oscillations obtained from such a MSM structure biased in the $V_{RT} < V < V_{FB}$ region.

1.4 Negative-mobility Diode

The electric field reaches high values inside the unipolar devices and therefore the carrier mobility, defined by $\mu = v/E$, cannot be assumed independent of field.

Several semiconductors like GaAs and InP exhibit (under normal conditions) a *negative* differential mobility. This can be seen in Figure 1.6a which shows the

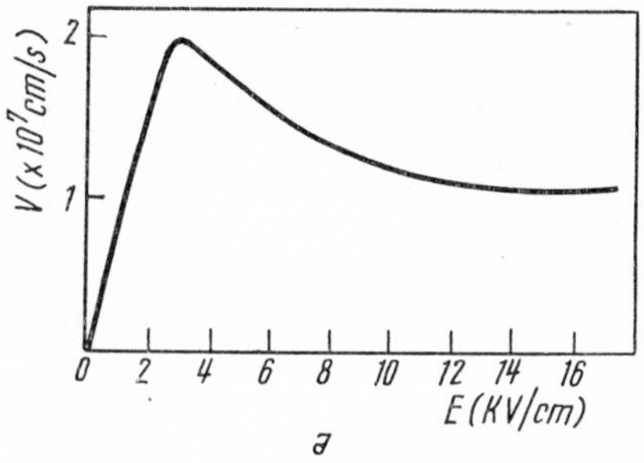

Figure 1.6a. Velocity field dependence of electrons in GaAs indicating that the carrier velocity decreases above a certain threshold field.

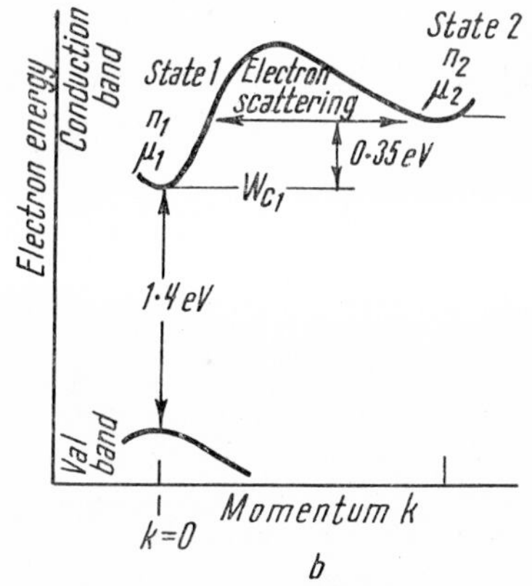

Figure 1.6b. Band structure of GaAs. The valence and conduction band energies are plotted as a function of the crystal momentum k. The lowest minimum in the conduction band is at $k = 0$ and a subsidiary minimum (valley) is located at the zone edge at an energy of 0.35 eV above the central minimum. When the electrons in the central minimum (lower valley) are accelerated by an electric field to an energy of 0.35 eV they can be scattered to the subsidiary minimum.

velocity-field dependence for electrons in GaAs: above a certain threshold field the electron velocity decreases with increasing electric field and thus the differential mobility dv/dE becomes negative.

This unusual $v - E$ dependence can be explained [23], [24] by considering the band structure shown in Figure 1.6b. The conduction band has two valleys: the electrons in the lower valley have a low effective mass and high mobility, the electrons in the upper valley are 'heavy' and have a low mobility. The measured drift velocity is a (weighted) average value. At low field intensities, almost all electrons lie in the lower valley and the low-field mobility is quite high. At larger field intensities, the electrons acquire sufficient energy to pass into the upper valley. This redistribution of the electrons takes place almost instantaneously (the intervalley scattering times are of the order of 10^{-11} s or shorter [25], [26]). Because

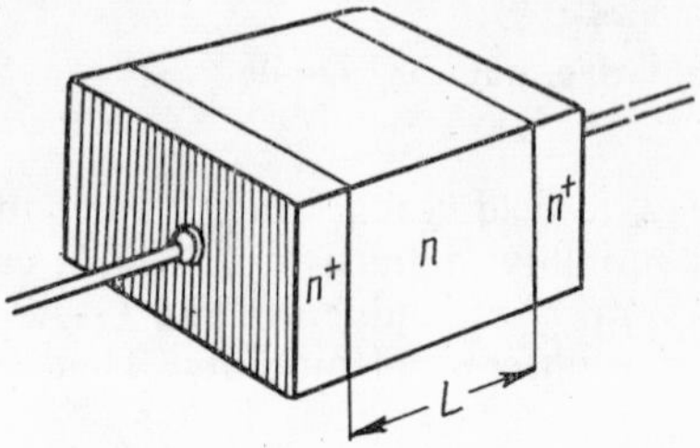

Figure 1.7. A n^+nn^+ GaAs diode, where N_D and L are the doping and the length, respectively, of the central n region.

the fraction of 'heavy' electrons increases very much with the electric field, the (average) drift velocity decreases as shown in Figure 1.6a. The shape of the $v - E$ characteristic appearing in Figure 1.6a also indicates the fact that at field intensities much higher than the threshold field, the electron velocity in the upper band does saturate. Because of the intervalley transfer described above, the electron devices using negative-mobility materials like GaAs are named 'transferred-electron devices' (or, sometimes, Gunn-effect devices).

The simplest transferred-electron device is the n^+nn^+ diode sketched in Figure 1.7. The characteristic parameters are the doping, N_D, and the thickness, L, of the central layer. At the first sight, this device is a simple semiconductor resistor and the current is given by

$$J = e N_D v \tag{1.3}$$

where the electron velocity, v should be determined from the $v - E$ characteristic, for a given bias field V/L. This implies a $J - V$ characteristic which reproduces the shape of the $v - E$ characteristic thus having a range of *negative* incremental resistance $R_i = \mathrm{d}V/\mathrm{d}J < 0$. However, such a steady-state $J - V$ characteristic does not actually occur. This is due to the effect of (non-uniform) space-charge inside the semiconductor bulk. Equation (1.3), which is written by assuming neutrality of the v region, is not actually valid.

Let us first discuss the behaviour of the n^+nn^+ resistor at thermal equilibrium (zero applied voltage). At the n^+n junctions a diffusion process takes place: electrons from the n^+ regions diffuse into the n region leaving uncompensated positively-charged donors. Therefore, a net space-charge and a non-zero electric field will appear in the vicinity of n^+n junctions, as shown in Figure 1.8. Mathematical computations [27] show that the thickness of the space-charge layer inside the n region should be comparable to the so-called extrinsic Debye length

$$L_{DE} = \left(\frac{k\,T\,\varepsilon}{e^2\,N_D}\right)^{1/2} \tag{1.4}$$

of the n region*⁾. The distribution indicated in Figure 1.8 is valid for $L \gg L_{DE}$, which means that the greatest part of the n region is practically neutral at thermal equilibrium.

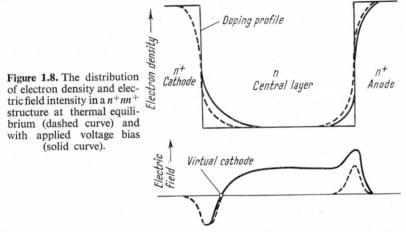

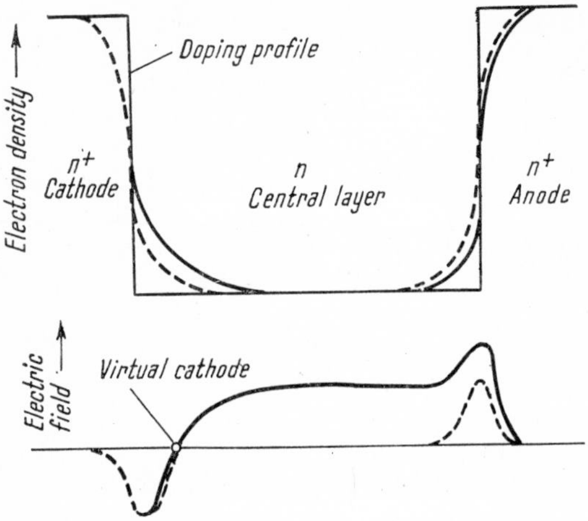

Figure 1.8. The distribution of electron density and electric field intensity in a n^+nn^+ structure at thermal equilibrium (dashed curve) and with applied voltage bias (solid curve).

Figure 1.8 also shows the electron density and electric field distributions when the n^+nn^+ structure is biased by an external voltage. Note the fact that the space-charge region of the negatively biased n^+ contact extends into the semiconductor bulk. The larger the applied bias, the larger the *non-uniformity* of the electric field in the n region. We can identify a 'virtual cathode', defined as in Section 1.2.

Because a net space-charge exists into the bulk we must assume the fact that the mechanism of dielectric relaxation is not efficient. This is because the mobile space-charge injected in excess passes very rapidly through the semiconductor under the influence of the applied electric field. This condition can be written mathematically

$$T_L \lesssim \tau_r, \quad \tau_r = \varepsilon / e N_D\, \mu \tag{1.5}$$

i.e. the carrier transit-time T_L is smaller than, or comparable to, the dielectric relaxation time τ_r.

Let us now derive the condition for approximate neutrality in a negative-mobility n^+nn^+ device biased at threshold field E_M (Figure 1.6a). The transit-time

---

*⁾ Sometimes this thickness is approximated by $3L_{DE}$. The space-charge layer developed in the n^+ region is much thinner.

is, approximately, $T_L = L/v_M$. The dielectric relaxation time may be approximated as having the same value as in equation (1.5) (see Figure 1.6a). For neutrality we require that $\tau_r \ll T_L$, and obtain

$$\varepsilon/e\,\mu\,N_D \ll L/v_M \quad \text{or} \quad N_D\,L \gg \frac{\varepsilon\, v_M}{e\,\mu}, \tag{1.6}$$

which means that the *doping-length product* N_DL should be sufficiently high. By introducing numerical data which are appropriate for GaAs at room temperature into equation (1.6), one obtains, approximately, $N_DL > 2 \times 10^{11}$ cm^{-2}. This particular value will be called *here* the critical doping-length product. We discuss below two cases:

(a) The doping-length product is above its critical value. The sample should be almost neutral (although large deviations from neutrality can occur in the vicinity of the n^+ contacts). However, this situation is not necessarily a stable one and the negative-mobility samples *are not* actually stable. This is a well-known experimental fact and can be understood as follows. The dynamics of small perturbations of the neutrality state (the cause of these perturbations does not concern us here) is governed by the *differential* relaxation time

$$\tau_d = \varepsilon/e\,N_D\,(\mathrm{d}v/\mathrm{d}\,E), \quad \mu_d = (\mathrm{d}v/\mathrm{d}\,E)\,. \tag{1.7}$$

This relaxation time is the time constant occurring in the exponential law which describes the decrease (or growth) of the bulk space-charge. If μ_d and τ_d are positive we encounter the common dielectric-relaxation mechanism. In negative mobility materials, if the sample is biased at high fields (above threshold), μ_d and τ_d will be negative. Any deviation from neutrality will *grow* in time ($\tau_d < 0$). This growing perturbation moves through the sample. If the transit-time is long compared to τ_d, the perturbation will have sufficient time to develop and can grow appreciably. We simply state here that such a perturbation can become a so-called *stable* high-field domain travelling inside the bulk and giving rise to a spike of the external current when it disappears at the positively-biased contact. The process of domain formation, propagation, and disappearance is repeated periodically leading to oscillations. The fundamental oscillation frequency is approximately equal to the 'transit-time frequency'

$$f_T = 1/T_L \tag{1.8}$$

and is of the order of several GHz, depending upon the device length, L.

(b) A subcritically doped negative-mobility diode is stable and can be used for small signal (high-frequency) amplification because it exhibits a negative resistance in the transit-time frequency region.

Depending upon the external circuit, the transferred-electron diode may have a variety of operation modes [24], [28], [29]. On the other hand, the diode can be constructed in planar form or have a non-ohmic injecting contact, etc. More complicated transferred-electron devices are studied in laboratories.

1.5 Conclusion

This chapter gives only a few examples of devices which can be studied by using the theory developed in this book.

These examples have introduced us to certain specific problems of unipolar currents in solids. We have seen that the current is carried by drift, the electric field is non-uniform (a bulk space-charge exists) and sufficiently intense to determine 'hot-carrier' effects (mobility field-dependence). We have also felt the importance of contact properties (by comparing diodes with ohmic and non-ohmic injecting contacts). The importance of such parameters as the relaxation time, the transit-time or the doping-length product is now evident. The previous Section indicated that interesting (and useful) effects could arise in simple structures if negative-mobility semiconductors are used.

In the next chapter we shall discuss briefly the equations which form the mathematical and phenomenological basis of our analysis of transit-time effects in unipolar solid-state devices.

Problems

1.1. Show that the punch-through voltage V_{PT} defined in Section 1.2 for the $p^+\nu p^+$ structure is given by

$$V_{PT} = eN_D L^2/2\varepsilon \qquad \text{(P. 1.1)}$$

where N_D is the effective donor concentration in the ν region (all donors are assumed to be ionized).

1.2. By using equation (1.2), show that the square-law dependence of the SCLC current should be of the form $J \propto \varepsilon\mu L^{-3}V^2$.

1.3. Prove the exact expression of the SCLC square-law

$$J = \frac{9}{8}\varepsilon\mu\, V^2/L^3 \qquad \text{(P. 1.2)}$$

Hint: Use the current-continuity equation, the Poisson equation and the boundary condition (1.1). Assume negligible diffusion, negligible ion density (high currents), and constant mobility.

1.4. Demonstrate that the carrier transit-time under SCL injection described by equation (P. 1.2), is equal to

$$T_L = \frac{4}{3}\frac{L^2}{\mu V} = 3R_i C_0 \qquad \text{(P. 1.3)}$$

where R_i is the incremental resistance (unit area device), $R_i = (\mathrm{d}J/\mathrm{d}V)^{-1}$ and C_0 the geometrical capacitance $C_0 = \varepsilon/L$.

1.5. Show that the flat-band voltage for a symmetrical MSM structure (Section 1.3) should be equal to V_{PT} given by equation (P. 1.1).

1.6. Prove the fact that the relation (1.5) (which requires a net space-charge inside a n^+nn^+ structure) can be approximately written

$$V \gtrsim \frac{kT}{e}\left(\frac{L}{L_{DE}}\right)^2 . \qquad \text{(P. 1.4)}$$

2.

Basic Equations

2.1 Maxwell's Equations

The basic formulae used throughout this book are Maxwell's equations and certain semiconductor equations.

Maxwell's equations are written down in their general form

$$\nabla \times \mathbf{E} = -\frac{\partial \mathbf{B}}{\partial t} \tag{2.1}$$

$$\nabla \times \mathbf{H} = \mathbf{j} + \frac{\partial \mathbf{D}}{\partial t} \tag{2.2}$$

$$\nabla \cdot \mathbf{D} = \rho \tag{2.3}$$

$$\nabla \cdot \mathbf{B} = 0 \tag{2.4}$$

(usual notation, see also List of Symbols). We may also write

$$\mathbf{D} = \varepsilon \mathbf{E} \tag{2.5}$$

$$\mathbf{B} = \mu \mathbf{H} \tag{2.6}$$

We shall simplify somewhat the above system of equations by making several assumptions. First, the medium will be considered isotropic and homogeneous and, thus, both the permeability, μ, and the permittivity, ε, should be scalar in nature and constant. Then, we assume that the minimum wavelength of the electrical oscillations studied is much longer than the (active) dimensions of the solid-state device. Such a regime is called quasi-stationary and is characterized by

$$\nabla \times \mathbf{E} \simeq 0 \tag{2.7}$$

As a result of this we can define an electric potential, Φ, such as

$$\mathbf{E} = -\nabla \cdot \Phi \,. \tag{2.8}$$

This is the first of the equations to be used directly in this book. The second one, the so-called Poisson equation, is the result of equations (2.3) and (2.5)

$$\varepsilon \nabla \cdot \mathbf{E} = \rho. \tag{2.9}$$

Finally, by using the vector identity $\nabla \cdot \nabla \times \mathbf{F} = 0$ and replacing **F** by **H** from equation (2.2) one obtains

$$\nabla \cdot \nabla \times \mathbf{H} = \nabla \cdot \left(\mathbf{j} + \varepsilon \frac{\partial \mathbf{E}}{\partial t} \right) = 0 \tag{2.10}$$

This is Maxwell's continuity equation, which states that the divergence of the total current

$$\mathbf{J} = \mathbf{j} + \varepsilon \frac{\partial \mathbf{E}}{\partial t} \tag{2.11}$$

is equal to zero.

In this book we consider the effect of an electric field upon the motion of charge carriers (either electrons, or holes) in solids. However, the magnetic field cannot be rigourously zero because $\nabla \times \mathbf{H}$ would be zero and thus the total current would be zero (see Eq. 2.2). This magnetic field is assumed very small and thus the particle current **j** appearing in equation (2.11) will be calculated taking into account only the electric field. This is a usual simplification in problems of this kind.

2.2 Semiconductor Equations

The semiconductor equations relate the electric charges and currents inside the semiconductor. These equatons were first formulated by Shockley [30] and remain until now [31] the basis of theoretical analysis of semiconductor devices. We repeat them below. The electron current and the hole current are given by respectively

$$\mathbf{j}_n = en\mu_n \mathbf{E} + eD_n \nabla \cdot n \tag{2.12}$$

$$\mathbf{j}_p = ep\mu_p \mathbf{E} - e\, D_p \nabla \cdot p \tag{2.13}$$

In the above equations μ and D denote the carrier mobility and the diffusion constant, respectively. *This notation will be used through the remainder of this book.* We may also write the continuity equations

$$\frac{\partial n}{\partial t} = - \frac{n - n_0}{\tau_n} + \frac{1}{e} \nabla \mathbf{j}_n \tag{2.14}$$

$$\frac{\partial p}{\partial t} = - \frac{p - p_0}{\tau_p} - \frac{1}{e} \nabla \mathbf{j}_p \,. \tag{2.15}$$

Here, the subscript zero denotes the thermal equilibrium value and τ_n, τ_p are time constants related to the generation-recombination process, which tends to restore the equilibrium.

The original set of five basic equations also contains the Poisson equation (2.9), written as follows

$$\varepsilon \nabla \mathbf{E} = e\,(p - n + N_D - N_A). \tag{2.16}$$

N_D, N_A are the densities of fully-ionized donors and acceptors, respectively.

The semiconductor equations must be written in a different form for the analysis of *unipolar* currents in solid-state devices. First of all, we have only one type of mobile charge carriers. The properties of unipolar currents and devices are 'symmetrical' for electrons and holes. Because it is easier to deal with positively-charged particles (this avoids confusing minus signs in equations) we shall chose *holes* for our theoretical calculations. Therefore, the electron density and the electron current will be assumed negligible. On the other hand, the generation of electron-hole pairs and their recombination will be considered insignificant. However, we must take into account another time-dependent process, which is trapping. Part of the mobile injected charge may be retained in defect centres (traps). As long as these charge carriers are trapped, they do not contribute to the particle current. The trapping process is described by a state equation which relates the free-hole and trapped-hole density

$$\mathcal{T}\,(p, p_t) = 0\cdot \tag{2.17}$$

If the deviations from equilibrium are not too large, the electric field does not occur explicitly in equation (2.17) [4].

The particle current is of the form

$$\mathbf{j} = \mathbf{j}_p = \mathbf{j}(p, \mathbf{E}) \tag{2.18}$$

whereas the Poisson equation (2.16) should be replaced by

$$\varepsilon\, \nabla\, \mathbf{E} = e\,(p + p_t + N_D - N_A) \tag{2.19}$$

(a hole trap is neutral when unoccupied). The above three equations are sufficient to determine the electric field as a function of the particle current. Proper boundary conditions should be also used (see the next chapter).

We shall discuss in detail the equation of the particle current (2.10)*. Equation (2.13) shows that the hole current has a drift component and a diffusion component

$$\mathbf{j} = \mathbf{j}_p = \mathbf{j}_{\text{drift}} + \mathbf{j}_{\text{diffusion}}\cdot \tag{2.20}$$

These components will be discussed in the next two sections.

---

*) Trapping effects are less important in usual semiconductors, and we postpone the discussion of these processes until Section 3.13.

2.3 Drift Current

The drift current is due to the flow of charged particles which move under the influence of the applied (external) field. The 'motion' equation of charged particle *in vacuum*, due to an electric field, **E**, is

$$m\frac{d\mathbf{v}}{dt} = q\mathbf{E}, \tag{2.21}$$

(m and q are the particle mass and electric charge respectively). A charge carrier (hole) moving in a solid, experiences 'collisions', which interrupt its motion under the influence of the electric field. An approximate equation which includes the effect of these collisions is

$$m\frac{d\mathbf{v}}{dt} + \nu_c m\mathbf{v} = e\mathbf{E}. \tag{2.22}$$

If these collisions are very frequent, the motion equation (2.22) will become approximately

$$\nu_c m\mathbf{v} = e\mathbf{E},\ \mathbf{v} = \mu\mathbf{E},\ \mu = e/\nu_c m, \tag{2.23}$$

where μ is the hole mobility. Therefore we have

$$\mathbf{j}_{\text{drift}} = ep\mathbf{v} = ep\mu\mathbf{E}, \tag{2.24}$$

which is exactly the expression used in writing equation (2.13).

Equation (2.24) is sufficiently accurate for studying the carrier drift in low-field regions. However, this relation breaks down for the very intense electric fields occurring in many unipolar devices. This occurs because the energy gained by the charge carrier is too high to be transferred to the lattice by the collision mechanism, the carrier energy increases much above its mean energy when it is in equilibrium with the lattice $\left(\frac{3}{2}kT\right)$ and the carrier becomes 'hot' (out of thermal equilibrium with the lattice). The mobility of a hot carrier is lower than the low-field mobility and depends upon the electric field intensity. Thus we should write

$$\mathbf{v} = \mu(E)\,\mathbf{E} \tag{2.25}$$

and the drift current becomes

$$\mathbf{j}_{\text{drift}} = e\,\mu\,(E)\,p\,\mathbf{E} = e\,p\,v\,(E)\,\frac{\mathbf{E}}{E}\,. \tag{2.26}$$

The relationship $v = v(E)$ should be determined by solving the transport (Boltzmann) equation and by taking into account various scattering mechanisms which characterize the carrier-lattice system. This is a difficult problem which requires a number of hypotheses and a detailed knowledge of the band structure. The theoretical velocity-field dependence cannot be obtained in analytical form (with rare exceptions)*). Therefore, the theory of solid-state devices should use for $v = v(E)$ either numerical data or a simple analytical approximation of experimental and/or theoretical data.

The time, t, does not appear explicitly in equation (2.26), i.e. the (average) carrier velocity is a local and *instantaneous* function of the electric field intensity. This is valid up to relatively high frequencies, as long as the mobility relaxation time may be considered negligible (for two-valley semiconductors the intervalley scattering time should be negligible, etc.).

2.4 Diffusion Current

Diffusion should be less important, in principle, in unipolar devices, because the electric field is high and the drift current is also high. However, the electric field is non-uniform and diffusion current might be important in certain regions of the unipolar device.

Equation (2.13) postulates the expression of the hole current as the sum of the drift current equation, (2.24), and the diffusion current

$$\mathbf{j}_{\text{dif}} = -\,e\,D_p\,\nabla p. \tag{2.27}$$

This superposition of field and diffusion currents is usual but it is not strictly correct. The drift current, equation (2.24), is derived for weak fields and low currents in homogeneous regions. The law (2.27) is justified for low concentration gradients and low currents in a field-free region. In a unipolar semiconductor device large electric fields and concentration gradients may exist †). Therefore, it is not obvious that the total particle current can be found by simply adding the drift current, equation (2.24), and the diffusion current (2.27).

We shall discuss below, for simplicity, the one-dimensional form of equation (2.13), which is

$$j = e\,\mu\,p\,E - e\,D\;\partial p/\partial x, \tag{2.28}$$

*) We also note the fact that the field-dependent mobility is not isotropic but depends upon the orientation of the electric field (related to the crystallographic planes).

†) A similar situation occurs for the carrier flow near a *pn* junction [32].

or, because of Einstein's relation

$$D = \mu\, kT/e, \tag{2.29}$$

one obtains

$$j = e\,\mu\, p\, E - \mu\, k\, T\, \partial p/\partial x. \tag{2.30}$$

A general expression for the current in the presence of electric fields and concentration gradients was derived by Blötekjaer [33]. For a one-dimensional flow of holes, this expression becomes [32]

$$i = e\,\mu\, p\, E - \mu \frac{\partial}{\partial x}\left(p \left\langle \frac{p_x^2}{m} \right\rangle \right), \tag{2.31}$$

where p_x is the x-component of the carrier momentum and the angular brackets $< \quad >$ indicate the average value. If the electric current is small the semiconductor will be close to thermal equilibrium and we can apply the principle of equipartition of energy, thus having

$$< p_x^2/2\,m > \; = \; < p_Y^2/2\,m > \; = \; < p_z^2/2\,m > \; = kT/2,$$

where T is the *carrier temperature*, defined by

$$< (p_x^2 + p_Y^2 + p_z^2)/2\,m > \; = \left(\frac{3}{2}\right) kT. \tag{2.32}$$

Therefore, at low currents, equation (2.31) becomes [32]

$$j = e\,\mu\, p\, E - \mu\,(\partial\, p\, kT/\partial\, x)\,. \tag{2.33}$$

If the carrier temperature T was taken constant, equation (2.33) would reduce to equation (2.30). Goldberg [32] indicated the fact that the carrier temperature cannot be assumed to be constant in a region with non-uniform electric field, even if the electric current is very small. This author [32] derived (for low currents) an equation of the form of equation (2.30) where the mobility should be replaced by an 'effective' value. The 'effective mobility' depends upon the gradient of the electric field and may be infinite or negative. This indicates the artificiality of the equation (2.30).

The problem becomes even more complicated at high electric fields. It was customary to replace the mobility and even the diffusion constant by a field-dependent value, such as

$$\mathbf{j} = e p\, \mu\,(E)\, \mathbf{E} - \mathrm{e}\, D\,(E)\, \nabla \cdot p\,. \tag{2.34}$$

Sometimes the diffusion term is written as shown below

$$\mathbf{j} = ep\,\mu(E)\,\mathbf{E} - e\nabla[D(E)\cdot p]. \tag{2.35}$$

Here $\mu = \mu(E)$ and $D = D(E)$ are material properties and should be calculated or measured. However, neither equation (2.34) nor (2.35) is correct because the mobility and the diffusion constant are not determined uniquely by the intensity of the local electric field [34]. Blötekjaer [35] demonstrated that at low frequencies and small field non-uniformities there exist a local and instantaneous relation between the lattice temperature and drift velocity. Because the carrier mobility (and diffusion constant) depend upon the carrier temperature, both these quantities should be considered to be velocity-dependent rather than field-dependent*).

Although incorrect, equations (2.34) and (2.35) were used by a number of workers for studying the Gunn (or transferred electron) devices. A correct description of the transport process in these devices should take into account the existence of (at least) two types of mobile charge carriers. For example, electron conduction in GaAs involves light electrons in the lower valley and heavy electrons in the upper valley. Here, the situation is even more complex and neither $\mu = \mu(E)$ nor $\mu = \mu(v)$ are acceptable. Blötekjaer derived [36] a set of transport equations for electrons in two-valley semiconductors. However, these new equations are too complicated to be used for an analytical theory of device behaviour.

2.5 Basic Equations for Unipolar Injection in Semiconductors

We shall summarize below the basic equations which should be used for the analysis of unipolar currents in semiconductors. These equations are: the current continuity equation (see equations (2.11) and (2.35))

$$\nabla\mathbf{J} = 0,\ \mathbf{J} = ep\,\mu(E)\,\mathbf{E} - e\nabla[D(E)\,p] + \varepsilon(\partial E/\partial t), \tag{2.36}$$

the Poisson equation

$$\varepsilon\nabla\mathbf{E} = e(p - N) \tag{2.37}$$

(where p is the free-hole density and $(-N)$ the density of immobile space charge) and the potential equation (2.8)

$$\mathbf{E} = -\nabla.\Phi. \tag{2.38}$$

*) If diffusion is not negligible there will be no local and instantaneous relation between $\mathbf{v}$ and $\mathbf{E}$, because $\mathbf{j} = ep\mathbf{v}$ where $\mathbf{v} = \mu\mathbf{E} + \dfrac{D}{p}\triangle p$ [35].

This set of equations should be completed with the dependence $\mu = \mu(E)$ and $D = D(E)$, which is specific to the particular semiconductor (and type of charge carrier) used, at a given temperature and doping. The doping profile should be given also. If incomplete ionization or trapping occurs, the immobile space-charge density is time-dependent. The solution can be found by postulating a set of equations describing the trapping (or ionization) kinetics.

2.6 High-frequency Electronic Conductivity

In this section, we shall discuss the effect of carrier collisions upon the high-frequency conductivity. Consider a semiconductor sample of uniform doping, N_A. This sample is bounded by two parallel metallic electrodes separated by the distance L. These electrodes are ohmic injecting contacts. The semiconductor resistivity is sufficiently low to provide the electrical neutrality, so that the mobile hole density is uniform and constant

$$p = N_A. \tag{2.39}$$

An alternating voltage

$$V(t) = V_1 \sin \omega t \tag{2.40}$$

is applied between these electrodes, thus determining an electric field

$$E(t) = V(t)/L = (V_1/L) \sin \omega\, t. \tag{2.41}$$

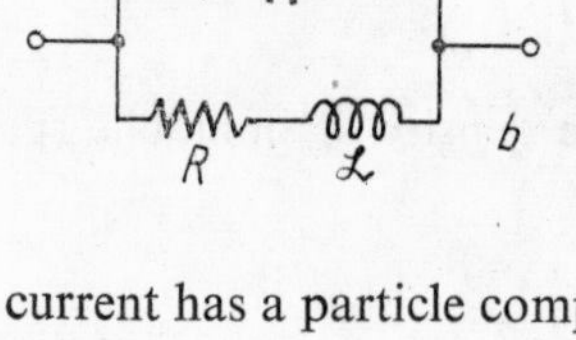

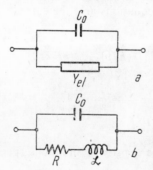

Figure 2.1. The high-frequency equivalent circuit of a semiconductor sample (C_0 is the geometrical capacitance, Y_{el} the electronic admittance and $\mathcal{L}$ an inductance indicating the effect of electron inertia).

The total current has a particle component and a displacement component and may be written

$$J(t) = e p v + \varepsilon \frac{\partial E}{\partial t} = e N_A v(t) + C_0 \frac{dV(t)}{dt}. \tag{2.42}$$

The equivalent a.c. circuit consists of a capacitance ($C_0 = \varepsilon/L$) shunted by an electronic admittance Y_{el} which 'conducts' the particle current (Figure 2.1a). Y_{el} can be evaluated by using the motion law, equation (2.22), rewritten here in one-dimensional form

$$\frac{\mathrm{d}v}{\mathrm{d}t} + \nu_c v = \frac{e}{m} E . \tag{2.43}$$

It can be found easily from equations (2.42) and (2.43) that the complex admittance Y_{el} is given by

$$Y_{el}(\omega) = \frac{e^2 N_A}{\nu_c m L} \frac{1}{1 + i\,\omega/\nu_c} . \tag{2.44}$$

The low frequency conductance (per unit of electrode area) is

$$G = Y_{el}(0) = \frac{e^2 N_A}{\nu_c m L} = \frac{e N_A \mu}{L} \tag{2.45}$$

(where μ is the mobility defined by equation (2.23)) and this is, of course, what we expect from a sample of resistivity $(eN_A\mu)^{-1}$ and length L. The reactive part of equation (2.44) corresponds to a *series* inductance $\mathcal{L}$ (see Figure 1.2b) having the value

$$\mathcal{L} = m\,L/e^2 N_A . \tag{2.46}$$

The parameter, $\mathcal{L}$, represents the inertial effect of carrier momentum between collisions ($\mathcal{L} \propto m$). This inductance can be neglected up to very high frequencies: the necessary condition is $\omega/\nu_c \ll 1$, where ν_c may be of the order of $10^{-11} - 10^{-12}\,\mathrm{s}^{-1}$.

The equivalent circuit to Figure 2.1b is characterized by two parameters: the resonance frequency ω_0 and the circuit Q, both depending only upon the bulk properties and not upon the length, L. We have

$$\omega_0 = \frac{1}{\sqrt{\mathcal{L} C_0}} = \left(\frac{e^2 N_A}{m\,\varepsilon,}\right)^{1/2} , \tag{2.47}$$

which is equal to the so-called *plasma-frequency**) [37]
The circuit Q is

$$Q_0 = \frac{\omega_0 \mathcal{L}}{R} = \frac{G}{\omega_0 C_0} = \omega_0 \frac{\mu m}{e} = \frac{\omega_0}{\nu_c} , \tag{2.48}$$

---

*) The plasma is formed by the mobile holes and the fixed ionized acceptors, which neutralize reciprocally.

which is exactly the ratio between the collision frequency and the plasma frequency. The plasma frequency may be also written as

$$\omega_0 = \left(\frac{\varepsilon\, m}{e^2\, N_A}\right)^{-1/2} = \left(\nu_c \frac{eN_A\mu}{\varepsilon}\right)^{1/2} = \left(\frac{\nu_c}{\tau_r}\right)^{1/2}, \tag{2.49}$$

where

$$\tau_r = \frac{\varepsilon}{eN_A\, \mu} \tag{2.50}$$

is the dielectric relaxation time. The circuit Q becomes

$$Q_0 = \frac{\omega_0}{\nu_c} = \left(\frac{\tau_c}{\tau_r}\right)^{1/2},\ \tau_c = \frac{1}{\nu_c}\,. \tag{2.51}$$

The sample impedance exhibits a sharp resonance (Q_0 large) only if the dielectric relaxation time is much shorter than the average time between collisions τ_c. Note the fact that if the dielectric relaxation time τ_r is very long (the resistivity is very high) the sample neutrality cannot be maintained. The effect of the bulk space charge will be discussed elsewhere.

2.7 Dynamic Bulk Negative Differential Conductivity in Semiconductors [38]

We shall now consider a 'two-valley' semiconductor*). By using subscripts 1 and 2 for the lower and upper valley, respectively, we may write the following conservation equation for the electron concentrations

$$\frac{dn_1}{dt} = -\frac{dn_2}{dt} = -\ [\nu_{12}(v_1)\, n_1 - \nu_{21}(v_2)\, n_2]. \tag{2.52}$$

Here ν_{12} and ν_{21} are the intervalley transfer rates (assumed velocity-dependent). Equation (2.43) should be rewritten for each valley i.e.

$$\frac{dv_1}{dt} + \nu_1\, v_1 = \frac{e}{m_1}\, E,\ \frac{dv_2}{dt} + \nu_2\, v_2 = \frac{e}{m_2}\, E. \tag{2.53}$$

We shall calculate below the a.c. conductivity by assuming that the sample is d.c. biased and the a.c. signal is very small as compared to the d.c. component. The a.c. part

*) An example of such a band structure was given in Figure 1.6b.

can be considered as a simple perturbation of the d.c. component and will be indicated by the letter δ

$$n_1(t) = n_1 + \delta n_1 \exp(i\omega t), \text{ etc.} \tag{2.54}$$

These expressions will be introduced in equations. The small quantities will be separated from the d.c. values*).

$$i\omega\,\delta n_1 = -i\omega\,\delta n_2 = -\left[\nu_{12}\,\delta n_1 - \nu_{21}\,\delta n_2 + n_1 \frac{d\nu_{12}}{dv_1}\,\delta v_1 - n_2 \frac{d\nu_{21}}{dv_2}\,\delta v_2\right] \tag{2.55}$$

$$(i\omega + \nu_1)\,\delta v_1 = \frac{e}{m_1}\,\delta E; \quad (i\omega + \nu_2)\,\delta v_2 = \frac{e}{m_2}\,\delta E. \tag{2.56}$$

The particle current should be

$$j = e(n_1 v_1 + n_2 v_2), \tag{2.57}$$

and its a.c. part is

$$\delta j = e\,[v_1\,\delta n_1 + v_2\,\delta n_2 + n_1\,\delta v_1 + n_2\,\delta v_2] = e\,[(v_1 - v_2)\,\delta n_1 + n_1 \delta v_1 + n_2 \delta v_2]. \tag{2.58}$$

Although the sample is neutral, the a.c. particle current also has a component due to density modulation. This is possible due to intervalley transfer of electrons (the same component was zero in the previous Section). We shall assume for simplicity $d\nu_{21}/dv_2 = 0$, $\nu_1 = \nu_2 = \nu_{12} + \nu_{21} = 1/\tau$.

The complex conductivity is [38]

$$\sigma(\omega) = \frac{\delta j}{\delta E} = \frac{1}{1 + i\omega\tau}\left[\sigma_0 + \frac{A\tau}{1 + i\omega\tau}\right] \tag{2.59}$$

where†)

$$\sigma_0 = e(n_1 \mu_1 + n_2 \mu_2), \; \mu_1 = \frac{e}{m_1 \nu_1}, \; \mu_2 = \frac{e}{m_2 \nu_2} \tag{2.60}$$

$$A = (v_2 - v_1)\,(e n_1 \mu_1)\, d\nu_{12}/dv_1. \tag{2.61}$$

The real part of the high-frequency conductivity is

$$\sigma_r(\omega) = \frac{1}{1 + \omega^2\tau^2}\left[\sigma_0 + A\tau\,\frac{1 - \omega^2\tau^2}{1 + \omega^2\tau^2}\right]. \tag{2.62}$$

---

*) This technique is explained in detail in Chapter 4.

†) Note the fact that the low-frequency conductivity is not equal to σ_0, except the case when both ν_{21} and ν_{12} are velocity independent.

We shall examine the possibility of a negative $\sigma_r(\omega)$. Here τ and σ_0 are positive, but A can be either positive or negative. By assuming that the probability of intervalley transfer from the lower valley to the upper valley, increases as the velocity v_1 increases (the carrier energy increases), we have $d\nu_{12}/dv_1 > 0$ in equation (2.61) and A has the sign of $v_2 - v_1$. In the case of GaAs and other semiconductors (Section 1.4) the electron mobility is lower in the upper valley and A is negative. The low-frequency ($\omega\tau \ll 1$) conductivity is $\sigma_0 + A\tau$ and is negative (Gunn effect) if

$$|A| > \sigma_0/\tau, \quad A < 0 \tag{2.63}$$

(or the rate of increase $d\nu_{12}/dv_1$ is sufficiently high). However, the a.c. conductivity of the Gunn-type semiconductor materials becomes *positive* at extremely high-frequencies. If $\omega\tau \gg 1$ in equation (2.62), we shall have

$$\sigma_1(\omega)\,|_{(\omega\gg 1/\tau)} \simeq \frac{\sigma_0 + |A|\,\tau}{\omega^2\,\tau}\,. \tag{2.64}$$

McGroddy and Guéret [38] investigated the possibility of the inverse situation: i.e. positive low-frequency conductivity and negative high-frequency conductivity. One of possibilities can be explained by using the previous formulae. If the upper valley has a higher electron mobility than the lower one (which is the opposite case of the Gunn effect), then A is positive. The condition

$$\tau A > \sigma_0, \quad A > 0 \tag{2.64'}$$

implies the fact the v.h.f. conductivity obtained from equation (2.62) for $\omega \to \infty$, should be negative. $\sigma_r(\omega)$ is positive at low frequencies and becomes negative as ω satisfies equation (2.65)

$$\omega\tau > \left(\frac{A\tau + \sigma_0}{A\tau - \sigma_0}\right)^{1/2}. \tag{2.65}$$

The above discussion is a useful introduction to negative-resistance effects studied later in this book. We examined the 'local' or 'bulk' conductivity and ignored the transport of transit-time processes. It was shown that the inertia due to carrier momentum, by itself, cannot provide a sufficient delay to determine a negative conductivity effect. In a multi-valley semiconductor, however, such a negative conductivity is possible. The frequency dependence of the a.c. conductivity depends upon the details of the band structure. The transfer of electron populations from one valley to another produces a delayed response at very high frequencies, i.e. frequencies which are comparable to the intervalley transfer rates. In certain conditions, the bulk conductivity can be negative at extremely high frequencies, with possible applications for generation and amplification of signals [38].

Problems

2.1 Show that the resonance frequency, equation (2.49), may be written

$$\omega_0 = \frac{1}{\sqrt{3}} \frac{v_T}{L_{DE}}, \tag{P.2.1}$$

where v_T is the carrier thermal velocity and L_{DE} the extrinsic Debye length (equation (1.4)).

2.2 Demonstrate that the quality factor, equation (2.48), may be expressed as

$$Q_0 = \sqrt{2\pi} \frac{\lambda_x}{L_{DE}} \tag{P.2.2}$$

where λ_x is the mean free path along the x axis.

3.

Physics of Unipolar Injection in Solids

3.1 Introduction

We shall start the analysis of transit-time effects in unipolar semiconductor devices by an investigation of the steady-state régime. This is, of course, necessary in order to clarify the physics of the device. On the other hand, the steady-state or direct-current (d.c.) behaviour forms the basis of the linear or small-signal operation: the latter is merely a 'perturbation' of the former.

There is still another reason to study the steady-state in detail. The transit-time effects do occur even in the stationary behaviour, but their importance is, very often, less evident. We shall adopt and develop here an analysis which introduces *the carrier transit-time as a key parameter*. The results would be then directly applicable to the small-signal calculations.

The steady-state analysis should emphasize also the importance of space-charge effects. The unipolar electric current is carried by charge carriers drifting in an electric field which is influenced by the space-charge of these carriers. A pronounced deviation from neutrality is peculiar to the unipolar devices studied here.

In this chapter we are not concerned with any special kind of device; we discuss some general properties instead. We do restrict, however, our discussion to a 'diode', i.e. a solid-state crystalline space bounded by two electrodes, both of them equipotential. The one called 'cathode' injects charge carriers of a certain sign, say holes, and these move under the influence of the external electric field until they reach the 'anode', which is properly biased with respect to the cathode (Figure 3.1a).

The first problem to be discussed here is that of the 'local' injection from the cathode surface and this discussion is correlated with the necessity for proper boundary conditions for the basic equations of semiconductor electronics.

Then, a simple analysis will show that the electric field should change monotonously with the distance from the cathode, at least in a one-dimensional device (plane-parallel electrodes of infinite extent, see Figure 3.1b). This simple 'geometrical' property will be used to develop a mathematical analysis for one-dimensional homogeneous semiconductor regions, where diffusion will be neglected. The possibility of a negative incremental conductance will be considered also.

A more detailed mathematical formulation follows. The normalized voltage and current are expressed in terms of the injection level and of the normalized transit-time. A space-charge factor is also defined and computed. Thus, a more clear evidence is obtained for both the transport of the injected mobile charge and the non-uniformity of the electric field associated with this space charge.

The remainder of this chapter is devoted to various phenomena which can occur in practical devices, i.e. trapping in the bulk, incomplete ionization of electronic impurities, non-uniform doping, large diffusion currents, etc.

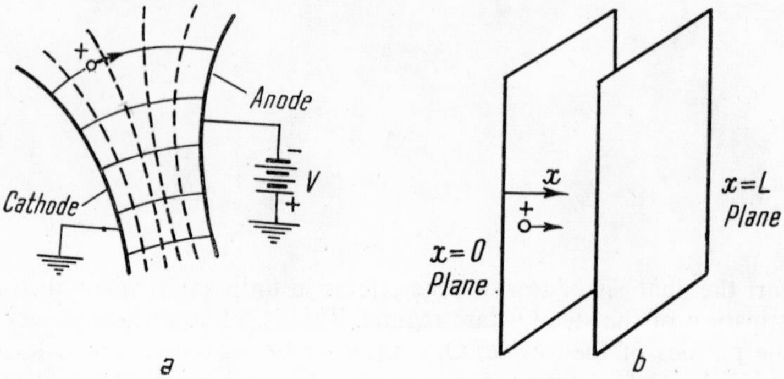

Figure 3.1. (a) Hole injection by a cathode of arbitrary shape. (b) One-dimensional space bounded by two plane-parallel electrodes of infinite extent: the cathode, at $x = 0$ and the anode $x = L$. All electrical quantities are function of x only (the direction x is perpendicular to the above planes).

3.2 Basic Equations

The basic equations of the steady-state régime of the device from Figure 3.1a are (see Chapter 2):

$$\mathbf{J} = ep\mathbf{v} - e\nabla(Dp) \tag{3.1}$$

$$\nabla \cdot \mathbf{J} = 0 \tag{3.2}$$

$$\varepsilon\nabla \cdot \mathbf{E} = e(p - N) \tag{3.3}$$

$$\mathbf{v} = \mathbf{v}(\mathbf{E}), \qquad D = D(E) \tag{3.4}$$

$$-\nabla\Phi = \mathbf{E} \tag{3.5}$$

where $(-N)$ denotes the density of fixed (immobile) space charge. Clearly, a differential equation in $\mathbf{E}$ should be solved. This equation will be obtained by replacing in equation (3.2) the current density $\mathbf{J}$ found from equations (3.1), (3.3) and (3.4), namely

$$\mathbf{J} = [\varepsilon\nabla\mathbf{E} + \mathrm{e}\mathbf{N}]\,\mathbf{v}(\mathbf{E}) - e\nabla(D\nabla\mathbf{E}) + e\nabla(DN) \tag{3.6}$$

For one-dimensional geometry, this equation reduces to

$$\left[\varepsilon \frac{\mathrm{d}E}{\mathrm{d}x} + eN(x)\right] v(E) - \varepsilon \frac{\mathrm{d}}{\mathrm{d}x}\left[D(E)\frac{\mathrm{d}E}{\mathrm{d}x}\right] + e\frac{\mathrm{d}}{\mathrm{d}x}[D(E)\,N(x)] = J \tag{3.7}$$

where J is constant and may be considered as given. Once $E = E(x)$ found, the voltage drop on the conduction space from $x = 0$ (cathode) to $x = L$ (anode) will be given by (see equation (3.5))

$$V = \int_0^L E(x)\,\mathrm{d}x \tag{3.8}$$

and thus we shall obtain $V = V(J)$

Equation (3.7) is second-order and two independent boundary conditions are necessary to select the particular solution suited to the device studied. *Only one* boundary condition is required if diffusion may be neglected, because equation (3.7) becomes first order ($D = 0$).

Equation (3.7) is non-linear and an analytical solution cannot be found, except for a few special cases. Thus, calculating the $J - V$ characteristic of unipolar injection is a very difficult task, even for devices which may be considered one-dimensional.

3.3 Mechanisms of Carrier Injection

The writing of boundary conditions is closely related to the physics of carrier injection into the semiconductor at the cathode and their extraction at the anode contact.

We discuss two general and opposite situations: the 'normal' contacts and the 'saturated' contacts.

A contact behaves in a normal way if the current density passing through it is much smaller than the saturation current (or the Richardson current) J_R. Assume for the sake of definiteness, a metal-semiconductor contact and a plane interface at $x = 0$. Let $p(0)$ be the random free-hole density at $x = 0$, at thermal equilibrium. For a 'perfect' metal-semiconductor contact (no interface trapped charges) the carrier density at the interface is given by

$$p(0) = N_v \exp(-e\Phi_p/kT), \tag{3.9}$$

where N_v is the effective density of states in the valence band, and (Figure 3.2)

$$\Phi_p = -e\,\Phi_M + \chi_s + W_c - W_v \tag{3.10}$$

is the potential barrier seen by holes looking from metal to semiconductor.

If a net current $J \neq 0$ flows, the contact is not at thermal equilibrium, but the carrier density at $x = 0$ can still be given by its thermal equilibrium value, provided that J is much smaller than the saturation current J_R defined as follows.

At thermal equilibrium, the random thermal current flowing from the metal into the semiconductor is equal, by the principle of detailed balance, to the current

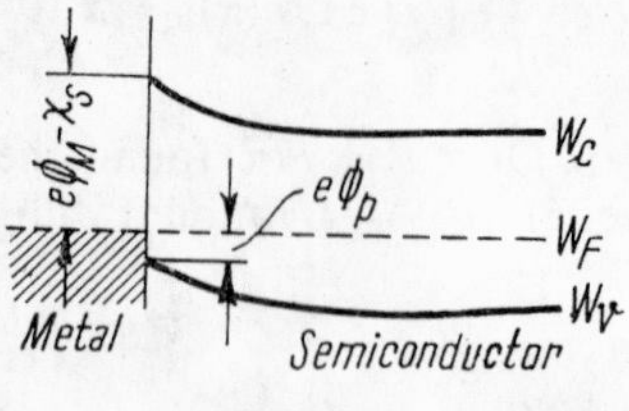

Figure 3.2. Thermionic emission of holes at a metal-semiconductor contact.

flowing from the semiconductor into the metal. The common value is the saturation (or temperature-limited) hole current which can be drawn from the cathode (metal)

$$J_R = RT^2 \exp(-e\Phi_p/kT), \tag{3.11}$$

where R is the Richardson constant. The theoretical value of R for electron emission from metal into vacuum is

$$R \to R_0 = 4\pi emk^2/h^3, \tag{3.12}$$

(where m is the electron mass, h is Planck constant and k is Boltzmann constant) and its numerical value is 120 A $cm^{-2}K^{-2}$. More detailed calculations for the metal-semiconductor interface should take into account the anisotropy of the effective

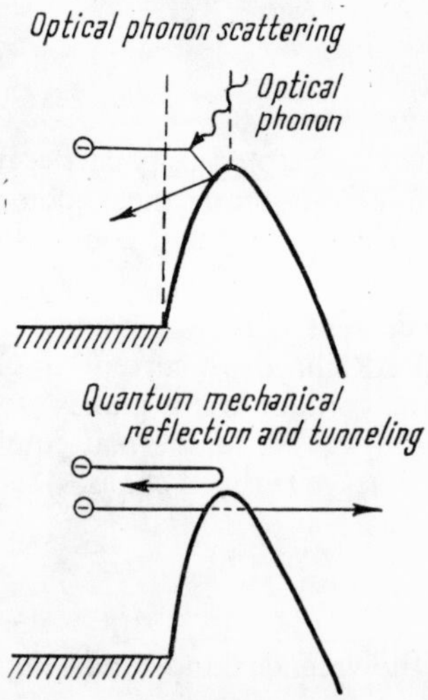

Figure 3.3. Schematic representation of the kinetic processes, existing at a metal-semiconductor interface, which modify the Richardson constant [39]. The barrier thickness is of the order of the inter-atomic distance.

mass of the carriers in the semiconductor, scattering by optical phonons into the high-field region near the metal-semiconductor interface, quantum-mechanical reflection and tunnelling at the potential barrier (Figure 3.3 [39]). When all these effects are incorporated into the model, one obtains an effective Richardson constant which depends upon the semiconductor, its doping and the crystallographic orientation of the interface; and upon the intensity of the electric field at the interface.

Andrews and Lepselter [39] indicated for R values of *ca.* 112 A cm^{-2} K^{-2} for electron injection into silicon (<111> orientation) and 32 A cm^{-2} K^{-2} for hole injection into silicon (the electric field at the contact is of the order of 10^4—10^5 V/cm and N_D, $N_A = 10^{16}$ cm^{-3}, $T = 300$ K).

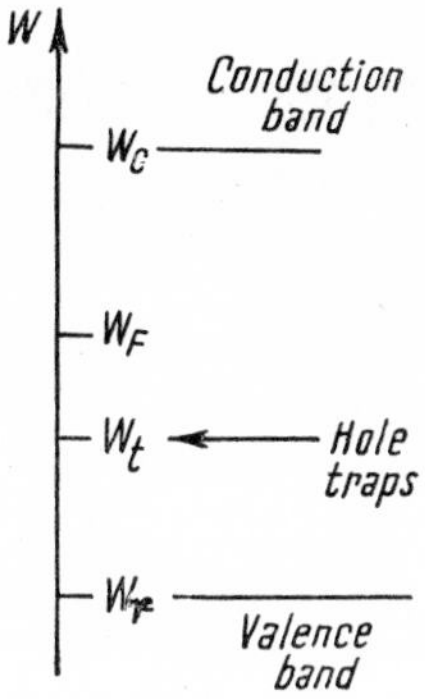

Figure 3.4. Energy levels at a metal-semiconductor interface with hole traps. W_v W_c, W_t, W_F are the top of the valence band, the bottom of the conduction band, the trap level and the quasi-Fermi level, respectively. The quasi-Fermi level W_F shall not approach W_v within less than at least several kT.

Assume now that *surface states* do exist at the metal-semiconductor interface and let Σ be their surface density and σ_t the surface density of trapped carriers. We may write [40]

$$\sigma_t = \Sigma\,[1 + \exp\,(W_F - W_t)/kT]^{-1} \tag{3.13}$$

where W_t is the energy level of hole traps (Figure 3.4). The hole concentration at the interface $p\,(0)$ is

$$p\,(0) = N_v \exp\left(\frac{W_v - W_F}{kT}\right) \tag{3.14}$$

and because of equation (3.13) we have

$$p\,(0) = \frac{N_v}{\beta}\left(\frac{\Sigma}{\sigma_t} - 1\right)^{-1}, \quad \beta = \exp\left(\frac{W_t - W_v}{kT}\right) \tag{3.15}$$

The interface concentration $p(0)$ may be calculated if the current density J is given (negligible diffusion)

$$J = ep\,(0)\,v\,(E_c)\,. \tag{3.16}$$

Here $E_c = E\,(0)$ is the cathode field. By Gauss' theorem [40] we have

$$E_c = e\sigma_t/\varepsilon \tag{3.17}$$

because the electric field at interface on the metal side may be assumed zero (negligible screening effect [41] — [43]). Now, $p\,(0)$ is given by

$$p\,(0) = \frac{N_v}{\beta\left(\dfrac{e\Sigma}{\varepsilon E_c} - 1\right)} \tag{3.18}$$

where E_c follows from equation (3.16).

For a 'high-low' junction (p^+p or n^+n) (Figure 3.5) or a heterojunction used as an emitter of charge carriers, a suitable boundary condition is of the type

$$p(0) = N_A^+ \tag{3.19}$$

where N_A^+ is the ionized-acceptor concentration on the heavily doped side (emitter).

The injection from a *pn* junction or a heterojunction will be examined in due course.

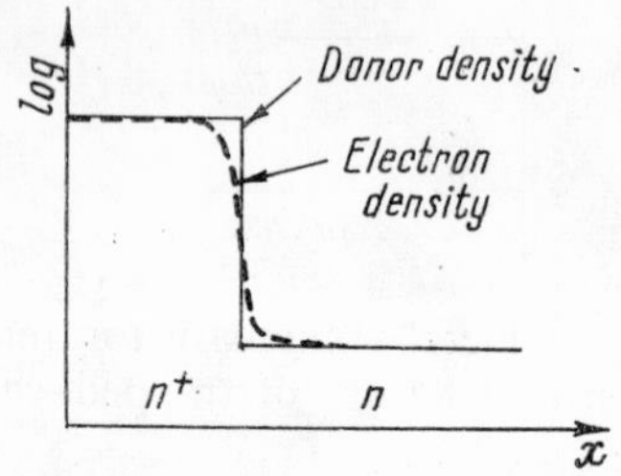

Figure. 3.5. Doping profile and electron density at a homojunction.

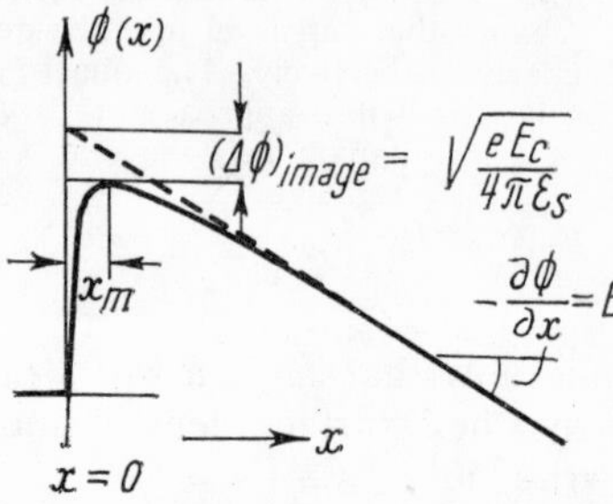

Figure 3.6. The image-force lowering of the barrier (Schottky effect) at a metal-semiconductor interface. The distance x_m is assumed to be very small as compared to the thickness of the semiconductor sample (see Problem 3.1).

A 'saturated' injection contact is characterized by the fact that the electric current passing through it is equal to the saturation current, J_{sat}. For insulators or slightly-doped semiconductors, at usual temperatures, this saturation current is purely a thermionic emission current and is given by an equation of the type (3.11) ($J_{sat} = J_R$). This formula predicts a perfect current saturation, at least as far as the effective Richardson constant does not depend upon the cathode field.

In practice, however, J_{sat} is not constant, because the potential barrier, Φ_p, is field-dependent.

First, the field-dependence of the barrier might arise from the image-force (or Schottky) effect. It is well-known the fact that the height of the barrier decreases by an amount [39], [44]

$$(\Delta\Phi)_{image} = \sqrt{\frac{eE_c}{4\pi\varepsilon_s}} \tag{3.20}$$

where E_c is the cathode field (see Figure 3.6) and ε_s is the dynamic (optical) dielectric constant of the semiconductor, which, for silicon, is close to the static value.

Therefore, the Schottky-emission current, J_{sat} may be written as

$$J_{sat} = J_R \exp \sqrt{E_c/E_R} \tag{3.21}$$

where E_R is a certain characteristic field

$$E_R = \frac{4\pi\, \varepsilon_s}{e} \left(\frac{kT}{e}\right)^2 \tag{3.22}$$

(its value is *ca.* 53 kV cm^{-1} for silicon, at room temperature).

It was shown [39], [41], [44] that the image-force lowering of the barrier is not always sufficient to explain the experimental data. A supplementary decrease of the barrier, by an amount which is proportional to the electric field E_c, is taken into account very often

$$(\Delta\Phi)_{suppl} = \alpha\, E_c, \qquad \alpha \simeq \text{const.} \tag{3.23}$$

Such a correction may arise from various physical causes. Sometimes one postulates the existence of a thin layer of oxide or other contaminant, intervening between a non-reacting metallic contact and the semiconductor. The oxide thickness t_{ox} is thin enough to allow carrier tunnelling*) but the static electric field sustained across such a layer contributes to the electric-field dependence of the effective barrier height ($\alpha = t_{ox}$ in equation (3.23)). If surface-states do exist, the field-dependence should be modified accordingly. In the case of a very large density of surface traps, the Fermi level at the surface is 'pinned' to a some value in the forbidden gap, the oxide field is constant and the corresponding correction $(\Delta\Phi)_{suppl}$ is no longer field dependent [39]

A similar barrier-lowering, equation (3.23), may occur due to the penetration of the electric field into the metal. For example, Mead *et al.* [41] reported a Thomas-Fermi screening distance in the metal of the order of 1 Å (for an Al-SiO_2-Si structure with t_{ox} = 550 Å)†).

For intimate metal-semiconductor contacts, such as for metallic silicide contacts††), the existence of an additional barrier lowering equation (3.23), cannot be attributed to an interface oxide layer. $(\Delta\Phi)_{suppl}$ may be explained as the effect of a dipole layer formed at the interface. It was shown that, at an intimate contact, 'the metal wave functions penetrate into the forbidden gap of the semiconductor' [44]. The resulting electronic states in the semiconductor are occupied by charge carriers and a space-charge of opposite sign occurs on the metal side of the contact, as shown

---

*) The magnitude of the injected current is determined by thermionic emission over the potential barrier which develops into the semiconductor in the immediate vicinity of the injecting contact.

†) The emitted current was light-stimulated.

††) For example, a Pt contact on Si was made, followed by a heat treatment which forms a Pd_2Si surface layer and almost removes the original oxide film [44].

in Figure 3.7 [44]. The same figure indicates the presence of a uniform and uncompensated fixed space-charge due to ionized donors (reverse-biased Schottky barrier contact). The potential profile is also shown (note the fact that in the absence of the dipolar layer, the same barrier would be parabolic). To a first approximation, the barrier lowering is linear in the cathode field. We stress the fact that in discussing the barrier-lowering-effects, the so-called cathode field is not the field at the interface but the (almost uniform) electric field on the semiconductor bulk side of the potential barrier.

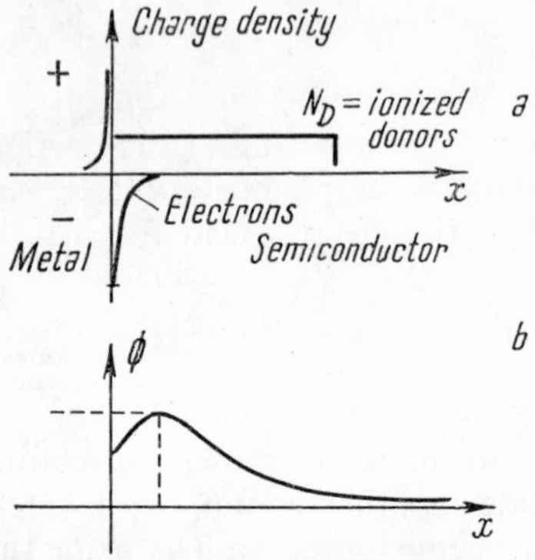

Figure 3.7. a) Schematic representation of the positive and negative charge distributions at the semiconductor surface. The uniform positive charge is caused by ionized donor atoms and the exponentially varying negative charge by electrons trapped in interface states. b) The potential profile resulting from the above charge distributions.

Generally speaking, the carrier emission at a metal-semiconductor contact can be both thermionic and quantic (field stimulated) in nature [45] — [47]. If the semiconductor doping is high and/or the temperature is low, the field emission can dominate the thermionic emission.

Field-emission into relatively lowly-doped semiconductors may be achieved by providing a highly-doped layer between the metal contact and the semiconductor bulk. This layer should be very thin in order to avoid carrier — multiplication effects. The field intensity inside such a layer may be sufficiently high to determine field-emission [48], [49]. A formula relating the injected (saturated) current and the cathode field E_c is [50]

$$J_F = a_F E_c^2 \frac{\theta_F}{\sin \theta_F} \exp\left(-\frac{E_F}{E_c} \right), \qquad \theta_F < \pi, \tag{3.24}$$

$$a_F = \frac{e^2}{8\,\pi h \Phi_p}, \quad \theta_F = \pi \frac{kT}{\Phi_p} \frac{E_F}{E_c}, \quad E_F = \frac{4}{3}\left(\frac{2m\Phi_p^{3/2}}{eh/2\pi} \right)^{1/2}.$$

To conclude, the boundary conditions for unipolar injection may be introduced by specifying either the carrier density at the contacts (for 'normal' contacts) or the injected (saturation) current (for 'saturated' injecting contacts).

A single boundary condition *at the injecting plane* is sufficient for problems where diffusion is neglected. The effect of diffusion and the corresponding boundary conditions will be discussed later in this chapter.

3.4 Accumulation or Depletion in the Bulk

We shall now demonstrate that, in a plane-parallel one-dimensional diode region, the electric field should change monotonously with x from $x = 0$ to $x = L$. This is a *bulk* property, independent of the boundary conditions at $x = 0$ and $x = L$. In a certain sense this is also a *geometrical* property since it may be not satisfied for geometries other than one-dimensional.

Assume, first, that the semiconductor is homogeneous, diffusion current is negligible and no trapping occurs. The Poisson equation (3.3) may be written as

$$\mathrm{d}E/\mathrm{d}x = (e/\varepsilon)(p - N), \quad N = \text{const.} \tag{3.25}$$

Clearly, $\mathrm{d}E/\mathrm{d}x$ is always positive for $N < 0$ (e.g. minority carrier injection [13] in a punched-through p^+np^+ structure, or injection in an insulator with deep traps). The question if $E = E(x)$ is a monotonous function or not, remains only for the case of majority carrier injection. Let us assume that $E = E(x)$ does exhibit a maximum (Figure 3.8a) at the $x = x_0$ plane. Thus

$$\left.\frac{\mathrm{d}E}{\mathrm{d}x}\right|_{x = x_0} = 0, \qquad 0 < x_0 < L. \tag{3.26}$$

Then, from equation (3.25) we have

$$p(x_0) = N > 0 \tag{3.27}$$

and $p(x)$ should cross the line $p = N$ as shown in Figure 3.8b. The semiconductor has to be depleted for $x > x_0$ $(\mathrm{d}E/\mathrm{d}x < 0)$ and accumulated for $x < x_0$ $(\mathrm{d}E/\mathrm{d}x > 0)$. In order to provide the current continuity $J = epv = \text{const}$, the carrier velocity should vary as shown in Figure 3.8c. But this is definitely not possible because Figure 3.8 (a and c) implies two distinct carrier velocities for the same value of the electric field. Thus, the starting hypothesis should be rejected. The electric field cannot have a maximum for $0 < x < L$. Of course, a similar demonstration proves that the electric field cannot have a minimum in the conduction space. Thus

$$\mathrm{d}E/\mathrm{d}x \neq 0 \text{ for } 0 < x < L \tag{3.28}$$

and the electric field should be a positive*) and monotonous function of x.

---

*) $E(x)$ cannot be negative in the absence of diffusion because the hole flow in the positive direction would not be possible. Note also that discontinuity of the electric field cannot be accepted as long as the charge density remains finite for $0 < x < L$. Later, we shall allow p to be infinite for $x = 0$ (and thus $E(0) = 0$, $v(0) = 0$) for space charge limited (SCL) injection in the absence of diffusion.

As a corollary, note the fact that, for majority carrier injection, the entire semiconductor space between $x = 0$ and $x = L$ should be either completely accumulated ($p(x) > N > 0$) or completely depleted ($p(x) < N > 0$); no other possibility exists (except the special case of complete neutrality).

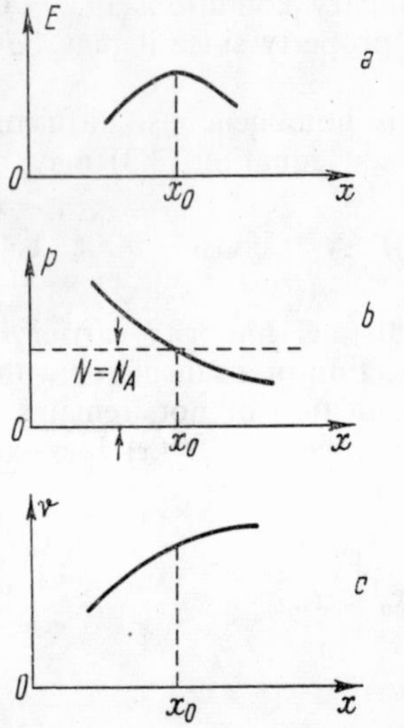

Figure 3.8. Figure used for a demonstration in the text.

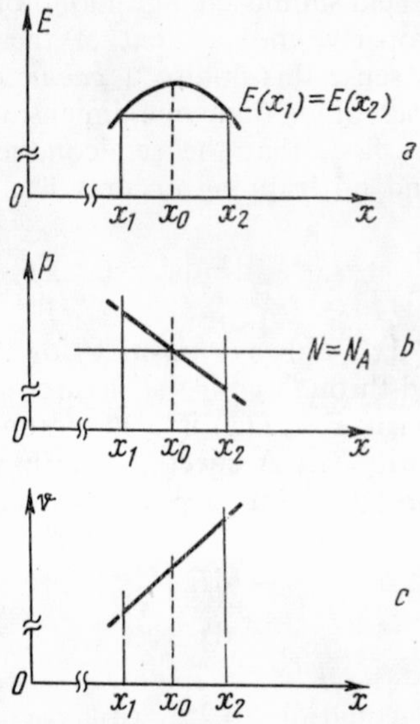

Figure 3.9. Figure used for a demonstration in the text.

Assume now that diffusion current is also taken into account and may be written

$$j_{\mathrm{dif}} = -eD\frac{\mathrm{d}p}{\mathrm{d}x} = J - j. \tag{3.29}$$

With reference to Figure 3.9 we consider a hypothetical neutral plane at $x = x_0$. We further assume only very small deviations from $x = x_0$, so that $E = E(x)$ may be approximated around $x = x_0$ by a parabola and $p = p(x)$ by a straight line (these approximations are consistent with Poisson equation (3.25). Let x_1 and x_2 be two values of x on both sides of x_0 such as $E(x_1) = E(x_2)$. Because the diffusion current (equation (3.29)), is almost constant and thus equal for $x = x_1$ and $x = x_2$ (D may be field-dependent) the particle current should be equal for $x = x_1$ and $x = x_2$, and thus $v(x_2) > v(x_1)$ because $p(x_2) < p(x_1)$. Then, because v must be a single-valued function of E, the possibility of existence of a neutral plane $x = x_0$ should be excluded. However, the above proof breaks down if the carrier velocity is saturated (see also [51]).

3.5 Analysis of Unipolar Injection in a Homogeneous Semiconductor by Neglecting Diffusion

Assume a homogeneous solid-state conduction-space bounded by two parallel planes of infinite extent: the injecting plane (cathode) at $x = 0$ and the collecting plane (anode) at $x = L$. The holes injected at $x = 0$ drift along the x-axis towards the anode (negatively biased with respect to the cathode). All (macroscopic) electric quantities (electric field and electric potential, hole density and velocity) should change only in the x direction. In this sense, our model is one-dimensional.

By neglecting the diffusion current in equation (3.1) we have

$$J = epv(E) \tag{3.30}$$

where the 'macroscopic' velocity v may be considered a uniform single-valued, function of E. The one-dimensional form of Poisson equation is (3.25), repeated here for convenience

$$\frac{dE}{dx} = \frac{e(p - N)}{\varepsilon}. \tag{3.31}$$

Here, $(-N)$ denotes the density of the fixed space charge, assumed uniform and independent upon the injection level. If we inject holes in a p-type semiconductor, $N = N_A > 0$ denotes the effective concentration of ionized acceptors. For minority carrier injection (holes introduced in an n-type semiconductor) N is numerically equal to the effective density of donors and should be negative. More generally, N in equation (3.4) has to be replaced by

$$N = N_{\text{ionized acceptors}} - N_{\text{ionized donors}} - N_{\text{deep traps}} \tag{3.32}$$

being either positive or negative*). Of course, the complete ionization of donors and the complete filling of traps can be provided only above a certain temperature. If charge carriers are injected in an insulator with traps, the traps may become deep (completely filled) only if the injection level (or the injected current) is sufficiently high [3], [5].

The total voltage drop across the conduction space (absolute value) is given by equation (3.8)

$$V = \int_0^L E\,dx. \tag{3.33}$$

*) We consider a trap (capture center) which is neutral when unoccupied. Such a localized level can retain *one* charge carrier, thus becoming charged with $+e$.

By combining equations (3.30) and (3.31) a first-order, non-linear equation in $E = E(x)$ is obtained. By using the boundary condition at $x = 0$, $E = E(x)$ will be obtained, in principle, as a function of J, $E = E(x; J)$. Then, E will be inserted in equation (3.33) to give $V = V(J)$. However such a calculation cannot be always analytically performed.

An alternative is to change the integration variable in equation (3.33). By eliminating dx and p from equations (3.30), (3.31) and (3.33) we obtain

$$V = \int_{E(0)}^{E(L)} \frac{\varepsilon E v(E)\,\mathrm{d}E}{J - eNv(E)}\,. \tag{3.34}$$

Another independent integral equation will be found from $L = \int_0^L \mathrm{d}x$. Thus we have

$$L = \int_{E(0)}^{E(L)} \frac{\varepsilon v\,(E)\,\mathrm{d}E}{J - eNv\,(E)}\,. \tag{3.35}$$

Then, the boundary condition

$$E(0) = \text{given} \tag{3.36}$$

should be specified. Therefore, the $J - V$ characteristic is given parametrically (and inexplicitly) by equations (3.34) and (3.35) (the parameter is $E(L)$). An analytical $J - V$ dependence or even an explicit parametric dependence is possible only in a few special cases.

3.6 The Control Characteristic

Equations (3.34) and (3.35) derived in the preceding Section take into account the bulk space charge (they loose their sense for strict neutrality, $J = eNv$) whereas the boundary condition, equation (3.36), specifies the properties of the $x = 0$ interface (i.e. the mechanism of carrier injection).

Let us assume the fact that the particle current at the cathode j_c may be written as a function of the cathode field $E_c = E(0)$

$$j_c = j_c(E_c) = J \tag{3.37}$$

thus defining the so-called *control characteristic* $j_c = j_c\,(E)$ [52]. Therefore, the boundary condition equation (3.37) (which implicitly determines E_c) can be replaced for equation (3.36).

From equations (3.34), (3.35) and (3.37) we have

$$V = \int_{E_c}^{E_a} \frac{\varepsilon E v(E)\,\mathrm{d}E}{j_c(E_c) - eNv(E)} \tag{3.38}$$

$$L = \int_{E_c}^{E_a} \frac{\varepsilon v(E)\,\mathrm{d}E}{j_c(E_c) - eNv(E)} \tag{3.39}$$

where $E_a = E(L)$. Equation (3.38) is similar to Kroemer's equation (11b) in reference [52]. The concept of control characteristic was introduced by the same author [52] to describe 'imperfect cathode boundary conditions' in Gunn devices. Kroemer has also shown that this concept is general and in no way restricted to the specific problem studied by him. The utility of this concept will be demonstrated here once again.

We shall now consider the existence of a control region between the emitting plane $x = 0$ and the cathode proper which is a metal or a semiconductor region of very high conductivity. We assume in most cases that the control region is very thin and a negligible voltage drops on it. Thus the length L and the voltage drop V from equations (3.38) and (3.39) are precisely known. In fact, these assumptions are more or less justified in practice as can be seen from a few characteristic examples. The control region may consist of:

(a) The interface between the metal and the semiconductor. This interface is actually a very thin layer of the order of a few interatomic distances and may contain an uncompensated electric charge and support a voltage drop. Sometimes a very thin insulator (oxide) layer occurs at the interface.

With reference to Figure 3.6 showing the variation of the macroscopic potential at the metal-semiconductor barrier which is modified by the image force, we note the fact that the control region extends from the metal beyond the potential maximum, more precisely until the region where the electric field is nearly constant; this region is usually very short (see Section 3.3 and problem 3.1).

(b) The region between the virtual cathode (defined as the plane where the potential is maximum) and the actual cathode. This region is not always short (in comparison with L) and constant, nor is the voltage drop on it negligible or fixed.

(c) A semiconductor region of different conductivity. This may be a highly doped and very thin semiconductor region introduced on purpose to enhance the field emission from the metal, still retaining the bulk resistivity at a relatively high value [48]. If this cathode region is too thick and has a non-negligible voltage drop, the whole device should be considered as made from two regions. This cathode region may be also a high resistivity region due to the imperfect formation of the cathode contact, as pointed out by Kroemer [52].

We shall now give some examples of control characteristic. This curve for the SCLC diode is just the vertical axis of the $j_c - E$ plane, since J may (theoretically) have any finite value for $E_c = 0$.

Assume a normal injecting contact (Section 3.2) where the mobile charge density at $x = 0$ is fixed by a boundary condition like (3.9) or (3.19). Therefore

$$J = j_c = ep(0)v(0) = ep(0)v(E_c) \tag{3.40}$$

and the shape of the control characteristic is precisely the shape of the $v - E$ characteristic. If the height of the barrier is modified by the field, E_c, then $j_c = j_c(E)$ deviates from the shape of $v = v(E)$. Kroemer gives as an example shallow barrier contacts, such as a heterojunction [52].

An interesting example is the effect of interface states at the metal-semiconductor contact [40]. From equations (3.18) and (3.40) we have

$$j_c = \frac{(eN_v/\beta)\, v\,(E_c)}{(e\Sigma/\varepsilon E_c) - 1} = j_c\,(E_c). \tag{3.41}$$

It can be shown easily that the cathode field cannot exceed the critical value $e\Sigma/\varepsilon$, since the current j_c tends to infinity due to $p(0) \to \infty$ ($W_F - W_t \gg kT$, all surface states are filled and $p(0)$ increases because W_F increases). For $E \to E_c = e\Sigma/\varepsilon$ there

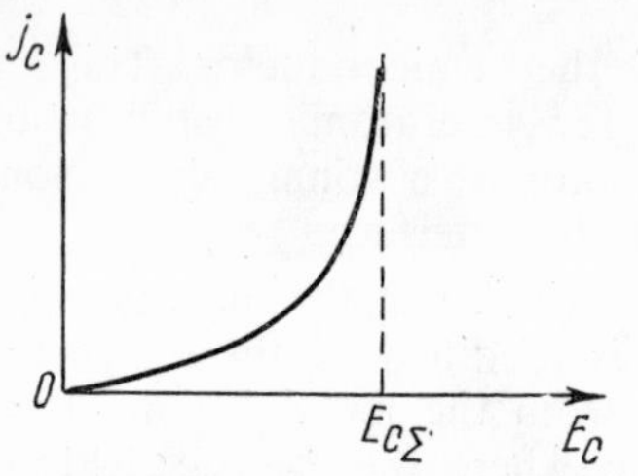

Figure 3.10. A possible form of the control characteristic, corresponding to a metal-semiconductor interface with traps.

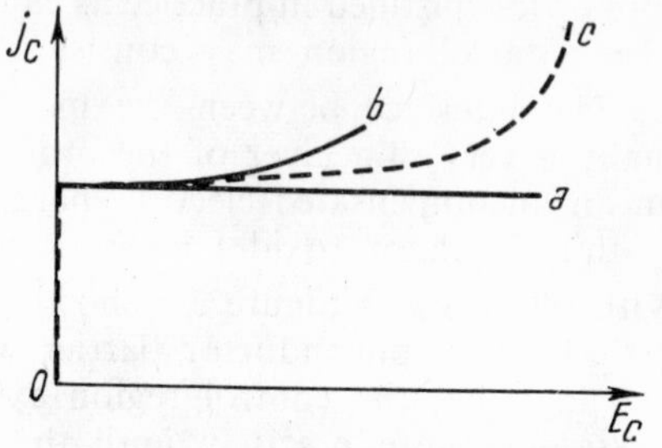

Figure 3.11. Examples of control characteristics for saturated contacts: (a) thermionic emission, (b) thermionic emission modified by the Schottky effect (image force lowering of the barrier); (c) field-emission.

is not, however, a true SCLC limitation, but E_c (finite) becomes very small as compared to V/L and may be neglected. We shall show later that the alternating-current boundary condition is approximately the same as for the SCLC injection, since the *variation* of the electric field E_c is negligible. A control characteristic for such a contact is shown in Figure 3.10, its detailed form depends upon the shape of the $v - E$ characteristic and the density of the interface traps.

Figure 3.11 shows examples of control characteristics for saturated contacts. If the barrier has a constant height and the emission is purely thermionic, the current j_c will be constant and equal to J_R (which is given by an equation of the form of equation (3.11)), and will be represented by an horizontal line in the $j_c - E$ plane. For $E = 0$, this characteristic should coincide with the vertical axis. The image-force lowering of the barrier gives an increase of the control characteristic (see equation (3.21)). We outline the fact that the semiconductor occurs here only through its dielectric constant.

The control characteristic for field-emission is given by equation (3.24). The exponential-like behaviour may lead to a sharp increase of the current at certain bias fields. A similar increase was noticed above for a 'normal' contact with interface traps. Kroemer [52] stated that 'similar control characteristics will lead to a similar electrical behaviour, despite any differences in physical origin', and that 'the only thing that matters is the shape of the control characteristic itself'.

3.7 The Current vs. Field Plane

We shall now introduce the concept of *neutral characteristic* [52] defined as

$$j_n = eNv(E) = j_n(E), \tag{3.42}$$

where N is positive for majority and negative for minority carrier injection. For the former case j_n is the actual 'neutral' current, i.e. the current flowing in a neutral semiconductor of doping N, biased at the static (uniform) field $V/L = E$.

The equations (3.37), (3.38) and (3.39) may be rewritten

$$V = \frac{\varepsilon}{eN} \int_{E_c}^{E_a} \frac{j_n(E)\, E\, \mathrm{d}E}{j_c(E_c) - j_n(E)} \tag{3.43}$$

$$\frac{eNL}{\varepsilon} = \int_{E_c}^{E_a} \frac{j_n(E)\, \mathrm{d}E}{j_c(E_c) - j_n(E)} \tag{3.44}$$

$$J = j_c(E_c). \tag{3.45}$$

Thus, the $J - V$ characteristic can be determined as follows. Assume that the bias current is given. The voltage drop may be calculated from the integral (3.43) where j_n depends upon the semiconductor material used (through the dependence $v = v(E)$) and is proportional to the doping N, whereas j_c depends upon the cathode interface. The anode and the cathode field must be known, however. The latter is implicitly given by equation (3.45), and the former should be evaluated from equation (3.44) and depends upon the length, L, of the device. All these computations may be performed graphically, numerically or analytically. A graphic representation in the $j - E$ plane is particularly useful and this is done for the following curves: the neutral (or bulk) characteristic $j_n = j_n(E)$, the control characteristic $j_c = j_c(E)$ and the $J - E$ steady-state characteristic of the device, where E is the bias field V/L. For the purpose of illustration, two cases are considered below. First, a silicon device with Schottky-barrier-emission cathode (Figure 3.12) and, secondly, a negative-mobility semiconductor with imperfect cathode [52] (Figure 3.13).

In the silicon device the carrier velocity does saturate at high electric fields. Let us denote by j_{sat} the 'saturated' neutral current

$$j_{\mathrm{sat}} = eNv_{\mathrm{sat}}. \tag{3.46}$$

If the emission current J_R is high (low barrier or high temperatures) and j_{sat} is small (low doping), then there is no intersection between j_n and j_c (Figure 3.12a). These curves do intersect each other in the opposite case (Figure 3.12b).

Figure 3.13 shows neutral characteristics j_n which are suitable for (a) GaAs and (b) Ge at low temperatures, respectively.

An interesting property of the crossover points curves j_n and j_c is demonstrated below.

Property 1. A crossover point of j_n and j_c is also a point of the steady-state characteristic $J-E$ (E is the bias field V/L) and corresponds to the neutrality of the semiconductor bulk.

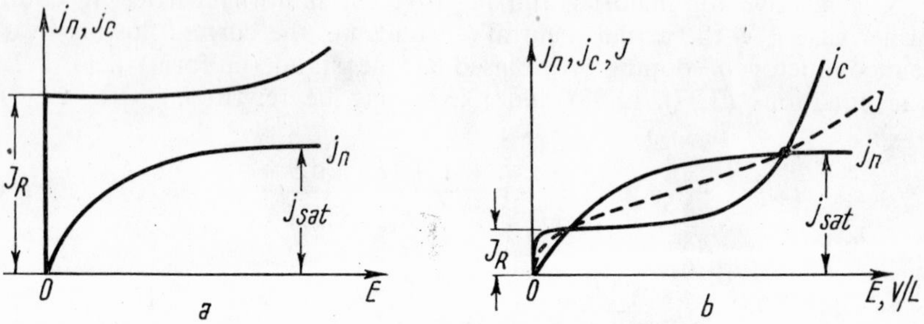

Figure 3.12. Control and neutral characteristics for a silicon diode with Schottky-barrier-emission cathode. (a) $J_R > j_{sat}$ (b) $J_R < j_{sat}$.

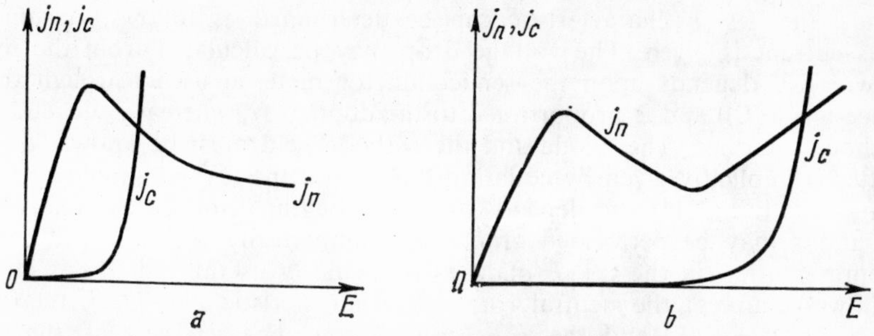

Figure 3.13. Control and neutral characteristics for a negative-mobility semiconductor with imperfect (non-ohmic) injecting contact. The neutral characteristic is typical for GaAs at room temperature (a) and Ge at low temperatures (b).

Let E_x be the abscissa of a crossover point, such as

$$j_c(E_x) = j_n(E_x). \quad (3.47$$

By use of equations (3.40) and (3.42), equation (3.47) becomes

$$J = eNv(E_x), \quad (3.48)$$

where J is the device current when the cathode field is equal to the crossover field. It is also evident that

$$J = ep(0)\,v(E_c) = ep(0)\,v(E_x), \tag{3.49}$$

such as, by comparison with equation (3.48), equation (3.50) follows

$$p(0) = N \tag{3.50}$$

and the injecting plane $x = 0$ is neutral. Moreover, the whole conduction space from $x = 0$ to $x = L$ should be neutral. This can be understood as follows: a deviation from neutrality near the cathode would require a variation of velocity from $v(0)$ and a corresponding variation of the electric field. But this electric field variation should be, according to Poisson equation (3.31), the result of a *cumulative* action of deviations from neutrality between the considered plane and the cathode. Thus the fact is plausible that these deviations from neutrality do not occur at all. This is proved mathematically in Section 3.9.

Because the entire semiconductor bulk is neutral, the electric field is uniform and equal to V/L, the bias field. Thus the crossover point is also a point of the $J - E$ steady-state characteristic*).

It was shown in Section 3.4 that the conduction space can be either an accumulation or a depletion region (simultaneous accumulation and depletion in the bulk is not possible). Whichever is the situation for a given device and bias current, it may be found by use of the $j - E$ characteristics as shown below.

Property 2. The semiconductor bulk is depleted of mobile carriers if $j_c(E_c) > j_n(E_c)$ and accumulated if $j_c(E_c) < j_n(E_c)$.

The proof of this theorem is very simple. The reader will find easily that $j_c(E_c) > j_n(E_c)$ determines $p(0) > N$ (see the demonstration of property 1) and since the emitting plane $x = 0$ is accumulated, the whole conduction space must be an accumulation region. The opposite situation may be verified similarly. The fact that the semiconductor is accumulated or depleted does not depend upon the sample length, L. If the semiconductor material (including doping) and the cathode interface properties are specified, then the current, J, is sufficient to determine if accumulation or depletion takes place†).

Property 3. The $j - E$ plane can be used in conjunction with equations (3.43) and (3.44), to determine the carrier and field distribution inside the device. This assertion is almost evident and it will be only exemplified.

Example (a). Highly doped silicon device with Schottky-barrier-emission cathode (Figure 3.14).

---

*) In Figure 3.12b there are three such crossover points (including the origin) and thus the steady-state $J - E$ characteristic may be free-hand interpolated as shown by the dashed line. Note the fact that j_n and j_c do not determine completely the $J - E$ characteristic. Its exact shape also depends upon the device length, L.

†) For minority carrier injection ($N < 0$ and $j_n < 0$) the semiconductor bulk is always an inversion layer and from Poisson equation (3.31) we have $E_a > E_c$.

Assume the device is operating with a current density, J, as shown in Figure 3.14a. According to the Property 2, the semiconductor bulk is depleted and thus $E_a < E_c$. The exact value of E_a can be calculated by use of equation (3.44). Clearly, the longer the device, the larger the difference $E_c - E_a$ ($N =$ given). However, the anode field E_a, cannot fall below the 'neutral' field E_N which is defined as

$$J = j_n(E_N) = eNv(E_N). \tag{3.51}$$

Assume the contrary, i.e. $E_a < E_N$. Therefore $eNv(E_a) < eNv(E_N) = J = ep(L)v(E_a)$ or $p(L) > N$ which is impossible because the entire semiconductor bulk should

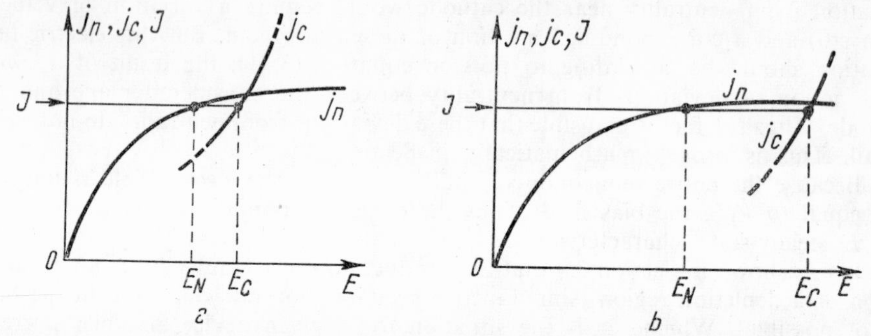

Figure 3.14. Evaluating the field non-uniformity in a silicon diode with Schottky-barrier-emission cathode (see the text).

be depleted. Thus, in the situation depicted in Figure 3.14a, the maximum field non-uniformity is relatively small. On the other hand, the case illustrated in Figure 3.14b is characterized by an almost constant drift velocity inside the bulk, which is an important simplification for calculations.

*Example (b). An n-*GaAs *device with imperfect cathode* (charge trapped at the interface or an embedded p layer, etc). The j_n and j_c characteristics are depicted in Figure 3.15 (see Section 3.6 and also Kroemer [52]). Assume that the injected bias current increases monotonously from zero to very high values above the cross-over current. J_x. We are interested here in the uniformity of the electric field. If J is sufficiently high (above the knee of the control characteristic j_c) then the cathode field will change only slightly with the increase in current. As far as $J < J_x$ the semiconductor is depleted of mobile carriers. Let J be above the knee of j_c but still very small as compared to J_x. Since $J = j_c(E_c)$ should be much smaller than j_n over almost the entire semiconductor space, equation (3.44) gives approximately

$$eNL/\varepsilon \simeq E_c - E_a \tag{3.52}$$

and since $E_c \simeq$ const, E_a should be also approximately constant. If the doping-length product is very small (or better, L is very short, because N also enters in $j_n = eNv(E)$) then the non-uniformity of the electric field will be very small.

As the current increases, the injected mobile charge increases and equation (3.52) should be modified. By the Gauss theorem we have

$$eNL - Q_{\text{mob}} = \varepsilon (E_c - E_a), \tag{3.53}$$

where Q_{mob} is the total injected mobile-charge per unit of electrode area, inside the semiconductor drift space. Thus, the non-uniformity of the electric field becomes even smaller as J increases. At $J = J_x$, the left hand side of equation (3.53)

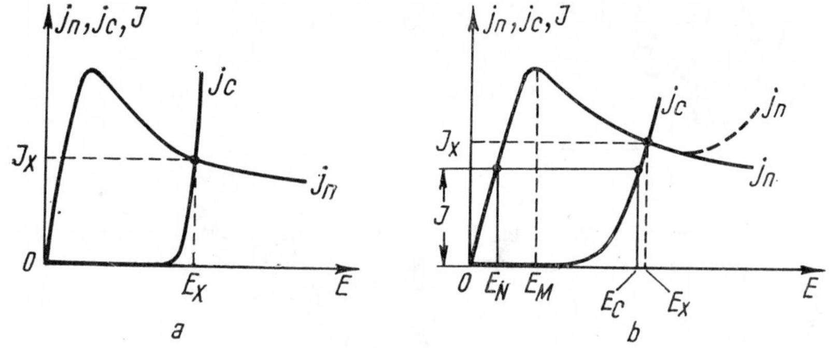

Figure 3.15. The field non-uniformity in an *n*-GaAs device with imperfect cathode (see also [52]).

cancels out and $E_c = E_a$ (neutrality), etc. This example shows that under imperfect cathode boundary conditions, the electric field may be almost uniform over a wide range of current intensities.

Example (*c*). The same situation as above, but we are now investigating the distribution of mobile charge. For the situation depicted in Figure 3.15 b assume an injected bias current below the crossover current J_x. The semiconductor is depleted of mobile carriers. If the device is sufficiently long, then the anode field, E_a, will be below the threshold and the carrier distribution looks like in Figure 3.16 [52]. The bias field, V/L, is given by equations (3.43) and (3.44). This bias field is, of course, restricted by $E_a < V/L < E_c$ (because E changes monotonously) and may be above or below the threshold field E_M (Figure 3.15b). Assume now that the injected current increases and the cathode field E_c also increases. Therefore, the quantity under the integral in equation (3.44) decreases and E_a increases (L = given) (curve b in Figure 3.16). If J increases further, the entire semiconductor will be in the negative mobility region (curve c in Figure 3.16). If the injected current exceeds the crossover current J_x the semiconductor will become accumulated. The anode field E_a does now exceed E_c. Depending upon the exact shape of the $v = v(E)$ characteristic at high fields (Figure 3.15b) and the value of the bias current (L is now assumed variable), E_a, may be limited or not (see example a). For a $v = v(E)$ characteristic such as the heavy-line in Figure 3.15b, the anode field, E_a, is not limited, and, if the device is sufficiently long E_a will increase beyond the critical value leading to carrier multiplication.

Following Kroemer [52] we note: 'the fact that a static solution to a space-charge problem exists does not guarantee that this solution is actually a stable one'. For a bias field (V/L) larger than the crossover field (E_x) the semiconductor is accumulated and this fact may lead to Gunn oscillations in above critically doped crystals [28], [29], [52].

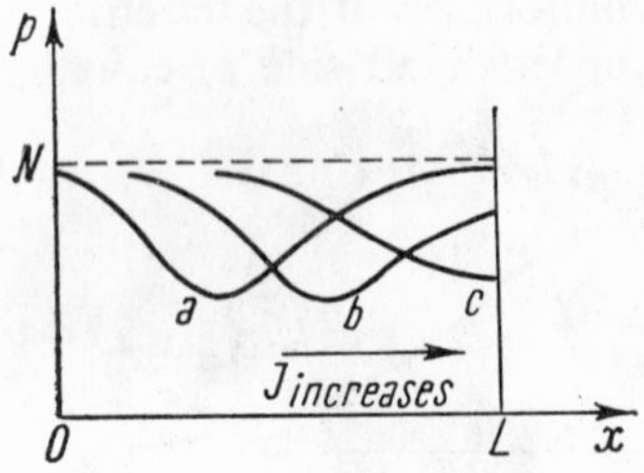

Figure 3.16. Investigating the carrier distribution in an n-GaAs device with imperfect cathode.

The stability problem requires, however, a small signal analysis which is beyond the scope of this introductory chapter and will be discussed later. We anticipate the fact that the slope of the control characteristic for a given bias current

$$\sigma_1(0) = \left.\frac{dj_c}{dE}\right|_{E=E_c} \tag{3.54}$$

will be a key parameter for the small-signal analysis. Here $\sigma_1(0)$ is a sort of a.c. conductivity of the injecting plane.

3.8 The Possibility of a Negative Conductance on the Static Characteristic

The existence of a negative conductivity, due to negative differential mobility ($dv/dE < 0$) in semiconductors like GaAs or InP raises, naturally, the problem of an external negative conductance ($dJ/dV < 0$). Consider a semiconductor diode consisting of a negative mobility material between injecting contacts. If the semiconductor were neutral, the $J - V$ characteristic would be given by

$$J = eNv(E) = j_n(E), \quad E = V/L \tag{3.55}$$

and the incremental conductance dJ/dE would be negative if the device is biased in the negative mobility region (see Figure 3.17 for a one-dimensional model).

However, as first shown by Shockley [14] the result will be different if proper boundary conditions are taken into account. We discuss below what is named in literature the 'Shockley's positive conductance theorem'.

We shall use equations (3.43) — (3.45) derived in Section 3.7 for a one-dimensional homogeneous semiconductor region (diffusion is neglected). As shown in the Appendix, the static (incremental) resistance may be written

$$\frac{\mathrm{d}V}{\mathrm{d}J} = \frac{\varepsilon}{eN} \int_{E_c}^{E_a} \frac{(E_a - E)\, j_n(E)\, \mathrm{d}E}{[j_c(E_c) - j_n(E)]^2} + \frac{\varepsilon}{eN} \frac{1}{\left(\frac{\mathrm{d}j_c}{\mathrm{d}E}\right)_{E=E_c}} \frac{E_a - E_c}{\left(\frac{j_c(E_c)}{j_n(E_c)} - 1\right)}. \tag{3.56}$$

We note in the expression of the second term in the right-hand side of equation (3.56), the cathode conductivity $\sigma_1(0)$ defined by equation (3.54). If the injection conditions were so that this conductivity is infinite (ohmic contact and space-charge-limited injection), then the second term in equation (3.56) would vanish. Therefore, the first term of the resistance in equation (3.56) may be considered the bulk space-charge contribution whereas the second one is due to the finite conductivity (or finite injection) of the cathode contact ($x = 0$ in Figure 3.1b).

The bulk resistance

$$\left(\frac{\mathrm{d}V}{\mathrm{d}J}\right)_{\text{bulk}} = \frac{\varepsilon}{eN} \int_{E_c}^{E_a} \frac{(E_a - E)\, j_n(E)\, \mathrm{d}E}{[j_c(E_c) - j_n(E)]^2}. \tag{3.57}$$

is *always positive*, irrespective of the sign of the differential mobility. Note that $j_n(E)/eN = v(E)$ is positive for both majority and minority carrier injection. If the semiconductor is accumulated or inverted then $E_c < E_a$, $E \leqslant E_a$ and the

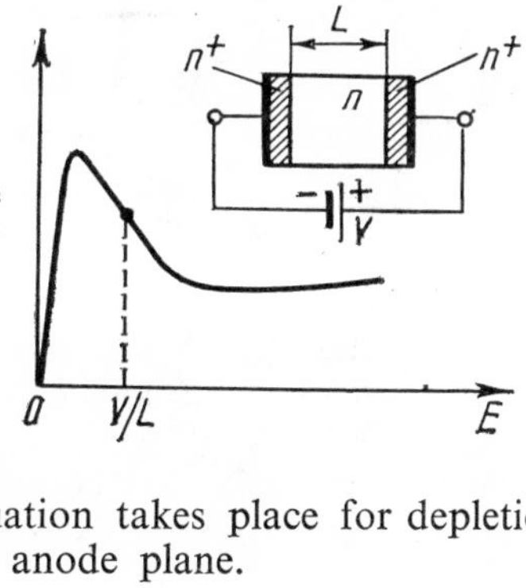

Figure 3.17. A negative-mobility diode with a bias field V/L within the negative-mobility region of the $v-E$ characteristic.

integral is obviously positive. The same situation takes place for depletion, when the minimum field in the device is at the anode plane.

The 'cathode' resistance

$$\left(\frac{\mathrm{d}V}{\mathrm{d}J}\right)_{\text{cath}} = \frac{\varepsilon}{eN} \frac{1}{\left(\frac{\mathrm{d}j_c}{\mathrm{d}E}\right)_{E=E_c}} \frac{E_a - E_c}{\frac{j_c(E_c)}{j_n(E_c)} - 1} \tag{3.58}$$

has the sign of the cathode conductivity (3.54), because $E_a > E_c$ for $j_c(E_c) > j_n(E_c)$ and $E_a < E_c$ for $j_c(E_c) < j_n(E_c)$ (Property 2 in Section 3.7).

Thus, a *necessary* condition for the external resistance (device of unit area)

$$\frac{dV}{dJ} = \left(\frac{dV}{dJ}\right)_{bulk} + \left(\frac{dV}{dJ}\right)_{cath} \tag{3.59}$$

to be negative is the negative conductivity of the cathode plane ($x = 0$). In other words, the control characteristic $j_c = j_c(E)$ should have a region of negative slope $\sigma_1(0) < 0$. By using Kroemers's terminology, a well-behaved cathode is characterized by a positive differential conductivity $\sigma_1(0)$. Our precedent discussion on the injection condition and the control characteristic (Sections 3.2 and 3.7) do reveal $\sigma_1(0) < 0$ only if the emitting plane is situated in a medium of negative mobility. The current density at the emitting plane is

$$J = ep(0)\, v(E_c) = j_c(E_c). \tag{3.60}$$

Here, $p(0)$ either is constant (equations (3.9) or (3.19)) or increases with the cathode field, whereas $v(E_c)$ decreases as E_c increases (negative mobility). By definition $(dj_c/dE)_{E=E_c} < 0$ and thus J decreases when E_c increases. However *it is not evident* that the cathode field increases as the bias field V/L increases*) and, consequently dV/dJ may be or may not be negative. This shows once again that $\sigma_1(0) <$ < 0 is necessary but not sufficient for a negative resistance to occur.

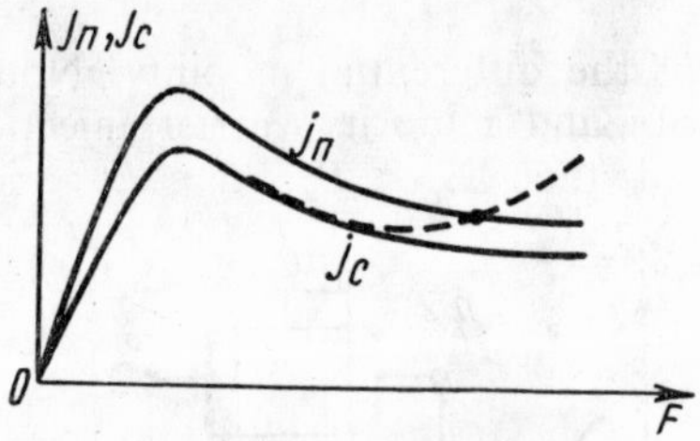

Figure 3.18. The possibility of a static negative resistance of a negative-mobility device with a shallow barrier injecting contact [52]. The dashed curve indicates the effect of the image-force lowering of the barrier.

A *sufficient* condition for a static negative resistance may be easily derived. If the semiconductor space is neutral, the cathode field will surely increase as the bias field increases. Thus, if there exists a crossover point of the neutral and the control characteristic in a region where the slope of the control characteristic is negative, the differential resistance dV/dJ must be negative if the device is properly biased.

Following Kroemer [52] we consider here the case of a shallow barrier injecting contact, when $p(0)$ = given and constant. We note the fact that

$$\frac{j_c(E)}{j_n(E)} = \frac{ep(0)\, v(E)}{eNv(E)} = \frac{p(0)}{N} = \text{const.} \tag{3.61}$$

---

*) We shall see shortly that the cathode field can decrease as the bias field increases (Section 3.10).

and the control characteristic is 'parallel' to the neutral characteristic and no crossover point exists. The sufficient condition for a static negative resistance is not satisfied. However, a crossover point may exist if the barrier is lowered by the electric field and the control characteristic looks like the dashed curve in Figure 3.18 [52].

A negative resistance, $\mathrm{d}V/\mathrm{d}J < 0$, may also occur if surface states exist at the cathode interface. This is discussed in the Appendix.

3.9 The Space-charge Factor

The results derived in the precedent Sections consist in some general properties of unipolar and one-dimensional currents carried by drift in a homogeneous semiconductor region. We stress the fact that the bulk and the injection effects were *separated* by using the neutral characteristic (depending upon the bulk properties) and the control characteristic (which is peculiar to the cathode-semiconductor interface).

Now we shall go further into the investigation of the bulk effects. This Section will outline the importance of transit-time effects on steady-state behaviour. We found that these effects may be clarified *without reference to the particular mechanism of carrier injection*. This will make the discussion more general. The interesting point is that using these general results in conjunction with a certain boundary condition, the steady-state characteristic may be obtained without difficulty, benefiting of a better understanding of the bulk processes.

Let us assume, first, that the carrier mobility is constant. The results will be generalized into the next Section, where a piecewise $v - E$ characteristic will be considered.

The basic equations used in this Section are

$$J = e \mu p E, \qquad (3.62)$$

(constant mobility, negligible diffusion), and

$$\frac{\mathrm{d}E}{\mathrm{d}x} = \frac{e(p - N)}{\varepsilon}, \qquad (3.63)$$

(N = constant, positive for majority and negative for minority carrier injection, see equation (3.32)). By eliminating the hole density one obtains

$$\varepsilon \frac{\mathrm{d}E}{\mathrm{d}x} = \frac{J}{\mu E} - eN. \qquad (3.64)$$

Now let us define T_x as the hole transit-time from $x = 0$ to x. This transit-time is, of course, the same for all carriers passing through the device. It is also evident that T_x is a uniform function of x and x is a uniform function of T_x. Thus T_x may be changed for x as an independent variable in equation (3.64). By replacing

$$\mathrm{d}x = v\,\mathrm{d}T_x = \mu E\,\mathrm{d}T_x, \tag{3.65}$$

and introducing

$$\omega_r = \frac{e\mu N}{\varepsilon}, \tag{3.66}$$

one obtains

$$\frac{\mathrm{d}E}{\mathrm{d}T_x} + \omega_r E = J/\varepsilon. \tag{3.67}$$

The replacing of T_x for x is justified by the fact that the electric field distribution is essentially determined by the electric charge *in motion*, hence a dependence $E = E(T_x)$ is natural. By integrating equation (3.67) one obtains

$$E(T_x) = E(0)\exp(-\omega_r T_x) + \frac{J}{\varepsilon\omega_r}[1 - \exp(-\omega_r T_x)]. \tag{3.68}$$

We shall introduce *the normalized transit-time*

$$\theta_r = \omega_r T_L \tag{3.69}$$

(T_L is the *total* transit-time, from $x = 0$ to $x = L$). Note the fact that for majority carrier injection, ω_r given by equation (3.66) is the reciprocal of the *dielectric relaxation time* τ_r. Thus, θ_r is the ratio of two characteristic time constants.

For the sake of definiteness, let us consider the *majority carrier injection* first ($N > 0$). Assume $\theta_r = T_L/\tau_r \gg 1$. Equation (3.68) yields

$$E_0(T_x) \simeq \frac{J}{\varepsilon\omega_r} = \frac{J}{e\mu N} = \text{const.} \tag{3.70}$$

which shows the fact that the semiconductor region is practically neutral and an ohmic current flows (irrespective of the injection mechanism). This is exactly what we expect because the dielectric relaxation time τ_r is much shorter than the carrier transit time and, due to the dielectric relaxation mechanism, the injected carriers cannot build up a space charge.

In the opposite case, $\theta_r \ll 1$, equation (3.68) becomes

$$E(T_x) \simeq E(0) + \frac{J_0 T_x}{\varepsilon}; \tag{3.71}$$

$J_0 T_x$ is exactly the total mobile charge (per unit area of electrodes) contained between the plane $x = 0$ and the plane x. By comparison of the above equation with the Gauss theorem, it follows that the fixed space charge is negligible.

An alternative boundary condition for equation (3.67) is to specify the so-called *injection level*

$$\Delta = \frac{p(0)}{N}. \tag{3.72}$$

Therefore, the electric field at $x = 0$ may be written

$$E(0) = \frac{J}{e\mu p(0)} = \frac{J}{\varepsilon\omega_r}\frac{1}{\Delta} \tag{3.73}$$

and equation (3.68) becomes

$$E(T_x) = \frac{J}{\varepsilon\omega_r}\left[1 + \left(\frac{1}{\Delta} - 1\right)\exp(-\omega_r T_x)\right]. \tag{3.74}$$

For majority carrier injection Δ and ω_r are positive. The electric field is constant for $\Delta = 1$. If $\Delta > 1$, E increases monotonously from $x = 0$ to $x = L$ and, according to the Poisson equation (3.63), $p(x) > N$ i.e. the entire semiconductor is an accumulation layer. If $\Delta < 1$, the entire conduction space is partially depleted of mobile charge. These results are not new. Similar properties were demonstrated in Section 3.4 under more general conditions.

The non-uniformity of the electric field inside the bulk, may be described quantitatively by *the space-charge factor**)

$$\zeta = 1 - \frac{E(0)}{E(L)}. \tag{3.75}$$

From equations (3.69), (3.74) and (3.75) one obtains

$$\zeta = [1 - \exp(-\theta_r)]\left[\frac{\Delta}{\Delta - 1} - \exp(-\theta_r)\right]^{-1} = \zeta(\Delta, \theta_r). \tag{3.76}$$

The space-charge factor ζ is zero for $\Delta = 1$ (neutrality, ohmic conduction) and equals unity for $\Delta \to \infty$ ($p(0) \to \infty$, $E(0) = 0$, SCL emission). The space-charge

*) This parameter is the ratio of the number of field lines starting from the space charge inside the conduction space and ending at the anode ($x = L$), to the total number of field lines ending at the anode (negatively biased with respect to the cathode).

factor ζ cannot exceed unity since the electric field cannot be negative (in the absence of diffusion such a negative field would prevent any hole transport towards the anode). Finally, ζ is positive ($0 < \zeta < 1$) for $\Delta > 1$ (accumulation) and negative for $\Delta < 1$ (depletion of the semiconductor).

Figure 3.19 shows $\zeta =$ const. curves plotted in the $\theta_r - \Delta$ plane (double logarithmic plot). The vertical axis ($\Delta = 1$) separates the enhancement (or accumulation) region at the right ($\Delta > 1$) from the depletion region at the left ($\Delta < 1$).

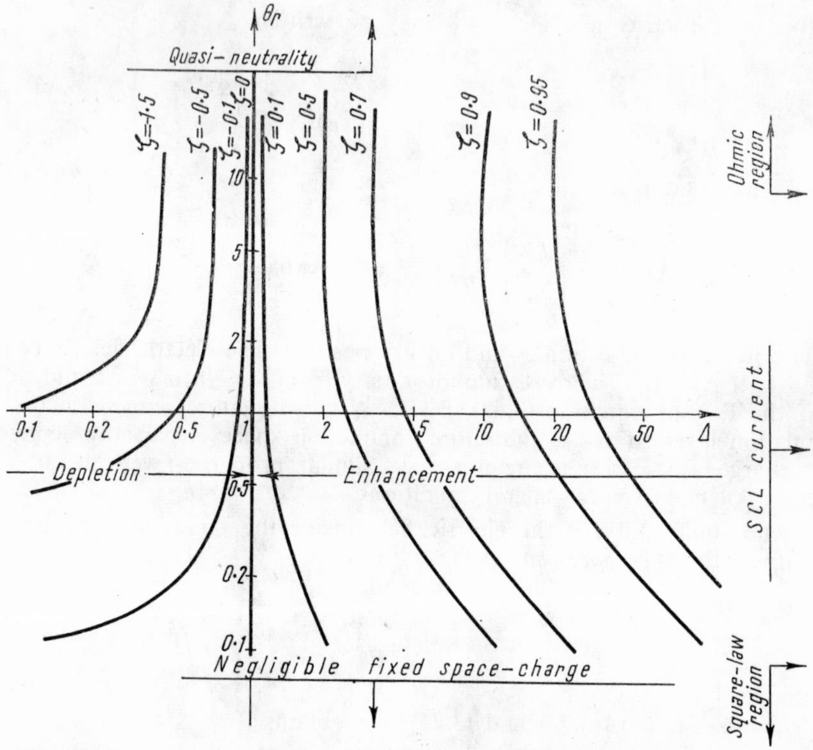

Figure 3.19. The $\zeta =$ const. curves in the $\theta_r - \Delta$ plane (majority carrier injection, constant mobility).

We stress the fact that for every physical device and at a given bias we have a representative point in the $\theta_r - \Delta$ plane (see below). Further, the upper region of the $\theta_r - \Delta$ plane, where $\theta_r \gg 1$ corresponds to quasi-neutrality, whereas for $\theta_r \ll 1$ the effect of the fixed space charge is negligible, as shown above. The space-charge-limited injection is described by the extreme right of this plane ($\Delta \gg 1$ or $\Delta \to \infty$).

The current density and the voltage drop, V, can be normalized and calculated as a function of Δ and θ_r. One obtains (see Appendix)

$$\bar{J} = \frac{J}{eNL\omega_r} = \bar{J}(\theta_r, \Delta) \tag{3.77}$$

$$\bar{V} = \frac{V}{eNL^2/\varepsilon} = \bar{V}(\theta_r, \Delta). \tag{3.78}$$

Several $\bar{J} =$ const. and $\bar{V} =$ const. curves are plotted in the $\Delta - \theta_r$ plane (Figure 3.20). These curves are 'universal' because they do not depend upon the injection mechanism. But their validity is restricted by the neglect of diffusion and the assumption of constant mobility.

Let E_l be a critical field beyond which the carrier mobility departs markedly from its low-field value. The results of Figure 3.20 are approximately valid for an 'average' field V/L below E_l, i.e.

$$V = \bar{V}(eNL^2/\varepsilon) < LE_l \tag{3.79}$$

or

$$\bar{V} < \frac{\varepsilon E_l}{eNL} = \bar{V}_{max}. \tag{3.80}$$

Then, for a given semiconductor and a given doping-length product NL the curves represented in Figure 3.20 are valid only in the region defined by equation (3.80).

The diffusion current is zero for $\zeta = 0$ ($\Delta = 1$) and it is small if the space-charge factor $\zeta = 1 - p(L)/p(0)$ is not very close to unity since the gradient of the carrier density is not too large and the diffusion is likely to be negligible, at least if the applied voltage is not very small. The diffusion effect may be important in the SCLC régime (Section 3.11), i.e.for $\zeta \lesssim 1$.

Now let us show how the $\theta_r - \Delta$ plane given in Figure 3.20 can be used to calculate the steady-state characteristics. Assume, first, cathode injection from a *shallow barrier contact*. $p(0)$ is given by a relationship of the form of equation (3.9) or (3.19) and thus the injection level, equation (3.72), is fixed to a particular value (at a given temperature). For a given $\Delta = \Delta_0$, every point on the vertical line $\Delta = \Delta_0$ corresponds to a point of the $J - V$ characteristic. By using the $\bar{J} =$ const. and $\bar{V} =$ const. curves the normalized $J - V$ characteristic can be (approximately) obtained. Such characteristics are shown in Figure 3.21. For $\Delta \to \infty$ we have $p(0) \to \infty$ and $v(0) = \mu E(0)$ must be zero in order to have a finite *drift* current. This is the case of the SCL current. The $J - V$ characteristic was first derived by Lampert [3] and then presented in normalized form by van der Ziel [53] and by Dascalu *et al.* [54] (see Problem 3.3). We note the gradual transition from the low-current ohmic region (θ_r large in Figure 3.20) to the square-law region at large voltages (small θ_r in Figure 3.20). However, at high applied voltages, the mobility becomes field-dependen (decreases) and the current, J, should be smaller than the value expected for $\mu =$ const. Note the fact that if the doping-length product NL has a large value, θ_r cannot become small (Figure 3.20) due to the restriction shown in equation (3.80) and thus the square-law region indicated on Figure 3.21 ($\Delta \to \infty$) will not actually exist. It was indeed confirmed by experiments that the square-law characteristic ($J \propto V^2$) occurs only for relatively low values of the doping-length product [55], [56].

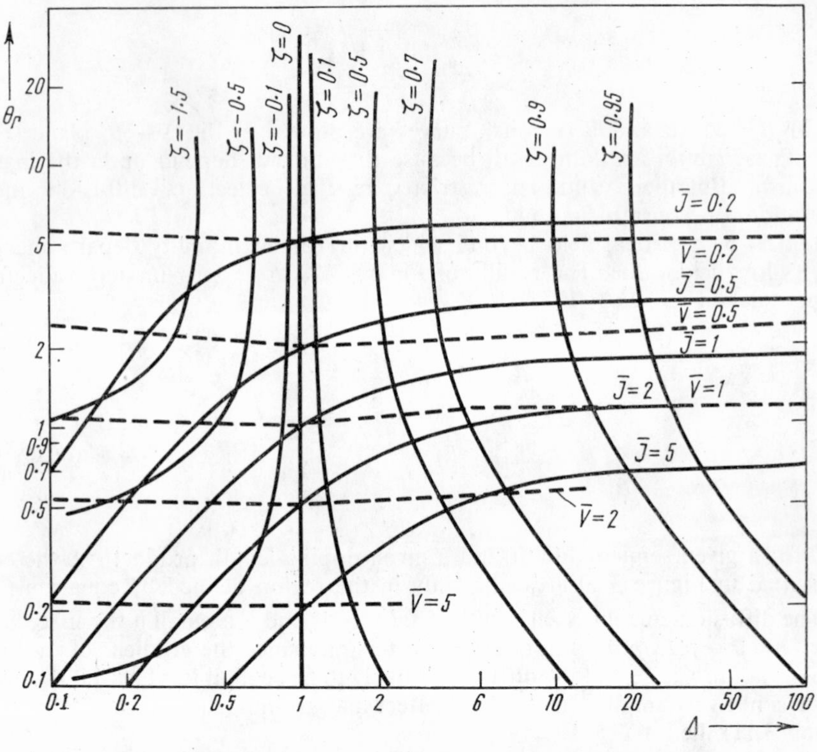

Figure 3.20. $\bar{J} =$ const. and $\bar{V} =$ const. curves in the $\theta_r - \Delta$ plane (majority carrier injection, constant mobility).

If Δ is finite and exceeds unity, the current is ohmic for low voltages. At higher voltages the characteristic remains below the SCL current (Figure 3.21). At very high voltages the $J - V$ characteristic becomes again linear (the carrier density is now almost uniform and equal to its value at the cathode plane). The reader will notice the fact that the finite emission at $x = 0$ leads to the same type of $J - V$ characteristic as the decrease of the carrier mobility under SCLC conditions (infinite injection).

Let us now consider another boundary condition which is very convenient to discuss in our approach. This is the pure thermionic emission without any effects of barrier-lowering. The injected current is constant and equal to the Richardson current, equation (3.11). In the $\theta_r - \Delta$ plane we simply follow the curve $\bar{J}_R = J_R/eNL\omega_r$=const. (see equation (3.77)). At each applied voltage we obtain directly the carrier transit time, the injection level and the space-charge factor, ζ, (Figure 3.20).

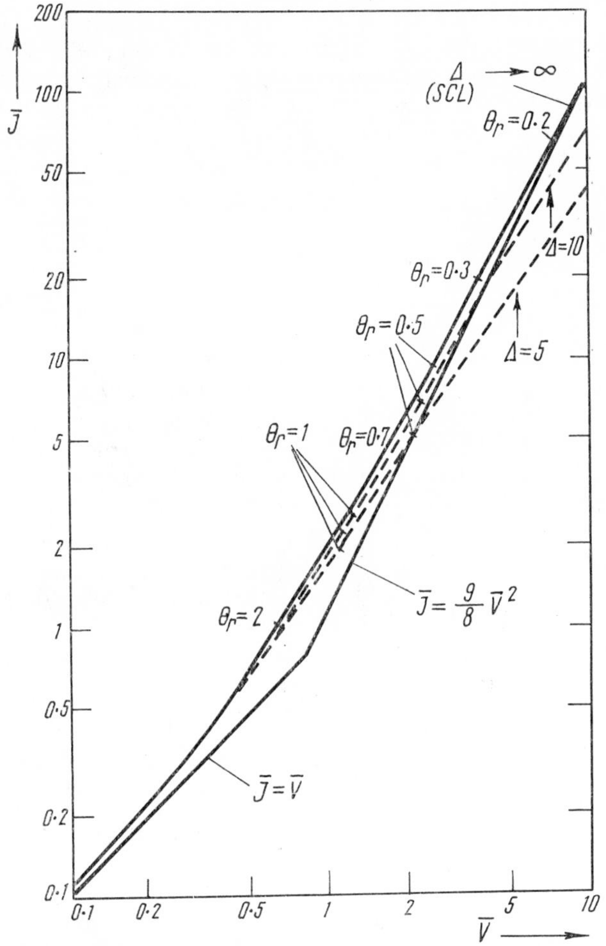

Figure 3.21. Normalized steady-state characteristics of a unipolar diode with majority-carrier injection from a shallow barrier contact (the carrier mobility is considered constant). The curve $\Delta \to \infty$ corresponds to SCL injection (see also [3], [53], [54]).

Note the fact that as the voltage increases, the semiconductor is first accumulated and then depleted of mobile carriers. The carrier transit-time decreases monotonously with increasing voltage, as expected. The applied voltage cannot be reduced below a certain value, which depends upon the injected current. In other words, a current equal to the emission current, J_R, cannot be sustained if the applied voltage is too low. The correct $J - V$ characteristic at low voltages must be calculated under SCLC conditions ($J < J_R$).

Now let us examine qualitatively the effect of changing J at a given voltage. For a given device, the current $J = J_R$ shows an exponential dependence upon temperature (ω_r changes very little with temperature and will be assumed constant). Thus the current and transit-time normalizations are not affected. It may be seen that θ_r (and thus T_L) changes very little with the injected current (Figure 3.20) and can be approximately evaluated for $\zeta = 0$, thus being of the order and a little longer than T_L for $\zeta = 0$ (Figure 3.20).

$$T_L \gtrsim T_{L,0} = \frac{L^2}{\mu V} \tag{3.81}$$

As the current increases from zero (V = const.) the following sequence of events takes place (Figure 3.20). For low current intensities (very low temperatures) the semiconductor is depleted because the emission is low and the injected charge carriers are swept out rapidly by the applied voltage. At higher currents, however, the semiconductor becomes accumulated. Finally, the current cannot exceed a certain maximum value (Figure 3.20) which is the space-charge-limited current for the particular voltage applied.

Another boundary condition will be now considered. Assume the fact that the pure thermionic emission is modified by the image-force (or Schottky) effect. According to equation (3.21), the cathode electric field may be written

$$E(0) = \frac{J}{\varepsilon\omega_r\Delta} = E_R\left[\ln\left(\frac{J}{J_R}\right)\right]^2, \tag{3.82}$$

and

$$\Delta = \frac{M\bar{J}}{(\ln \bar{J}/\bar{J}_R)^2}, \tag{3.83}$$

where

$$M = \frac{eNL}{\varepsilon E_R} \propto NL, \tag{3.84}$$

and

$$\bar{J}_R = J_R/eNL\omega_r, \tag{3.85}$$

are two normalized parameters. The $J - V$ characteristic may be evaluated as follows. Assume $\bar{J}_R =$ given. Clearly, $\bar{J} > \bar{J}_R$, a condition which restricts the possible position of the representative point in the $\theta_r - \Delta$ plane (Figure 3.22). The allowed region is further restricted by condition (3.80). Inside the remainder of the $\theta_r - \Delta$ plane, we have to draw curves $M = \text{const.}$ ($\bar{J}_R = \text{given}$). Each of these curves correspond to a normalized $\bar{J} - \bar{V}$ characteristic. Note the fact that the $M = \text{const.}$ lines tend to coincide with $\bar{J} = \bar{J}_R$ for small values of the applied voltage. In fact, the *deviation* from $\bar{J} = \bar{J}_R$ represents just the image-force effect. For a given semi-

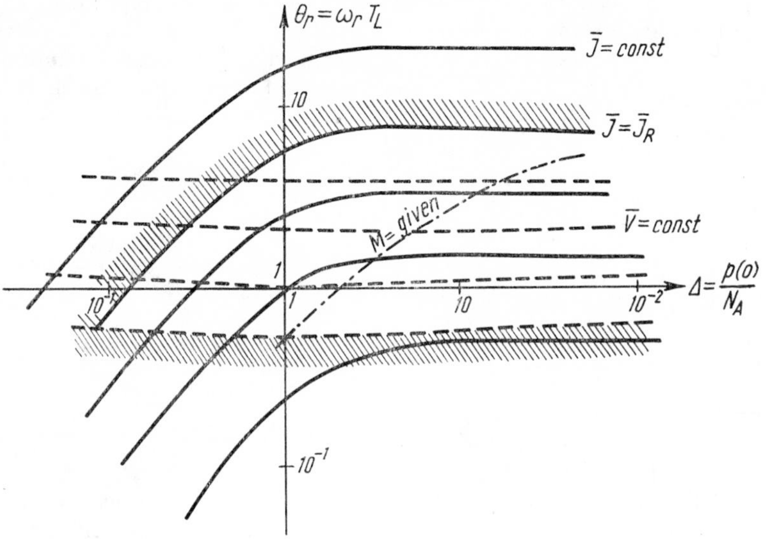

Figure 3.22. Diagram indicating the graphic evaluation of the normalized static characteristics of a unipolar diode with Schottky-type injection.

conductor material and a given emission current J_R, the shape of the normalized steady-state characteristic should depend upon the parameter M given by equation (3.84) and thus the doping-length product will be taken as a parameter.

To conclude, the $\Delta - \theta_r$ plane of Figure 3.20 is universal (constant mobility, negligible diffusion) and takes into account the bulk effects. If the injection mechanism is specified, then it is possible to draw into the $\theta_r - \Delta$ plane certain trajectories which correspond to the steady-state $J - V$ characteristics.

It is true that a direct calculation of the $J - V$ characteristic may be preferable in one case or another*⁾. However, the method of the $\theta_r - \Delta$ plane still retains several advantages. These are:

(a) A suitable selection of the normalization constants and undimensional parameters, as shown above for Schottky emission. These parameters may be then used for computer calculations.

---

*⁾ Only a few very simple cases can be studied analytically, but the $J - V$ characteristic can, of course, be computed numerically.

(b) A better understanding of the physical situation in the semiconductor bulk. As an example, we indicate the fact that it is possible to show directly if the electric field inside the device may be considered almost uniform ($\zeta \simeq 0$) or not. We have already seen that the occurrence of hot carrier effects can be predicted. In general, the simplifying assumptions can be handled now with less difficulties.

c) The small-signal characteristics may be conveniently expressed and computed using the parameters Δ and θ_r. Thus, the $\bar{J}=$ const. and $\bar{V}=$ const. curves in the $\Delta - \theta_r$ plane allow us a direct calculation of the small-signal response as a function of the applied bias, for any given device. It is often a difficult problem to make this connection between the steady-state solution and the small-signal solution (which strongly depends upon the former). The frequent (and often unjustified) approximations in evaluating the bias effect upon the device small-signal characteristics prove that such a problem does really exist.

The discussion of minority carrier injection follows the same course and is left to the reader.

The following Section represents a generalization of the present results for majority and minority carrier injection into the positive and negative-mobility semiconductor regions. This will prepare the ground for the general small-signal analysis developed in the next chapter.

3.10 A Generalized Theory of Space-charge Effects

For most of the unipolar semiconductor devices, which concern us here, the carrier mobility cannot be approximated by a constant. An arbitrary velocity-field dependence has to be taken into account in equation (3.30). Various analytic approximations may be used for $v = v(E)$. There is, however, little chance to get an analytical expression, even in parametric form, for the $J - V$ characteristic.

We shall assume here the fact that the differential mobility is approximated by a constant

$$\frac{\mathrm{d}v}{\mathrm{d}E} = \mu_d = \text{const.} \tag{3.86}$$

By integration, one obtains

$$v = v_d + \mu_d E = \mu_d (E + E_d) \tag{3.87}$$

where $v_d = \mu_d E_d$ is also a constant. Of course, a straight-line approximation like equation (3.87) may be acceptable only for a certain range of field intensities, or for a certain range of carrier velocities, which is the same thing. Such an approximation may be used between the peak and the valley electron velocity in a negative mobility semiconductor like InP (Figure 3.23a). Another example is the high-field low-mobility region of the $v = v(E)$ dependence for holes in silicon (Figure 3.23b).

A piecewise straight-line approximation is familiar (see for example [29]). If the electric field inside the semiconductor does not change too much, a single line approximation like equation (3.87) may be sufficient. This is the case studied here and it is a natural extension of the quasi-neutrality approximation (uniform carrier velocity), which was very often used. Both majority and minority carrier injections may be studied. The calculations may be developed analytically up to a certain point. Clearly, these results do not depend upon the particular semiconductor under study. For a given semiconductor, suitable numerical values should be used for E_d and v_d in equation (3.87).

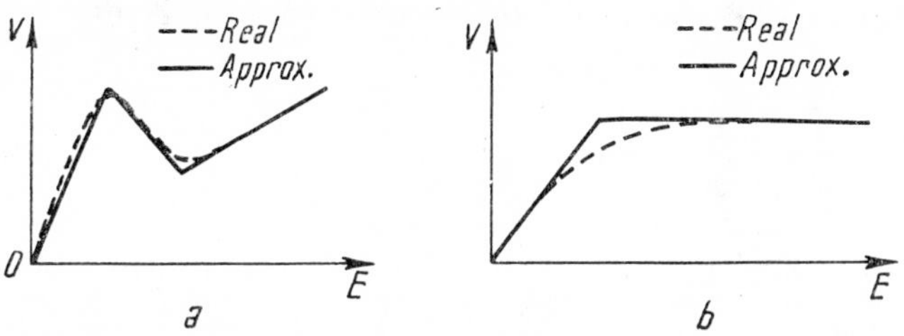

Figure 3.23. A straight-line approximation for a region of the velocity-field characteristic. (a) electrons in In P, (b) holes in silicon.

Equation (3.87) will be used in conjunction with

$$J = epv \tag{3.88}$$

and the Poisson equation

$$\frac{\mathrm{d}E}{\mathrm{d}x} = \frac{e(p - N)}{\varepsilon}. \tag{3.89}$$

From the last two equations we obtain

$$\varepsilon \frac{\mathrm{d}E}{\mathrm{d}x} = \frac{J}{v} - eN \tag{3.90}$$

where $\mathrm{d}x = v\mathrm{d}T_x$ and, from equation (3.87) $\mathrm{d}E = \mathrm{d}v/\mu_d$. Thus

$$\mathrm{d}T_x = \frac{\varepsilon}{\mu_d} \frac{\mathrm{d}v}{J - eNv} \tag{3.91}$$

which can be readily integrated and yields

$$v = v(0)\,[\Delta - (\Delta - 1)\exp(-\,\theta_d\xi)] \tag{3.92}$$

where

$$\theta_d = \frac{T_L}{\tau_d}. \tag{3.93}$$

and τ_d is a sort of differential relaxation time

$$\tau_d = \frac{\varepsilon}{eN\,\mu_d}. \tag{3.94}$$

Then, $v(0)$ is the cathode velocity, Δ the injection level (3.72) and ξ the normalized transit-time

$$\xi = \frac{T_x}{T_L} \tag{3.95}$$

We stress the fact that Δ and θ_d may be either positive or negative, depending upon the sign of N and μ_d. The electric field distribution is given by equations (3.87) and (3.92). The hole density distribution may then be deduced from the Poisson equation. The space charge effects can be discussed as shown in the preceding Section.

There is another problem which concerns us here. Because equation (3.87) is valid only for a certain range of velocities the same thing is true for equation (3.92) and Δ, θ_d cannot take any values.

From equation (3.92) the anode carrier-velocity $v(L)$ is given by ($\xi = 1$)

$$v(L) = v(0)\,[\Delta - (\Delta - 1)\exp(-\theta_d)]. \tag{3.96}$$

$v(0)$ can be found by integrating

$$L = \int_0^L \mathrm{d}x = \int_0^{T_L} v\,\mathrm{d}T_x = \int_0^1 T_L v(\xi)\,\mathrm{d}\xi, \qquad \xi = \frac{T_x}{T_L} \tag{3.97}$$

where $v(\xi)$ is given by equation (3.92). The result is

$$v(0) = \frac{L}{\tau_d}\,[\theta_d\Delta + (\Delta - 1)(-1 + \exp - \theta_d)]^{-1}. \tag{3.98}$$

We shall define two normalized parameters

$$\varphi_0 = \frac{v(0)}{L/\tau_d} = [\theta_d\Delta + (\Delta - 1)\,(-1 + \exp - \theta_d)]^{-1} = \varphi_0(\theta_d, \Delta) \tag{3.99}$$

$$\varphi_L = \frac{v(L)}{L/\tau_d} = \frac{\Delta - (\Delta - 1)\exp - \theta_d}{\theta_d\Delta + (\Delta - 1)(-1 + \exp - \theta_d)} = \varphi_L(\theta_d, \Delta) \tag{3.100}$$

which are proportional to the cathode and anode carrier-velocities, respectively. Let us consider the example of a negative-mobility region such as shown in Figure 3.23a. The approximation (3.87) (where $\mu_d < 0$, $E_d < 0$, $v_d = \mu_d E_d > 0$) is valid throughout the entire conduction space only if

$$v_m < [v(0),\ v(L)] < v_M. \tag{3.101}$$

Assume, for convenience, majority carrier injection and accumulation ($\Delta > 1$). The electric field increases with x whereas the carrier velocity must decrease. Thus $v(0)$ must be below v_M, whereas $v(L)$ must be above v_m. From equations (3.99) and (3.100) we obtain

$$|\varphi_0| = |\varphi_0(\theta_d, \Delta)| < \frac{v_M\,|\tau_d|}{L} = \left(\frac{v_M \varepsilon}{e|\mu_d|}\right)\frac{1}{NL} \tag{3.102}$$

$$|\varphi_L| = |\varphi_L(\theta_d, \Delta)| > \frac{v_m|\tau_d|}{L} = \left(\frac{v_m \varepsilon}{e|\mu_d|}\right)\frac{1}{NL}. \tag{3.103}$$

The above relationships restrict the values which Δ and θ_d can have simultaneously*).

The current and voltage may be written as follows (Appendix)

$$\bar{J} = \frac{J}{eNL/\tau_d} = \bar{J}(\theta_d, \Delta) \tag{3.104}$$

$$\bar{V}' = \frac{V'}{eNL^2/\varepsilon} = \frac{V + E_d L}{eNL^2/\varepsilon} = \bar{V}'(\theta_d, \Delta). \tag{3.105}$$

We use again a representation in the $\theta_d - \Delta$ plane. There are four distinct situations indicated in Table 3.1.

Table 3.1

	majority carrier		minority carrier	
	pos. mob.	neg. mob.	pos. mob.	neg. mob.
N	>0	>0	<0	<0
μ_d	>0	<0	>0	<0
Δ	>0	>0	<0	<0
θ_d	>0	<0	<0	>0
φ_0	>0	<0	<0	>0
φ_L	>0	<0	<0	>0
$\bar{J}$	>0	<0	>0	<0
eNL^2/ε	>0	>0	<0	<0

*) For majority carrier injection and negative differential mobility Δ is positive ($N > 0$) and θ_d is negative ($\mu_d < 0$).

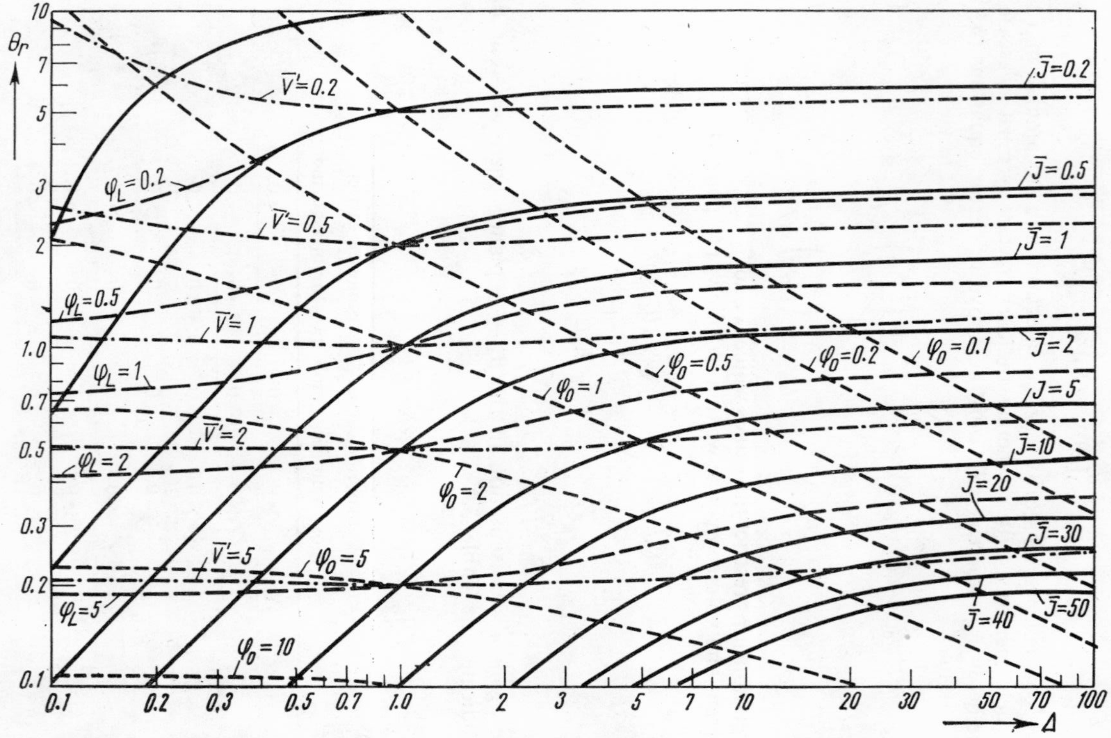

Figure 3.24. The $\theta_d - \Delta$ plane for *majority* carrier injection in semiconductor regions biased in the positive-mobility range.

All these situations are represented separately in Figures 3.24 — 3.27. The following curves appear in the θ_d — Δ plane

$$\bar{J} = \text{const.}$$

$$\bar{V}' = \text{const.}$$

$$\varphi_0 = \text{const.}$$

$$\varphi_L = \text{const.}$$

For simplicity, the *absolute values* of $\bar{J}$, φ_0, φ_L, θ_d, Δ are indicated.

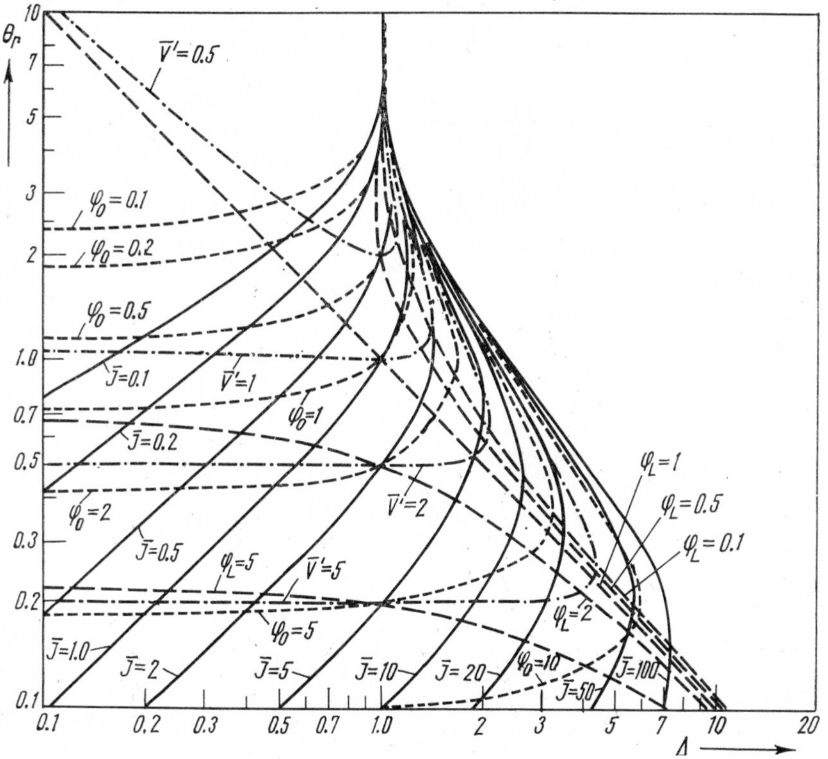

Figure 3.25. The θ_d — Δ plane for *majority* carrier injection in semiconductor regions biased in the *negative-mobility* range (all numbers indicate absolute values, see also Table 3.1).

Figure 3.24 shows the θ_d — Δ plane for majority carrier injection and positive mobility. Assume that a relation of the form of equation (3.87) holds for $E \geqslant E_l$.

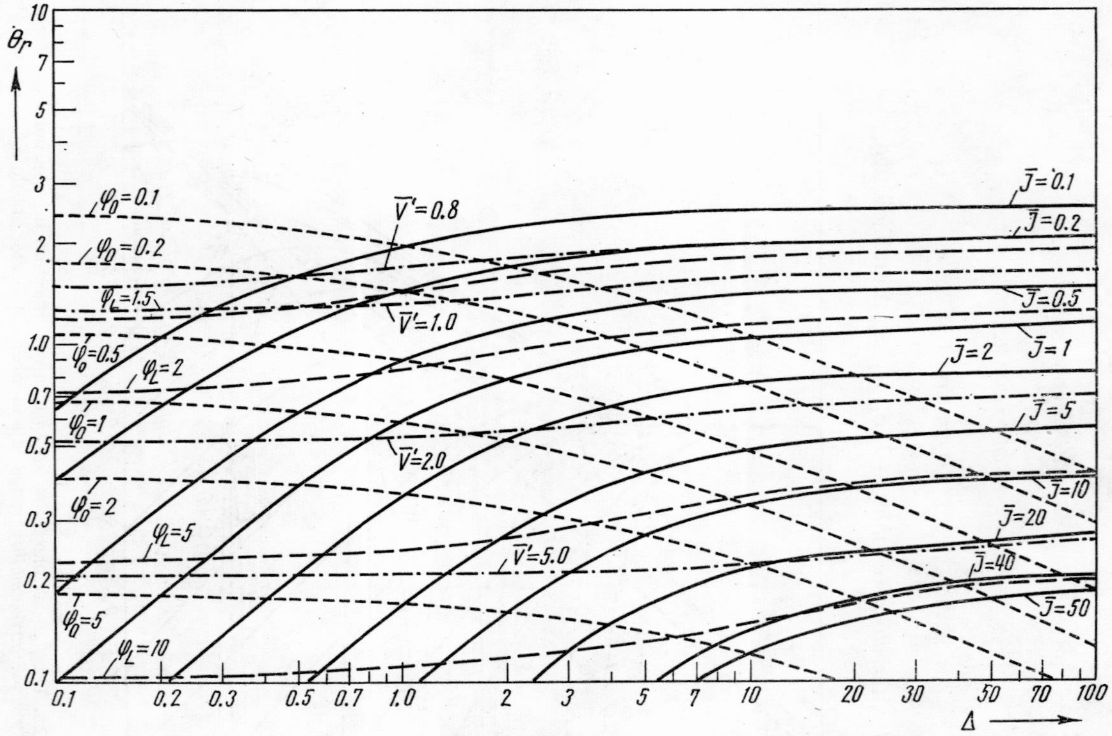

Figure 3.26. The $\theta_d - \Delta$ plane for *minority* carrier injection in semiconductor regions biased in the *positive-mobility* range (all numbers indicate absolute values, see also Table 3.1).

If both $v(0)$ and $v(L)$ exceed $v = v_d + \mu_d E_l = v_c$ this approximation will be satisfied. Consequently we must have

$$\varphi_0, \varphi_L > \frac{v_c}{L/\tau_d} = \left(\frac{v_c \varepsilon}{e\mu_d}\right)\frac{1}{NL} = \varphi_c. \qquad (3.106)$$

The region in the $\Delta - \theta_d$ plane which satisfies the above condition is given by (see Figure 3.24)

$$\varphi_0 > \varphi_c \text{ for } \Delta > 1 \text{ (accumulation)} \qquad (3.107)$$

$$\varphi_L > \varphi_c \text{ for } \Delta < 1 \text{ (depletion).} \qquad (3.108)$$

Once the injection mechanism specified, the steady-state characteristic can be determined as shown in the preceding Section.

Figure 3.25 shows the $\theta_d - \Delta$ plane for majority carrier injection and negative differential mobility. There is an absolute limit given by $\varphi_L = 0$ which almost coincides with $\varphi_L = 0.1$ (absolute value). Only that part of $\theta_d - \Delta$ plane which is situated on the left of the $\varphi_L = 0$ curve may be used for calculations.

An interesting problem to be discussed here is the possibility of a negative resistance on the static characteristic. Assume injection at a shallow barrier contact and thus a boundary condition of the form $\Delta =$ given. The trajectory followed by the representative point in the $\theta_d - \Delta$ plane is simply a vertical line. By use of Figure 3.25, we see that as the voltage increases, the current increases and no negative resistance can occur*). However, the inexistence of a negative resistance was not yet rigorously demonstrated, because the electric field was restricted to the negative-mobility region.

Figures 3.26 and 3.27 show similar diagrams for *minority* carrier injection. Their interpretation and discussion are again left to the reader.

Finally, we shall stress again that these diagrams take into account the bulk effects and do not depend in any way upon the device studied. Then, for a given semiconductor and a given doping-length product there exist a precise delimitation of the region in the $\theta_d - \Delta$ plane, which can be used for our calculations.

If we wish to find the $J - V$ dependence, the particular injection mechanism will be specified. Thus, taking into account the boundary condition, one or more characteristic lines may be drawn in the $\theta_d - \Delta$ plane. Each of them coresponds to a normalized steady state characteristic. Simultaneously, we can obtain information on the space charge inside the bulk and the carrier transit-time.

If the $J - V$ characteristic is known†) the $\theta_d - \Delta$ plane will be used to found θ_d, Δ, $v(0)$ and $v(L)$ which are necessary for small-signal calculations, at a given bias. The fact that certain parameters of the equivalent small-signal circuit may be plotted directly in the $\theta_d - \Delta$ plane will be shown later.

*) Alternatively we can see that the cathode velocity (which is proportional to $|\varphi_0|$), increases (i.e. the cathode field decreases) as the bias voltage increases.

†) The direct calculations are unaffected, of course, by graphical and interpolation errors, but for most situations a computer must be used.

If the $v = v(E)$ characteristic must be approximated by two or more linear 'pieces', the task of calculating the $J - V$ characteristic will be carried out only by means of a computer, but the same system of parameters may be used. Such a programme was actually written and used for calculations [57].

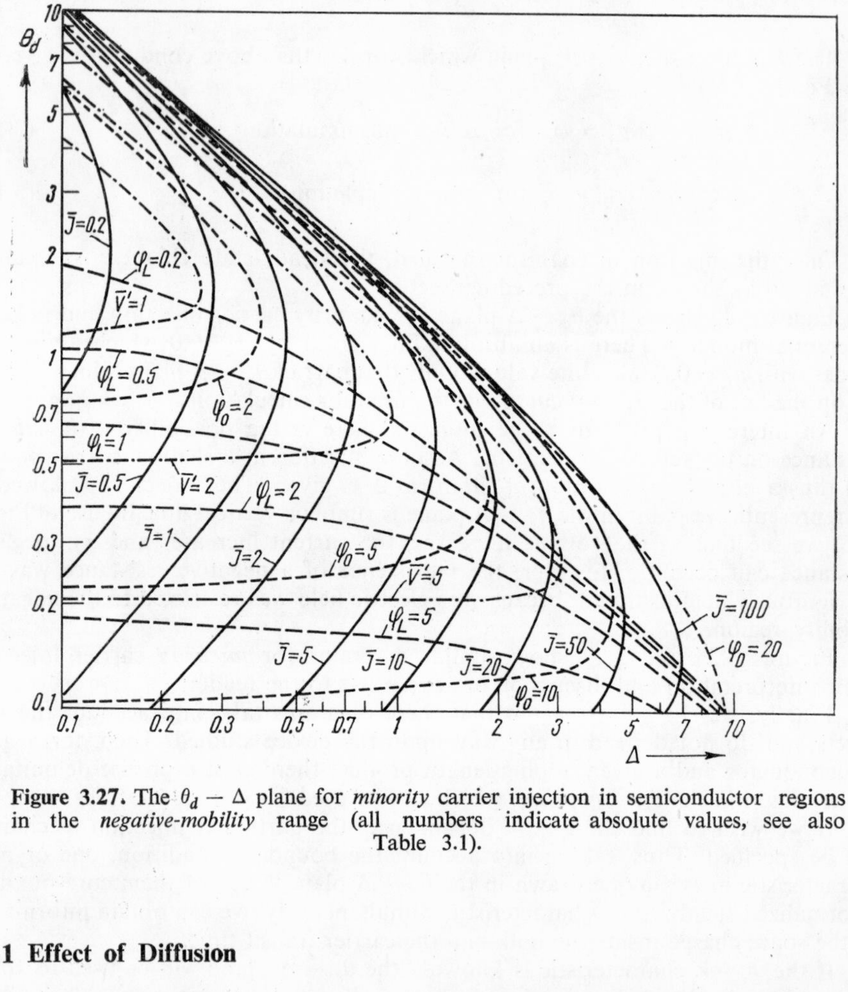

Figure 3.27. The $\theta_d - \Delta$ plane for *minority* carrier injection in semiconductor regions in the *negative-mobility* range (all numbers indicate absolute values, see also Table 3.1).

3.11 Effect of Diffusion

The effect of the diffusion current was neglected so far. In principle, this neglect is possible if the electric field is sufficiently intense and the drift current dominates over the diffusion current. Nevertheless, if a copious carrier injection takes place and the carrier-density gradient is very large, the diffusion current cannot be neglected in the vicinity of the injecting cathode, even for very intense fields.

Let us consider the case of the insulator (dielectric) diode with zero fixed space charge and constant mobility. The total direct current is given by equation (3.7) (D is also constant), and, by also using the Einstein relation, one obtains

$$J = \varepsilon\mu E \frac{\mathrm{d}E}{\mathrm{d}x} - \varepsilon D \frac{\mathrm{d}^2E}{\mathrm{d}x^3} = \varepsilon\mu \left(E \frac{\mathrm{d}E}{\mathrm{d}x} - \frac{kT}{e} \frac{\mathrm{d}^2E}{\mathrm{d}x^2} \right). \tag{3.109}$$

Mott and Gurney have shown [1] that the diffusion current may be neglected if the applied voltage, V, satisfies the following inequality*)

$$V \gg kT/e. \tag{3.110}$$

The fact that no parameter characterizing the contact is taken into account shows this condition cannot be correct.

The exact solution of equation (3.109) was searched for by a number of workers [9], [13], [58], [59], etc. The particular solution suited to our problem could be found by imposing proper boundary conditions. If the diffusion current is taken equal to zero, the carrier density at the cathode and the total voltage drop are sufficient to determine the electric field intensity and the current (Problem 3.8). If diffusion is taken into account, one more boundary condition is necessary and the carrier density at the anode will be also specified. One assumes that the current passing through the crystal is much below the saturation current of both electrodes and 'the electron atmospheres on each side of the contact will remain in thermal equilibrium' [9]. Boundary conditions of the type shown in equation (3.9) may be used.

Unfortunately, even for this simple case ($N = 0$, $\mu =$ const., $D =$ const.), the analytical solution is complicated and difficult to discuss in simple physical terms. We illustrate below the physical conditions inside the device by using the computer solution obtained by Page [60].

Figure 3.28 shows the $J - V$ characteristic computed for a particular CdS device. The current is proportional to the square of the applied voltage in the high current range (forward bias). Figure 3.29 shows the potential distribution inside the device for various current levels. As the bias current increases from negative values passing through zero, a potential minimum occurs and then moves towards the cathode and decreases in magnitude. At current densities corresponding to the square-law region in Figure 3.28, the potential minimum or virtual cathode is very close to the actual cathode and its magnitude is almost negligible as compared to the total voltage drop.

Lampert and Edelman [61] have taken into account the fixed space charge ($N =$ const. < 0) and also obtained a computer solution; no analytical solution was possible. If the mobility-field dependence is taken into account, the calculations become even more complicated and their generality is restricted to a certain semiconductor.

---

*) The diffusion current is evaluated and compared to the drift current by use of the electric field distribution calculated in the absence of diffusion.

An interesting problem is the validity of the 'Shockley positive conductance theorem' (Section 3.8). Kroemer [62] stated that the presence of diffusion cannot invalidate this theorem (which he proved for an arbitrary geometry, non-uniform doping and SCLC conditions). Later, Hauge [63] and Döhler [64] discussed the theoretical possibility of a negative resistance on the static characteristic of Gunn devices, by assuming a field-dependent diffusion constant.

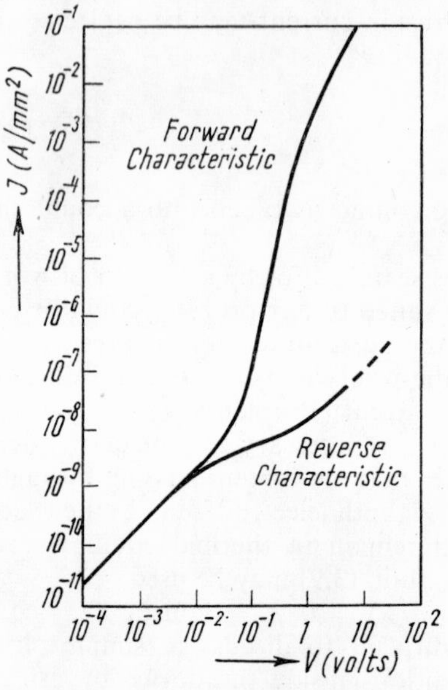

Figure 3.28. The forward and reverse $J - V$ characteristic of a SCLC CdS diode, computed by Page [60] by including the effect of the diffusion current (cathode barrier of 0.2 V, anode barrier of 0.7 eV, $\varepsilon = 10^{-10}$ F/m, $\mu_e = 0.021$ m²/Vs, $L = 10$ μm, $T = 300$ K).

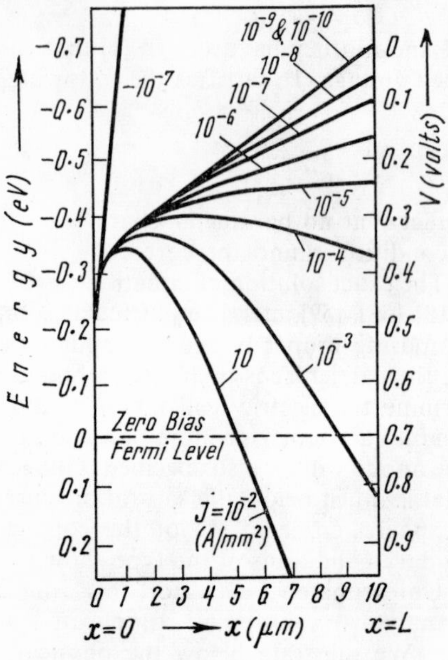

Figure 3.29. The potential distribution in the device indicated in Figure 3.28, at various current levels (computed by Page [60]).

3.12 Non-uniformly Doped Semiconductor Region

We shall consider now a semiconductor region which is non-uniformly doped with donor (or acceptor) impurities. The problem of calculating the unipolar current flow, in the presence of the space charge cannot be solved, generally speaking, without using a computer. However, this problem can be simplified to a certain extent if the doping profile can be approximated by successive regions of uniform doping.

Therefore, we shall consider a semiconductor sample of length L which consists in a number of successive regions characterized by a uniform doping $N_1, N_2, \ldots$ $\ldots$, respectively (Figure 3.30) and a thickness $L_1, L_2, \ldots\ldots$, where $L_1 + L_2 +$ $+ \ldots = L$. Let us denote by $E_c, E_{12}, E_{23}, \ldots\ldots$, etc., the field at the emitting plane ($x = 0$), at the interface between region 1 and region 2 ($x = L_1$), $\ldots\ldots$, respectively, and by $j_{n1}, j_{n2}, \ldots\ldots$ the neutral characteristics of regions 1,2, $\ldots$. We stress the

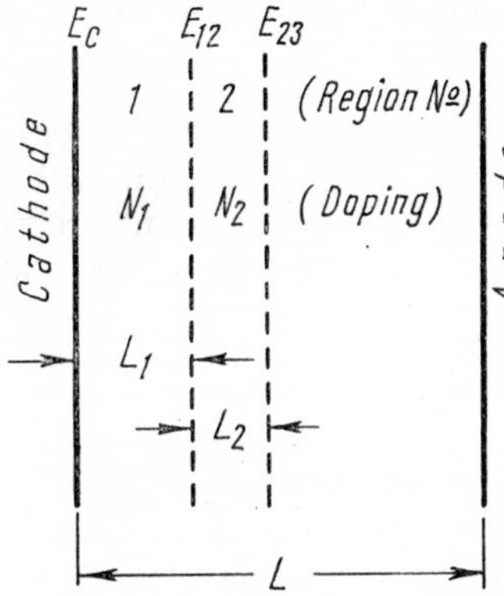

Figure 3.30. A semiconductor sample consisting of several semiconductors regions separated by plane-parallel boundaries and having uniform (but different) dopings.

fact that N_i ($i = 1,2,3, \ldots\ldots$) can be either positive or negative: the structure may comprise *p-n* junctions or even heterojunctions.

The steady-state behaviour of such a device can be studied by use of equations of the form (3.43)—(3.45). The cathode boundary condition is

$$J = j_{c1}(E_c) \tag{3.111}$$

The transition field E_{12} can be determined by using

$$\frac{eN_1L_1}{\varepsilon} = \int_{E_c}^{E_{12}} \frac{j_{n1}(E)\,\mathrm{d}E}{j_{c1}(E) - j_{n1}(E)} \tag{3.112}$$

which is equation (3.44) written for the region 1. We note the fact that equations (3.111) and (3.112) should determine implicitly a dependence $J = J(E_{12})$. Because J is also the particle current injected in region 2, the above characteristic may be considered as the control characteristic of the region 2.

The transition field E_{23} is given by

$$\frac{eN_2L_2}{\varepsilon} = \int_{E_{12}}^{E_{23}} \frac{jn_2(E)\,\mathrm{d}E}{J - j_{n2}(E)}\,. \tag{3.113}$$

etc. The total voltage drop is the sum of the individual voltages on each region in part, and are given by equations of the form (3.43).

The electric field should vary monotonously in each region in part, but does not necessarily do so in the whole structure. We suggest that the reader shows that the demonstration from Section 3.4 is no longer valid if the doping is non-uniform. Examples of non-uniformly doped semiconductor regions are given in the following chapters.

3.13 Incomplete Donor Ionization and Trapping

Until now we have assumed the fact that the fixed space charge is constant. We shall consider now a semiconductor with a single energetic level in the forbidden gap. The defect centres are uniformly distributed in space but their occupancy, at a given temperature, depends through 'a state equation' upon the local concentration of mobile carriers.

We shall discuss below, according to Wright [65], the behaviour of a *pnp* structure. The extreme sides are heavily doped as compared to the central region. Figure 3.31 shows a simple model for the energetic diagram of the central region. A single donor level, situated at energy W_D can exchange carriers with both conduction band and valence band. Direct band-to-band transitions have a low probability for a semiconductor like silicon [65] and are neglected. The rate of increase of mobile

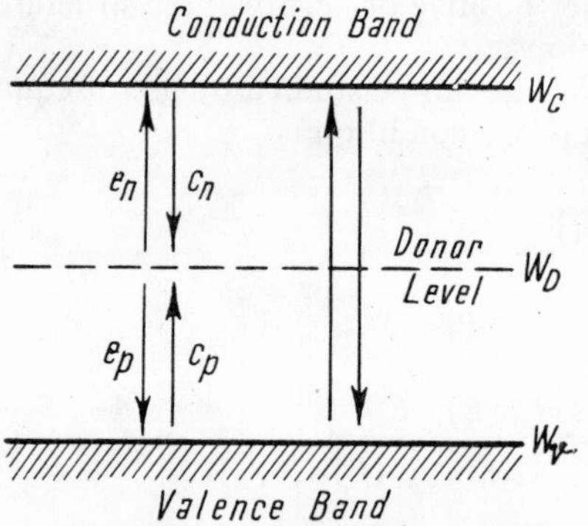

Figure 3.31. Simple energetic model of a semiconductor, with a single donor level exchanging carriers with both the conduction band and the valence band. The band-to-band transitions have a very low probability (see also [65]).

electron density is the difference between the rate of emission (e_n) of electrons from the centres less the rate of capture (c_n) of electrons by the centres (Figure 3.31). Therefore

$$\frac{\mathrm{d}n}{\mathrm{d}t} = n_D \nu_n \exp\left(-\frac{(W_c - W_D)}{kT}\right) - n\,\sigma_n v_{th(n)}\, p_D. \tag{3.114}$$

Similarly, the rate of increase of the mobile holes in the valence levels may be written

$$\frac{\mathrm{d}p}{\mathrm{d}t} = p_D \nu_p \exp\left(-\frac{W_D - W_v}{kT}\right) - p\sigma_p v_{th(p)}\, n_D. \tag{3.115}$$

In the above equations n_D and p_D represent the densities of electrons and holes respectively, occupying the donor level. Clearly, we must have

$$p_D + n_D = N_D \tag{3.116}$$

where N_D is the density of donor centers (uniform). The capture cross section of the empty centre is denoted by σ, the thermal velocity of the mobile carrier is v_{th} and the attempt-to-escape frequency is ν. Subscripts 'n' and 'p' represent electron and hole parameters, respectively.

Assume, first, the zero-current thermodynamic equilibrium case. By the principle of detailed balance we must have $e_n = c_n$ and $e_p = c_p$ or $\frac{dn}{dt} = 0$ and $\frac{dp}{dt} = 0$. Then from the above equations, one obtains

$$n_{D,0} = N_D \left[1 + \frac{\nu_n}{v_{th(n)}\,\sigma_n n} \exp\left(-\frac{W_c - W_D}{kT}\right)\right]^{-1} \tag{3.117}$$

$$p_{D,0} = N_D \left[1 + \frac{\nu_p}{v_{th(p)}\,\sigma_p p} \exp\left(-\frac{W_D - W_v}{kT}\right)\right]^{-1}. \tag{3.118}$$

Further, n_D and p_D can be eliminated between equations (3.116) — (3.118) and one obtains a relation between n and p (which determines the Fermi level).

Then, assume a relatively small (a few kT/e) bias voltage applied to the structure. A depleted region will occur in the vicinity of the negatively biased p^+ region. In this part of the central region, the mobile carrier densities will be considered to be zero. This is not strictly true, but it can be still a good approximation. As a matter of fact, there will be a very small minority electron current injected into the depleted region from the negatively biased p-type contact and there will be a small current carried by holes from the extremity of the depleted region which is in contact with the neutral n region. However, both these currents should be very small and the carriers injected in the depletion region will have a very low concentration because they are swept very rapidly out of this region [65].

The density of positively charged (ionized) donor centres in the depleted region (p_D) can be found by introducing $dp/dt = 0$, $dn/dt = 0$, $n = 0$, $p = 0$ in equations (3.114)–(3.116). When the depletion region just extends over the entire n region, the applied voltage is equal to the so-called 'reach-through' voltage V_{RT}, which may be obtained by integrating twice the Poisson equation

$$V_{RT} = \int_0^L dx \int_0^L \frac{e p_{D,0}\, dx}{\varepsilon} = \frac{e p_{D,0} L^2}{2\varepsilon}. \tag{3.119}$$

If the donor levels are shallow (i.e. $W_c - W_D$ is small and $p_{D,0} \simeq N_D$), then V_{RT} is equal to the punch-through voltage

$$V_{PT} = \frac{eN_D L^2}{2\varepsilon} \tag{3.120}$$

(see the punch-through diode discussed in Section 1.2). The opposite case is $W_D - W_v$ very small and $p_{D,0} \simeq 0$ so that $V_{RT} \simeq 0$.

By introducing, as Wright [65] did, $\nu_p = \nu_n$, one obtains

$$\frac{V_{RT}}{V_{PT}} = \frac{p_{D,0}}{N_D} = \left[1 + \exp 2 \, \frac{(W_c - W_v)/2 - W_D}{kT}, \right]^{-1}, \tag{3.121}$$

and the ratio V_{RT}/V_{PT} depends upon the position of the donor level with respect to the middle of the forbidden gap. If the donor level is a few kT/e above, the reach-through voltage will be practically equal to V_{PT}. If W_D is a few kT/e below, V_{RT} is very small as compared to V_{PT}. In the second case, we must have space-charge injection with trapping, because the deep lying donors will act as traps for mobile injected holes.

The steady-state characteristics of such a *pnp* structure have been calculated by Wright [65], by neglecting electron flow and diffusion current, assuming constant mobility and using the boundary condition for SCLC injection (zero field at the positively biased contact)*⁾. The curves plotted in Figure 3.32 have been obtained for

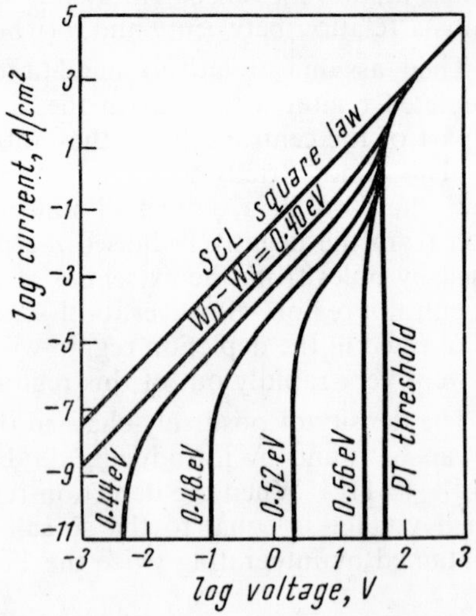

Figure 3.32. The current-voltage characteristics computed by Wright [65] for a silicon *pnp* structure.

*⁾ By using similar hypotheses, Timan and Fesenko [66] have calculated the analytical $J - V$ dependence for a $n^+ nn^+$ structure with incomplete ionization of donors in the central region.

a *pnp* silicon sample having a central n-type region 25 μm thick, containing a density of 10^{14}cm^{-3} donors, at various energetic levels, as shown on each curve ($\mu_p = 500$ cm^2Vs, $\varepsilon = 10^{-10}$ F m^{-1}, $W_c - W_v = 1.12$ eV, $T = 300$ K, $\nu_n = \nu_p$).

If the donor level is a few kT/e above the mid-band position, the device will behave like a normal punch-through structure (Section 1.2). The current rises very steeply and then follows the square-law of the SCLC current (Figure 3.32).

If $W_D - W_v$ is very small, the donor levels 'are swamped by these (valence) levels and have no effects' [65]. The $J - V$ characteristic should coincide with the ideal square law (Figure 3.32).

If the donor levels lie in the lower half of the forbidden gap but sufficiently far from the valence band, these levels will act as hole traps. At low currents these traps are *shallow* (i.e. practically unoccupied by holes, $p_D \simeq 0$, $n_D \leqslant N_D$). At higher currents almost all traps will be filled ($p_D \leqslant N_D$) and are named *deep*. The transition from one situation to another is indicated by a steep increase of the current, which takes place at approximately $V \leqslant V_{PT}$.

We shall discuss below the case of shallow trapping which was first studied by Rose [2] and Lampert [3]. We assume the fact that V_{RT} is extremely small as compared to V_{PT} and the voltage bias satisfies the equation

$$V_{RT} \ll V \ll V_{PT}. \tag{3.122}$$

Consequently, we must have

$$p_{D,0} \ll p_D \ll N_D. \tag{3.123}$$

Therefore, equation (3.115) yields

$$\frac{\partial p}{\partial t} = -\frac{p_D}{\tau_t} + \frac{p}{\tau}, \tag{3.124}$$

where

$$\tau_t = \nu_p^{-1} \exp \frac{W_D - W_v}{kT}, \tag{3.125}$$

is the lifetime of a trapped hole, and

$$\tau_f = \frac{1}{\sigma_p v_{th(p)} n_D} \simeq \frac{1}{\sigma_p v_{th(p)} N_D} \tag{3.126}$$

is the lifetime of a free hole, and is practically constant because $n_D = N_D - p_D \simeq$ $\simeq N_D$, according to equation (3.123). Note the fact that almost all donors are not ionized. The Poisson equation is

$$\frac{\partial E}{\partial x} = \frac{e(p + p_D)}{\varepsilon}. \tag{3.127}$$

The steady-state characteristic will be obtained by use of equations (3.124) (where $\mathrm{d}p/\mathrm{d}t = 0$) and (3.127). Because $J = e\mu pE$, one obtains

$$J = \frac{9}{8}\,\delta\varepsilon\mu\,\frac{V^2}{L^3}, \tag{3.128}$$

where

$$(\delta = 1 + \tau_t/\tau_f)^{-1} \tag{3.129}$$

denotes the ratio of free hole density to the total (free and trapped) hole density. This ratio is the same at every distance x from the injecting electrode. We note the fact that (over a voltage range) the $J - V$ characteristic has the same parabolic shape predicted for the ideal SCL current flow in solid-state diodes (see equation P.1.2) but the current is reduced at the fraction $\delta < 1$ from the trap-free value. The reader may also note the fact that δ decreases (and can become extremely small as compared to unity) if the trap density and the energetic distance from the valence band increases. A set of shallow levels should produce the same effect. However, the 'shallow' or 'deep' quality of a trap level depends upon the injection level*). The form of a $J - V$ characteristic dominated by trapping can be very complex. Sometimes, this SCLC characteristic is used to determine some properties of impurities and defects in solids [5], [67], [68].

Finally, we note the fact that there is no difference in equations describing the punch-through diode and the (insulator) diode with deep traps [3], [13], [61].

Problems

3.1 Consider the image-force lowering of the barrier (Section 3.3). Show that the potential maximum (hole injection) occurs at the distance $x_m = \frac{1}{2}(e/4\pi\,\varepsilon_s E_c)^{1/2}$ from the metal-semiconductor interface (E_c is the electric field in semiconductor). Calculate the distance x_m for a metal-silicon contact and discuss the result.

3.2 Demonstrate that if the control characteristic is biased in a saturation region, the $J - V$ dependence will tend to saturate.

3.3 Derive the parametric equations of the SCL current, starting from the more general equations given in Section 3.9. Discuss separately the majority and minority carrier injection. Show that if the bias current is sufficiently high, the $J - V$ dependence given by equation (P.1.2) will be obtained (the carrier mobility is kept constant).

3.4 Find, approximately, how large should be the resistivity of silicon used to fabricate a *square-law* n^+nn^+ resistor.

---

*) In other words, it depends upon the position of the Fermi level in the forbidden gap of the semiconductor. If the Fermi level is a few kT/e below the trap level, the level is shallow, etc.

3.5 Calculate and plot the $J - V$ characteristics of a Schottky-emission cathode diode.
Hint: Use the theory of Section 3.9.

3.6 Show mathematically, starting from the equations of Section 3.9, that the punch-through voltage for minority-carrier injection does not depend upon the emission at the cathode ($V_{PT} = eNL^2/2\varepsilon$).

3.7 How is it possible to apply the theory developed in Section 3.10 to a non-uniformly doped semiconductor region?
Hint: The doping profile of this semiconductor region will be approximated by segments of constant doping.

3.8 Calculate the $J - V$ characteristics of a semi-insulating diode, by assuming negligible fixed space-charge, negligible diffusion, constant mobility, and by use of $p(0) =$ given, as the cathode boundary condition. Show that the asymptotic dependence for small bias currents is the SCLC square-law dependence ($I \propto V^2$).

Part 2

Linear Theory: Small-signal Operation Stability and Noise

4

Transport Coefficients and General Linear Analysis

4.1 Introduction

This chapter will put forward a general linear theory for unipolar and one-dimensional currents in solid-state devices.

Here, the small-signal behaviour will be analysed by assuming the fact that the applied signal merely causes small perturbations of the steady state. Every electrical quantity, such as the current density or the carrier concentration, consists of a large steady state value plus a very small time-dependent perturbation. Such a form of solution is introduced into the differential equations and solved for the perturbation, by taking into account the fact that this perturbation is very small and its presence does not modify the steady-state component.

The mathematical technique is straightforward. The small-signal behaviour is, however, very complex and rich in interesting effects. Certain internal phenomena which are less evident in the steady state become dominant in the small-signal operation.

Assume a small sinusoidal voltage superimposed on a relatively large steady-state bias and applied to a unipolar semiconductor diode. If the frequency of the sinusoidal signal is very low, the device behaves for alternating current as a simple resistance, which is equal to the incremental resistance (for unit area)

$$R_i = \left(\frac{\mathrm{d}J}{\mathrm{d}V}\right)^{-1}$$

defined as the reciprocal of the slope of the steady-state characteristic for a given d.c. bias. The device behaviour is said to be quasi-stationary.

As the frequency increases, we expect the diode capacitance to shunt the resistance (series parasitic resistances and inductances are neglected). It really does so but this effect is accompanied or even obscured, by several 'dramatic' changes in the device behaviour.

First, at relatively low frequencies ($10^2 - 10^6$ Hz), when the signal period is still much larger than the diode RC time constant, a modification of the device resistance, accompanied by large capacitive effects may indicate the effect of trapping inside the bulk or at interface [8], [69]. This occurs simply because the time constants of carrier trapping and releasing are finite and the readjustment of the injected charge cannot follow the signal. An analysis for the SCL diode with shallow traps shows that the capacitive effects disappear at higher frequencies but the device conductance is smaller than the low-frequency conductance.

At high frequencies (of the order of $10^9 - 10^{10}$ Hz) the diode a.c. behaviour is modified by the transit-time effect: the charge carriers which are injected into the semiconductor experience a change of the a.c. electric field distribution during the time they travel from one electrode to another. Due to their inertia in transit, an inductive effect will accompany the response of the particle current and the device will exhibit a capacitance which is smaller than the space-charge steady-state capacitance. At even higher frequencies the device conductance and parallel capacitance depend in an oscillatory fashion upon the signal frequency.

Of paramount importance is the fact that certain devices may exhibit a high-frequency *negative* resistance, even when the incremental resistance is positive. The transit-time effects play a major rôle in the process of *internal power generation* which leads to such a negative resistance. The linear theory should predict high-frequency instability, and a.c. power amplification or generation due to such a negative resistance.

Recent theories have shown that the processes taking place at the emitting contact may have considerable influence upon the behaviour in the transit-time frequency region. We shall discuss this later.

At very high frequencies the transit-time effects 'quieten-down' and the diode behaves essentially as a capacitor whose capacitance is equal to the geometrical capacitance of the electrodes.

At still higher frequencies ($10^{11} - 10^{12}$ Hz) electron relaxation phenomena occur. These were discussed already in Chapter 2 for a semiconductor resistance (without internal space-charge).

The small-signal theory should give:

(1) The frequency characteristics and the high-frequency equivalent circuits of amplifying devices (like the field-effect devices), and of negative-resistance structures.

(2) The stability for a given d.c. bias (this is important for Gunn effect devices, etc.).

(3) Some noise properties.

Furthermore, the small-signal analysis clarifies some aspects of the internal mechanism of device behaviour. Finally, the linear theory is a good start for certain methods of *non-linear* analysis.

The small-signal theory may be conveniently developed by using the so-called *transport coefficients*. These coefficients take full account of the *bulk* phenomena for a one-dimensional semiconductor region. Used in conjunction with proper boundary conditions, the transport coefficients should give the small-signal response for every semiconductor device made from one or more (cascaded) semiconductor regions. Such a separation of bulk and interface processes was also introduced in the steady-state analysis (Chapter 3) and is very convenient for a systematic study of a wide group of electronic devices.

4.2 Basic Equations

We consider in this chapter a homogeneous semiconductor region bounded by two parallel planes (Figure 3.1b): $x = 0$ is the emitting plane, or, shortly, the cathode and $x = L$ is the collecting plane (anode). These two planes do not necessarily coincide with actual electrodes (see, for example, Section 3.6). The mobile holes are injected at $x = 0$ and *drift* towards the anode due to an electric field parallel to the x direction.

The one-dimensional form of Maxwell's continuity equation is

$$J(t) = j + \varepsilon \frac{\partial E}{\partial t}, \qquad \frac{\partial J}{\partial x} = 0. \tag{4.1}$$

Here ε is constant and real, and thus we ignore possible high-frequency relaxation effects. The convection or particle current is denoted by j. We assume that the hole flow is carried only by drift, with field-dependent mobility

$$j = epv(E). \tag{4.2}$$

The fact should be stressed that neglection of diffusion in the calculation of steady-state characteristics does not imply that the alternating current component of the diffusion current is also negligible. We shall return to this problem later.

The hole velocity is assumed to be a local and instantaneous function upon the electric field intensity, i.e.

$$v = v(E) \tag{4.3}$$

where the time t does not occur explicitly. Therefore, for a.c. operation, the period of the applied signal should be much longer than the mobility relaxation time or than the inter-valley scattering time. This will set an upper limit for our results at values of several tens of GHz or higher [25], [26].

The divergence of the electric field is given by the Poisson equation

$$\frac{\partial E}{\partial x} = \frac{e(p - N)}{\varepsilon}, \tag{4.4}$$

where $(-N)$ is the density of the fixed (immobile) space charge. As shown in Section 3.5.

$$N = N_A - N_D - p_t \tag{4.5}$$

where p_t is the density of trapped holes. An equation of the form

$$\mathcal{T}(p, p_t) = 0, \tag{4.6}$$

should be also given [4]. Equation (4.6) describes the dynamic trapping-releasing process (see also Section 3.13) [8], [70].

Finally, if the oscillation wavelength is much longer than L the total potential drop is given by (see equation (2.38)):

$$V(t) = \int_0^L E(x,t)\,\mathrm{d}x. \tag{4.7}$$

From equations (4.1) – (4.4) we obtain

$$\left(\varepsilon\frac{\partial E}{\partial x} + eN\right) v(E) + \varepsilon\frac{\partial E}{\partial t} = J(t) \tag{4.8}$$

where, in general $N = N(x, t)$. If $J(t)$ is assumed to be given, the above relation is a first-order non-linear equation in $E = E(x, t)$ with partial derivatives. Once $E = E(x, t)$ found, the voltage drop will be calculated from equation (4.7). Unfortunately, in most problems $V(t)$ is given and $J(t)$ should be calculated. However, for a small-signal analysis this 'inversion' does not represent a problem.

Equation (4.8) is quasi-linear, i.e. linear with respect to its derivatives. The general solution of such an equation should depend upon an arbitrary function of $k-1$ variables, where k is the number of independent variables in the differential equation (here $k = 2$). Thus, the general solution of equation (4.8) may be determined by imposing *the boundary condition*

$$E(0, t) = \text{given function of } t. \tag{4.9}$$

Some possible boundary conditions were discussed in Chapter 3, with regard to the steady-state behaviour. The emission processes discussed there may be considered, for our purposes, practically instantaneous. Thus, the boundary condition may be written in a general (and implicit form)

$$j(0, t) = F[E(0, t)] \tag{4.10}$$

where j is $j(x, t)$ and F does not explicitly contain the time t.

4.3 Method of Linear Analysis

There are two main approaches to the small-signal (linear) problem.

(A) The small-signal response may be calculated as the first-order approximation in a series of successive approximations which attempt to solve the general equations under the assumption of a small, *but not infinite small*, magnitude of the applied signal. Such an approach is more general than a linear analysis because it can also yield non-linear effects (second and higher order approximations).

An example is the analysis of high-frequency behaviour of planar vacuum devices, developed successively by Benham [71], Müller [72], and Llewellyn [73]. This analysis was reviewed recently by Birdsall and Bridges [74]. This method, usually called the 'Llewellyn method' was also applied to the small-signal be haviour of unipolar solid-state devices [10], [17], [75]. A similar approach is used by ourselves in Chapter 10, for evaluation of non-linear effects.

(B) The small-signal effects are calculated by assuming vanishingly small perturbations for all quantities of interest, by introducing these quantities (current, field, etc.) *directly* into the basic equations and linearizing them with respect to these perturbations. This method is simpler than the first one. No non-linear effects can be calculated in this manner. The method was first used in conjunction with semiconductor devices.

The 'impulsive impedance' method suggested by Shockley [14], [16] belongs to this category. However, the more complicated Llewellyn method was still used for more than a decade. The 'pure' linear method was reinvented in a different form*) by McCumber and Chynoweth [76], Draganescu [77], van der Ziel and Hsu [78], etc. It was later shown that the vacuum diode can be studied in the same manner [79], [80]. The linear method is superior to the Llewellyn method because it is simpler and can be successfully applied to more complicated problems. We shall exemplify this. The Llewellyn method was ineffective for two problems, namely: (a) the simultaneous evaluation of trapping and transit-time effects [8], and, (b) the effect of mobility field-dependence upon the high-frequency behaviour of SCLC diodes [81]. Both these problems were later solved by Dascalu [70] and, respectively, by Dascalu [82] and Kroemer [83], by using the linear method.

It is this linear method [76], [77], [78] that we shall use throughout Chapters 4 – 9. The non-linear problems will be solved by specific methods (Chapters 10 – 14).

The method will be exemplified later. In this section we shall indicate only the main steps of this approach.

We consider the alternating current operation at small-signals. The voltage drop which is applied on the semiconductor region is denoted by

$$V(t) = V_0 + \underline{\mathbf{V}}_1 \exp i\omega t, \quad |\underline{\mathbf{V}}_1| \ll V_0. \tag{4.11}$$

Here V_0 is the steady-state bias. The second term in the right-hand part is the applied signal. $\underline{\mathbf{V}}_1$ is a complex number called the complex amplitude and $i = \sqrt{-1}$ is the imaginary unit. Altogether $\underline{\mathbf{V}}_1 \exp i\omega t$ is a rotating phasor, familiar in electrical engineering. The actual signal is either the real or the imaginary part of $\underline{\mathbf{V}}_1 \exp i\omega t$.

The sum of a real number V_0 and a complex quantity in equation (4.11) is quite unusual (in fact $V(t)$ is not a real function of time, as expected). However, this does not affect the validity of our calculations, since the two components of $V(t)$ will be separated during calculations.

*) Shockley's analysis [14] takes place in the time domain. Later, the analysis in the frequency domain was used almost exclusively [76] – [78], etc.

The signal is much smaller than the d.c. bias and the response should be, to a first approximation, of the same form, i.e.

$$J(t) = J_0 + \mathbf{\underline{J}}_1 \exp i\omega t, \quad |\mathbf{\underline{J}}_1| \ll J_0. \tag{4.12}$$

(a sinusoidal variation superimposed on a d.c. component of relatively large amplitude). $J(t)$ from equation (4.12) will be introduced in equation (4.8). The solution $E(x, t)$ of the latter should be of the same form

$$E(x, t) = E_0(x) + \mathbf{\underline{E}}_1(x) \exp i\omega t, \quad |\mathbf{\underline{E}}_1| \ll E_0. \tag{4.13}$$

We note at this point that the separation of alternating and direct currents components not only of current and voltage drop, but also of electron velocity, charge density and electric-field intensity, is known even since Llewellyn [73]. What is new in the method described here is the fact that $E(x, t)$ given by equation (4.13) is introduced directly into the differential equation of the problem.

Then, because boldface symbols correspond to vanishingly small quantities, the above equation should lead to two separate equalities: one for the d.c. or zero-order (large) quantities, and another for a.c. or first order (small) quantities. Second order quantities are neglected. Thus, exp $i\omega t$ which multiplies all the first order quantities will disappear and the time, t, is eliminated.

There are now two ordinary differential equations to be solved. The first one is in $E_0(x)$ and is exactly the one met in the steady-state problem. The second one has $\mathbf{\underline{E}}_1 = \mathbf{\underline{E}}_1(x)$ as a dependent variable and must, of course, be linear. Once $\mathbf{\underline{E}}_1(x)$ is found, the signal $\mathbf{\underline{V}}_1$ will be obtained from equations (4.7), (4.11) and (4.13). Because $\mathbf{\underline{E}}_1(x)$ should be proportional to $\mathbf{\underline{J}}_1$ from the a.c. part of equation (4.8), one obtains immediately the ratio $\mathbf{\underline{V}}_1/\mathbf{\underline{J}}_1$, which is the (complex) impedance of the semiconductor region (for unit-area electrodes) *).

4.4 Transport Coefficients

The a.c. transport coefficients do completely characterize the a.c. non-stationary behaviour of a semiconductor region (defined as in Section 4.2).

Consider a diode consisting of a semiconductor region and an injecting cathode at one boundary of this region. The a.c. behaviour (i.e. the impedance) of this diode may be calculated easily as a function of the transport coefficients and a proper boundary condition at the cathode.

---

*) An a.c. boundary condition should be also used to determine $\mathbf{\underline{E}}_1 = \mathbf{\underline{E}}_1(x)$. Now it is clear that this method is exactly the one currently used in the theory of the bipolar transistor (after 1950). We try to imagine why this method was not used for unipolar semiconductor devices and Llewellyn's method was preferred. We note the fact that there is an analogy between the space-charge unipolar solid-state devices and the space-charge vacuum devices. On the other hand there is an important difference between the analysis of unipolar and bipolar currents, the *displacement current* is taken into account in the former and neglected in the latter case.

Assume now a more complicated device which may be dissected in several regions. The a.c. properties of this device will be found by using the transport coefficients for each region and also boundary and continuity conditions at the interfaces.

For a semiconductor region described by equations (4.1) — (4.7) we shall define four transport coefficients, as follows:

$$\underline{\mathbf{V}}_1 = K_{11}\underline{\mathbf{J}}_1 + K_{12}\underline{\mathbf{E}}_1(0) \tag{4.14}$$

$$\underline{\mathbf{E}}_1(L) = K_{21}\underline{\mathbf{J}}_1 + K_{22}\underline{\mathbf{E}}_1(0). \tag{4.15}$$

Here, the total current and the cathode field $\underline{\mathbf{E}}_1(0)$ are taken as the 'input' variables, and the voltage drop and the anode field as the 'output' variables.

These coefficients are analogous to the Llewellyn-Peterson coefficients [84] for planar vacuum regions. Similar coefficients were introduced by Yoshimura [17] for semiconductor regions (see below).

Equations of the form of equations (4.14) and (4.15) can be readily obtained as shown in the preceding paragraph. Detailed calculations are given below. We anticipate the fact that K_{ij} are complex numbers which depend upon the frequency of the applied signal and also upon the steady-state bias of the semiconductor region.

Now we assume the fact that the cathode plane $x = 0$ is an actual emitting electrode which is characterized by a boundary condition of the form (4.10). The a.c. component of the particle current at $x = 0$ is

$$\underline{\mathrm{j}}_1(0) = \sigma_1(0)\underline{\mathbf{E}}_1(0), \quad \sigma_1(0) = \left(\frac{\mathrm{d}j(0, t)}{\mathrm{d}E(0, t)}\right)_{\mathrm{d.c}}. \tag{4.16}$$

The above relation was obtained by introducing $\underline{\mathbf{E}}(0, t)$ given by equation (4.13) into equation (4.10), using Taylor series expansion and equalizing the first-order quantities. We re-obtain $\sigma_1(0)$ which is the slope of the control characteristic (Section 3.6).

The a.c. part of the continuity equation (4.1) written for $x = 0$ yields

$$\underline{\mathbf{J}}_1 = \underline{\mathrm{j}}_1 + i\omega\varepsilon\,\underline{\mathbf{E}}_1(0). \tag{4.17}$$

The impedance of the diode considered above is

$$Z = \frac{\underline{\mathbf{V}}_1}{\underline{\mathbf{J}}_1} = K_{11} + \frac{K_{12}}{\sigma_1(0) + i\omega\varepsilon} \tag{4.18}$$

and was obtained by use of equations (4.14), (4.16) and (4.17). We identify two special cases. If the current is completely space-charge limited then $\sigma_1(0) \to \infty$ (vertical control characteristic, see Section 3.6) and $Z = K_{11}$. The opposite case is: complete electrode limitation, constant injected current, horizontal control characteristic and $\sigma_1(0) = 0$, etc.

If we consider several cascaded semiconductor regions which are not separated by electrodes but merely differ in their properties (doping), then equations (4.14) and (4.15) should be applied for each region. Equation (4.15) will be used to determine the electric field at the $x = L$ interface, which is the boundary condition for the next region, etc.

For other physical situations the particle current should be continuous at the interface of two regions. It is then more convenient to use the C_{ij} coefficients defined by

$$\underline{\mathbf{V}}_1 = C_{11}\,\underline{\mathbf{J}}_1 + C_{12}\,\underline{\mathbf{j}}_1\,(0) \tag{4.19}$$

$$\mathbf{j}_1\,(L) = C_{21}\,\underline{\mathbf{J}}_1 + C_{22}\,\mathbf{j}_1\,(0). \tag{4.20}$$

These were the original coefficients introduced by Yoshimura [17]. We have

$$C_{11} = K_{11} + K_{12}/i\omega\varepsilon \tag{4.21}$$

$$C_{12} = -\,K_{12}/i\omega\varepsilon \tag{4.22}$$

$$C_{21} = 1 - i\omega\varepsilon\,K_{21} - K_{22} \tag{4.23}$$

$$C_{22} = K_{22}. \tag{4.24}$$

Sometimes, a boundary condition of the form

$$\underline{\mathbf{j}}_1\,(0) = \beta\,\underline{\mathbf{J}}_1, \tag{4.25}$$

is used. β may be a complex quantity, defined, for example, as a transport factor which takes into account the delay of carriers in a region situated at $x < 0$ (examples in Chapter 6). From equations (4.19) and (4.25) the impedance of the semiconductor region is

$$\underline{\mathbf{V}}_1/\underline{\mathbf{J}}_1 = C_{11} + \beta C_{12}. \tag{4.26}$$

We shall outline the fact that the transport coefficients are calculated for a certain semiconductor region with *specific* properties. The form of these coefficients depends, for example, upon the velocity field dependence. This dependence is different, of course, for holes in silicon and for electrons in InP. A complete different set of transport coefficients should be used if one has to take trapping into account.

There is a peculiar property of *solid-state* regions, the fact that they cannot be described by 'universal' transport coefficients as the vacuum regions are described by Llewellyn-Peterson coefficients [84].

In the next Section we shall calculate the transport coefficients for semiconductor with negligible trapping.

4.5 Derivation of Transport Coefficients for Semiconductors

The model to be considered here is given in Section 4.2, but we assume that

$$N(x, t) = N = \text{const.} \tag{4.27}$$

i.e. only completely ionized donors and deep traps are taken into account. We recall the fact that diffusion is neglected and the velocity-field dependence is still arbitrary.

The basic equation of the problem is given by equation (4.8). Here we assume a solution of the form

$$E(x, t) = E_0(x) + E_1(x, t), \ |E_1| \ll E_0. \tag{4.28}$$

The perturbation, $E_1(x, t)$, which is a very small quantity as compared to $E_0(x)$, is the result of a small perturbation $V_1(t)$ of the d.c. bias V_0, i.e.

$$V(t) = V_0 + V_1(t), \ |V_1| \ll V_0 \tag{4.29}$$

where $V_1(t)$ may have an arbitrary time-dependence. From equations (4.7), (4.28) and (4.29) one finds

$$V_0 = \int_0^L E_0(x)\,\mathrm{d}x, \ V_1(t) = \int_0^L E_1(x)\,\mathrm{d}x. \tag{4.30}$$

The analytical procedure follows closely the method sketched in Section 4.3 for a sinusoidal time dependence. We shall determine the average values (subscript 0) by neglecting the small quantities. Then, the small quantities (denoted with subscript 1) will be found by using the above determined d.c. quantities and neglecting any term containing a product of two small quantities. This is *a typical perturbation method* which is familiar in physics, for example in solving Schrödinger's equation.

The hole velocity, v, may be written successively as

$$v = v(E) = v(E_0 + E_1) = v(E_0) + \left(\frac{\mathrm{d}v}{\mathrm{d}E}\right)_{E=E_0} \times E_1 + \ldots, \tag{4.31}$$

or, approximately

$$\left.\begin{aligned} &v = v_0(x) + v_1(x, t), \ |v_1| \ll v_0, \\ &\text{where} \\ &v_0 = v(E_0), \ v_1(x, t) = \left(\frac{\mathrm{d}v}{\mathrm{d}E}\right)_{E=E_0} \times E_1(x, t). \end{aligned}\right\} \tag{4.32}$$

The total current density, equation (4.1), should be also of the form

$$J(t) = J_0 + J_1(t), \ |J_1| \ll J_0 \tag{4.33}$$

By introducing equations (4.28), (4.31), (4.32) into equation (4.8) and neglecting all small quantities we obtain

$$J_0 = \left(\varepsilon \frac{\mathrm{d}E_0}{\mathrm{d}x} + eN\right) v_0(E_0) \tag{4.34}$$

which is the equation of the steady-state regime (compare with Section 3.5).

Then we solve equation (4.8) for small quantities by taking equation (4.34) into account. Equation (4.8) then becomes

$$J_1(t) = \varepsilon v_0(E_0) \frac{\partial E_1}{\partial x} + \left(\varepsilon \frac{\mathrm{d}E_0}{\mathrm{d}x} + eN\right) v_1 + \varepsilon \frac{\partial E_1}{\partial t} \tag{4.35}$$

or

$$\varepsilon v_0 \frac{\partial E_1(x, t)}{\partial x} + \varepsilon \frac{\partial E_1(x, t)}{\partial t} + \frac{J_0}{v_0}\left(\frac{\mathrm{d}v}{\mathrm{d}E}\right)_{E=E_0} \times E_1(x, t) = J_1(t). \tag{4.36}$$

This is a first-order *linear* equation with partial derivatives. Thus the assumption of very small perturbations of the physical quantities allowed the linearization of equation (4.8) with respect to these perturbations.

The second important step is the transformation of equation (4.36) into an *ordinary* equation by applying the Laplace transform. By denoting

$$\mathcal{L}\{E_1(x, t)\} = \mathcal{E}_1(x; s) \tag{4.37}$$

$$\mathcal{L}\{J_1(t)\} = \mathcal{J}_1(s) \tag{4.38}$$

$$\mathcal{L}\{V_1(t)\} = \mathcal{V}_1(s), \tag{4.39}$$

where s is the complex variable in the Laplace transform, equation (4.36) becomes

$$\begin{aligned} &\varepsilon v_0 \frac{\mathrm{d}\mathcal{E}_1(x; s)}{\mathrm{d}x} + \varepsilon\left[s\,\mathcal{E}_1(x; s) - E_1(x, 0)\right] + \\ &+ \frac{J_0}{v_0}\left(\frac{\mathrm{d}v}{\mathrm{d}E}\right)_{E=E_0} \times \mathcal{E}_1(x; s) = \mathcal{J}_1(s). \end{aligned} \tag{4.40}$$

We recall the fact that the Laplace transform is not valid for $t < 0$. The signal $V_1(t)$ is applied at $t = 0$. The *initial condition* $E_1(x, 0)$ = given, should be specified. $E_1(x, 0)$ does not depend, however, upon x. The divergence of $E_1(x, t)$ is determined by the supplementary mobile charge injected by the applied signal (Poisson equation written for small first-order quantities). At $t = 0$ this charge does not yet exist inside the semiconductor because it requires a finite time for propagation. Thus, at $t = 0$ the supplementary field should be uniform for $x > 0$. Therefore

$$\mathcal{J}'(s) = \mathcal{J}_1(s) + \varepsilon E_1(x, 0) \tag{4.41}$$

does not depend on x and equation (4.40) may be rewritten as

$$\varepsilon v_0 \frac{\mathrm{d}\,\mathcal{E}_1(x; s)}{\mathrm{d}x} + \left[\frac{J_0}{v_0}\left(\frac{\mathrm{d}v}{\mathrm{d}E}\right)_{E=E_0} + \varepsilon s\right]\mathcal{E}_1(x; s) = \mathcal{J}_1'(s). \tag{4.42}$$

Here, s may be regarded as a parameter. Thus, equation (4.42) is an ordinary differential equation. This is the basic equation for small-signal operation. In principle, the small-signal analysis should be simpler than the steady-state analysis because equation (4.42) is linear, whereas the differential equation for the steady state (4.34) is non-linear. However, the small signal calculations require the distribution of the steady-state field, i.e. $E_0 = E_0(x)$.

It is interesting to note that there are two particular cases when equation (4.42) has *constant coefficients* and its solution is very simple. These are:

(a) Quasi-neutrality; the d.c. electric field is uniform and v_0 and $(\mathrm{d}v/\mathrm{d}E)_{E=E_0}$ are independent of x. If the d.c. field is not actually uniform, we may consider an average value, as McCumber and Chynoweth did in their paper [76]. The averaging of d.c. values (i.e. the neglection of d.c. space-charge) is permissible in only a few cases*), otherwise the small-signal calculation is in error.

(b) Saturated carrier velocity; therefore $v = v_l = \text{const.}$ and $(\mathrm{d}v/\mathrm{d}E) = 0$. Consequently the d.c. bias can affect the small-signal operation only through the boundary condition at $x = 0$.

We return now to the general case. Equation (4.42) will be integrated by use of a change of independent variable, i.e. the transit time

$$T_x = \int_0^L \frac{\mathrm{d}x}{v_0} \tag{4.43}$$

will be replaced for x (see Section 3.9). By taking into account $\mathrm{d}x = v_0\,\mathrm{d}T_x$ and also equation (4.34), we obtain

$$\frac{\mathrm{d}\,\mathcal{E}_1}{\mathrm{d}T_x} + P(T_x)\,\mathcal{E}_1 = \mathcal{J}_1'(s)/\varepsilon \tag{4.44}$$

*) This averaging is possible in semiconductor diodes with majority carrier injection, high doping-length product and positive differential mobility (the last condition is required for stability).

where

$$P(T_x) = \frac{J_0}{\varepsilon v_0}\frac{dv_0}{dE_0} + s = \frac{dv_0}{dT_x}\left\{v_0(T_x)\left[1 - \alpha v_0(T_x)\right]\right\}^{-1} + s \tag{4.45}$$

and

$$\alpha = \frac{eN}{J_0}. \tag{4.46}$$

$v_0 = v_0(T_x)$ in equation (4.45) is implicitly given by

$$T_x = \frac{\varepsilon}{J_0}\int_{E_0(0)}^{E_0(x)} \frac{dE_0}{1 - \alpha v_0(E_0)} \tag{4.47}$$

which was obtained from equation (4.34).

The general solution of equation (4.44) should be of the form

$$\mathcal{E}_1(T_x) = \exp\left(-\int P\,dT_x\right)\left[\mathcal{C} + \frac{\mathcal{J}'(s)}{\varepsilon}\int \exp\left(\int P\,dT_x\right)dT_x\right] \tag{4.48}$$

$\mathcal{C}$ being an integration constant. By introducing $P(T_x)$ from equation (4.45) into the above equation, one finds

$$\mathcal{E}_1(T_x) = \frac{1 - \alpha v_0}{v_0}\left[\exp(-sT_x)\right]\left[\mathcal{C} + \frac{\mathcal{J}'(s)}{\varepsilon}\int_0^{T_x}\frac{v_0(\eta)}{1 - \alpha v_0(\eta)}\exp(sT_x)\,dT_x\right] \tag{4.49}$$

or

$$\mathcal{E}_1(T_x) = \frac{\mathcal{J}'(s)}{\varepsilon}\left(\frac{1 - \alpha v_0(T_x)}{v_0(T_x)}\right)\int_0^{T_x}\frac{v_0(\eta)}{1 - \alpha v_0(\eta)}\left[\exp s(\eta - T_x)\right]d\eta + \left[\exp(-sT_x)\right]\frac{1 - \alpha v_0(T_x)}{v_0(T_x)}\,\frac{v_0(0)}{1 - \alpha v_0(0)}\,\mathcal{E}_1(0). \tag{4.50}$$

Then, from equation (4.30) we obtain

$$V_1 = \int_0^L E_1\,dx = \int_0^{T_L} v_0 E_1\,dT_x \tag{4.51}$$

and

$$\mathcal{V}_1(s) = \int_0^{T_L}\mathcal{E}_1(T_x)\,v_0(T_x)\,dT_x \tag{4.52}$$

where $\mathcal{E}_1(T_x)$ is given by equation (4.50). Thus $\mathcal{V}_1(s)$ may be written as

$$\mathcal{V}_1(s) = \mathcal{K}_{11}(s)\mathcal{J}_1'(s) + \mathcal{K}_{12}(s)\mathcal{E}_1(0; s) \tag{4.53}$$

and $\mathcal{E}_1(T_x)$ becomes for $x = L$

$$\mathcal{E}_1(T_L; s) = \mathcal{K}_{21}(s)\mathcal{J}_1'(s) + \mathcal{K}_{22}(s)\mathcal{E}_1(0; s). \tag{4.54}$$

Note the fact that $\mathcal{K}_{ij}$ are a sort of *transport coefficients* (Section 4.4) written in operational form. From equations (4.50) and (4.54) there follows

$$\mathcal{K}_{21}(s) = \frac{1}{\varepsilon}\frac{1-\alpha v_0(T_L)}{v_0(T_L)}\int_0^{T_L}\frac{v_0(\eta)}{1-\alpha v_0(\eta)}\exp s(\eta - T_L)\,d\eta \tag{4.55}$$

$$\mathcal{K}_{22}(s) = \exp(-sT_L)\left[\frac{1-\alpha v_0(T_L)}{v_0(T_L)}\right]\left[\frac{v_0(0)}{1-\alpha v_0(0)}\right]. \tag{4.56}$$

By taking into account equations (4.50), (4.52) and (4.53) one obtains

$$\mathcal{K}_{11}(s) = \frac{1}{\varepsilon}\int_0^{T_L}dy\left[1-\alpha v_0(y)\right]\exp(-sy)\int_0^y\frac{v_0(\eta)}{1-\alpha v_0(\eta)}\exp s\eta\,d\eta \tag{4.57}$$

$$\mathcal{K}_{12}(s) = \frac{v_0(0)}{1-\alpha v_0(0)}\int_0^{T_L}\left[1-\alpha v_0(\eta)\right]\exp(-s\eta)\,d\eta. \tag{4.58}$$

Once the transport coefficients in equations (4.55) — (4.58) are found, the small-signal problem is, in principle, solved. There are, however, three main difficulties in getting the actual response of a certain device to a given excitation.

First, v_0 should be found as a function of T_x from the implicit relationship (4.47). Only for a few very simple analytical approximations of $v = v(E)$, can be $v = v(T_x)$ found explicitly.

Secondly, the integrals appearing in equations (4.55) — (4.58) cannot be, generally, analytically performed.

Thirdly, taking the inverse Laplace transform of $\mathcal{J}_1'(s)$ to get $\mathcal{J}_1(t)$ is a formidable task, since $\mathcal{J}_1'(s)$ may have a very complicated transcedental expression.

In the next section we shall consider a sinusoidal excitation and the application of the Laplace transform is no longer necessary.

4.6 Transport Coefficients for the Alternating Current Regime

Consider now that the voltage applied to the semiconductor region is of the form (4.11). The transport coefficients will be calculated exactly as shown in the preceding Section with the only difference that the complex quantities of the form

$\underline{V}_1 \exp i\omega t$ will replace the Laplace transforms of the form $\mathcal{L}\{V_1(t)\}$. We note that instead of

$$\mathcal{L}\left\{\frac{\partial E_1}{\partial t}\right\} = s\mathcal{E}_1 - E_1(x, 0) \tag{4.59}$$

we have (see equation (4.13))

$$\frac{\partial}{\partial t}\left[\underline{\mathbf{E}}_1(x) \exp i\omega t\right] = i\omega \underline{\mathbf{E}}_1(x) \exp i\omega t. \tag{4.60}$$

The exponential factor in equation (4.60) is not important, because it is common to all terms in an equation of the form (4.36) and thus, it may be simplified. By comparing the above two relations we conclude that an equation of the form (4.42) is also valid for $E_1(x)$, if the following changes are made

$$\mathcal{E}_1 \to \underline{\mathbf{E}}_1,\ \mathcal{J}'(s) \to \underline{\mathbf{J}}_1 \tag{4.61}$$

$$s \to i\omega. \tag{4.62}$$

Thus, the transport coefficients defined by equations (4.14) and (4.15) should be given by the relationships (4.55) — (4.58) where $i\omega$ should be replaced for s, i.e.

$$K_{ij} = \mathcal{K}_{ij}\big|_{s=i\omega}. \tag{4.63}$$

We suggest below a more elegant and concise form for the K_{ij} coefficients. First we obtain

$$\underline{\mathbf{E}}_1(T_x) = K_{21x}\,\underline{\mathbf{J}}_1 + K_{22x}\,\underline{\mathbf{E}}_1(0) \tag{4.64}$$

where

$$K_{22x} = \frac{p_0(T_x) - N}{p_0(0) - N} \exp(-\,i\omega T_x) = K_{22x}(T_x) \tag{4.65}$$

and

$$K_{21x} = \frac{K_{22x}}{\varepsilon} \int_0^{T_x} \frac{\mathrm{d}\eta}{K_{22x}(\eta)}. \tag{4.66}$$

Then, the transport coefficients will be

$$K_{11} = \int_0^{T_L} v_0(T_x)\, K_{21x}\, \mathrm{d}\, T_x \tag{4.67}$$

$$K_{12} = \int_0^{T_L} v_0 (T_x) K_{22x} \, \mathrm{d} T_x \tag{4.68}$$

$$K_{21} = K_{21x} (T_L) \tag{4.69}$$

$$K_{22} = K_{22x} (T_L). \tag{4.70}$$

The proof of the above relations is not difficult, as the reader may convince himself.

By using the relationships (4.66) — (4.70), all four transport coefficients can be calculated as a function of K_{22x}. The definition of K_{22x} from equation (4.65) is particularly simple and suggestive. The exponential factor corresponds to the delay caused by the finite time T_x required for charge transport from the injecting plane $x = 0$ to the distance x. The ratio $[p_0 (T_x) - N] [p_0 (0) - N]^{-1}$ from equation (4.65) is related to the bulk space charge.

There are also other possible forms for the transport coefficients. Instead of discussing them we prefer to calculate these coefficients for one or two particular, but highly relevant, cases.

4.7 An Example

The theory will be illustrated for a situation similar to that studied by Llewellyn and Peterson [84].

We assume a semiconductor space which is free of fixed space-charge (i.e. $N = 0$ and α defined by equation (4.46) is also zero). Let the carrier mobility be constant i.e.

$$v = \mu E, \ \mu = \text{const.} \tag{4.71}$$

We shall apply the results derived in Section 4.5. From equation (4.47) where $\alpha = 0$, we obtain

$$T_x = \frac{\varepsilon}{J_0} [E_0 (x) - E_0 (0)] \tag{4.72}$$

and

$$v_0 = \mu E_0 = \mu \left[\frac{J_0 T_x}{\varepsilon} + E_0 (0) \right] \cdot \tag{4.73}$$

Let us introduce the space-charge factor, equation (3.75), repeated here for convenience

$$\zeta = 1 - E_0 (0)/E_0 (L). \tag{4.74}$$

After some algebraic manipulation, equations (4.72) – (4.74) yield

$$v_0(T_x) = v_0(L)\left[\zeta\left(\frac{T_x}{T_L}\right) + 1 - \zeta\right]. \tag{4.75}$$

This expression of v_0 will be now replaced in equations (4.55) – (4.58) ($s \to i\omega$) and the result is (see also [85]):

$$K_{11} = \frac{v_0(L)\,T_L^2}{\varepsilon}[\zeta\,P(i\theta) + (1-\zeta)\,Q(i\theta)] \tag{4.76}$$

$$K_{12} = v_0(L)\,T_L(1-\zeta)\,R(i\theta) \tag{4.77}$$

$$K_{21} = \frac{T_L}{\varepsilon}[\zeta Q(i\theta) + (1-\zeta)\,R(i\theta)] \tag{4.78}$$

$$K_{22} = (1-\zeta)\exp(-i\theta) \tag{4.79}$$

where

$$P(i\theta) = \frac{1}{(i\theta)^3}\left[1 - i\theta + \frac{(i\theta)^2}{2} - \exp(-i\theta)\right] \tag{4.80}$$

$$Q(i\theta) = \frac{1}{(i\theta)^2}[\exp(-i\theta) + i\theta - 1] \tag{4.81}$$

$$R(i\theta) = \frac{1}{i\theta}[1 - \exp(-i\theta)] \tag{4.82}$$

and

$$\theta = \omega T_L \tag{4.83}$$

is the transit angle. For the so-called transit-time frequency

$$f_T = \frac{1}{T_L} \tag{4.84}$$

(the signal period is equal to the carrier transit time) the transit angle is $2\pi f_T T_L = 2\pi$.

Due to the presence of the transit angle and the space-charge factor, ζ, (which is zero for uniform electric field and unity for complete space charge limitation), the *form* of the transport coefficients, equations (4.76) – (4.79), recalls the aspect of the Llewellyn-Peterson coefficients for planar vacuum regions [84].

Clearly, the K_{ij} coefficients depend upon: (a) the properties of the semiconductor space itself (ε, μ, L) (b) the steady-state bias, V_0; and (c) the signal angular-frequency ω (called frequency in the following discusion).

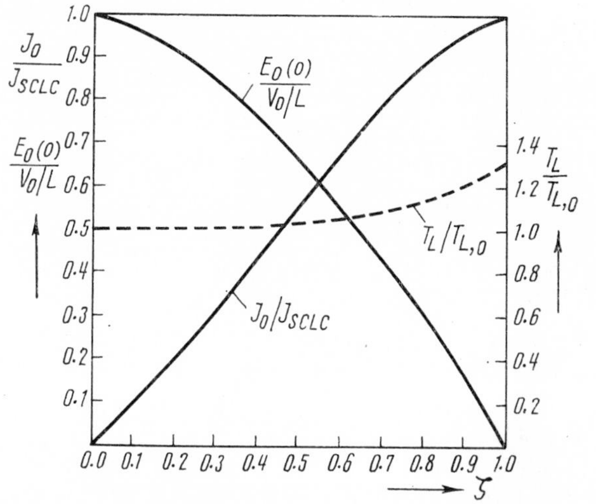

Figure 4.1. The normalized current, the normalized field and the normalized transit-time represented *vs.* the space-charge factor (zero fixed space-charge and constant mobility).

The space-charge factor, ζ, may be calculated as a function of the bias current. The result is plotted in Figure 4.1 (see problem (4.1)). The injected current J_0 is normalized to the SCL current

$$J_{\mathrm{SCLC}} = \frac{9}{8}\,\varepsilon\mu\, V_0^2/L^3 \tag{4.85}$$

which is the maximum current which can flow for a given applied voltage V_0*). Also represented as a function of ζ are the cathode field normalized to the bias field V_0/L, and the transit-time normalized to its zero space charge-value

$$T_{L,0} = \frac{L}{v_0} = \frac{L}{\mu E_0} = \frac{L^2}{\mu V_0}. \tag{4.86}$$

These curves may be used to calculate the steady-state characteristics (as soon as the cathode boundary condition is specified, see Chapter 3) and the K_{ij} coefficients.

*) The concept of maximum current which can be injected in a semiconductor space was discussed in Section 3.9, in connection with majority carrier injection ($N > 0$). Here we are in the limit case $N = 0$.

Now let us discuss the special case of the SCL injection. This is defined by

$$E_0(0) = 0, \ \zeta = 1. \tag{4.87}$$

From equations starting with (4.76)

$$\underline{\mathbf{V}}_1 = K_{11}\,\underline{\mathbf{J}}_1 = \frac{v_0(L)\,T_L^2}{\varepsilon} P(i\theta)\,\underline{\mathbf{J}}_1 \tag{4.88}$$

$$\underline{\mathbf{E}}_1(L) = K_{21}\,\underline{\mathbf{J}}_1 = \frac{T_L}{\varepsilon} Q(i\theta)\,\underline{\mathbf{J}}_1. \tag{4.89}$$

The diode impedance is (for unit area electrodes)

$$Z_{\mathrm{SCLC}} = \frac{\underline{\mathbf{V}}_1}{\underline{\mathbf{J}}_1} = \frac{v_0(L)\,T_L^2}{\varepsilon}\,\frac{1}{(i\theta)^3}\left[1 - i\theta + \frac{(i\theta)^2}{2} - \exp(-i\theta)\right] \tag{4.90}$$

and was derived successively by Shockley [14], Shao and Wright [10], Yoshimura [17], Drăgănescu [77] and van der Ziel and Hsu [78]. It is convenient and customary to normalize the impedance to its very low-frequency value which must coincide with the incremental resistance, R_i, derived from the $J - V$ steady-state characteristic

$$R_i = Z_{\mathrm{SCLC}}\big|_{\theta\to 0} = \frac{v_0(L)\,T_L^2}{6\varepsilon}. \tag{4.91}$$

The normalized impedance is

$$\overline{Z}_{\mathrm{SCLC}} = \frac{Z_{\mathrm{SCLC}}}{R_i} = \frac{6}{\theta^2}\left(1 - \frac{\sin\theta}{\theta}\right) - i\,\frac{3}{\theta}\left[1 - \frac{2}{\theta^2}(1 - \cos\theta)\right] \tag{4.92}$$

and depends solely upon the normalized frequency θ (transit-angle). The SCLC impedance ($\mu = \text{const.}$, $N = 0$, negligible diffusion) is shown in Figure 4.2a in a polar plot. The impedance of the SCLC planar vacuum diode [73], [84] is also plotted for comparison (Figure 4.2b).

Now let us consider the opposite case, namely

$$\zeta \simeq 0 \tag{4.93}$$

or negligible space-charge (nearly uniform electric field). Such a situation can be achieved if the injection conditions are such that $J_0 \ll J_{\mathrm{SCLC}}$ (Figure 4.1). It is permissible to write

$$v_0(T_x) \equiv v_0(L) = \mu V_0/L,\ T_L = L/v_0(L) = \frac{L^2}{\mu V_0} = T_{L,0}$$

and from equations (4.14), (4.15), (4.76) — (4.79) one obtains ($C_0 = \varepsilon/L$)

$$\underline{\mathbf{V}}_1 = \frac{T_L}{C_0} Q(i\theta)\,\underline{\mathbf{J}}_1 + LR(i\theta)\,\underline{\mathbf{E}}_1(0) \tag{4.94}$$

$$\underline{\mathbf{E}}_1(L) = \frac{T_L}{\varepsilon} R(i\theta)\,\underline{\mathbf{J}}_1 + \exp(-i\theta)\,\underline{\mathbf{E}}_1(0)\,. \tag{4.95}$$

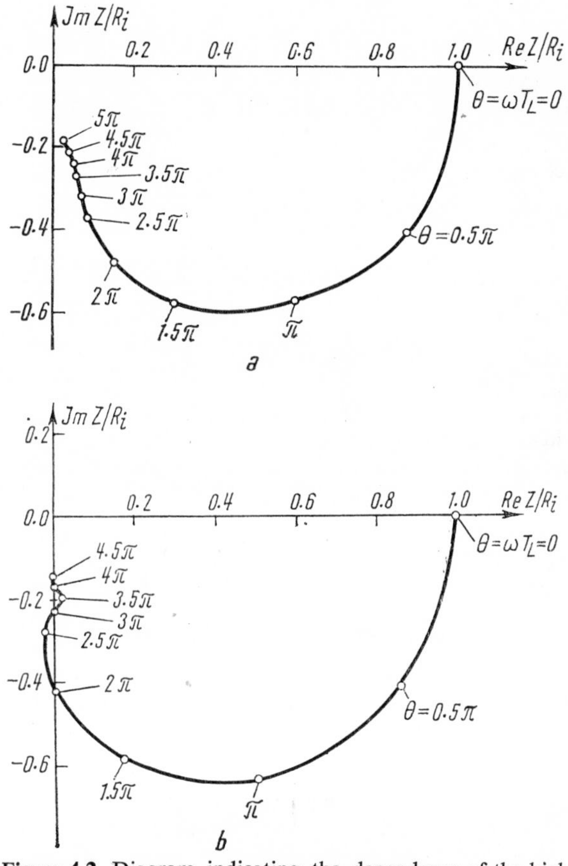

Figure 4.2. Diagram indicating the dependence of the high-frequency small-signal impedance (normalized to the incremental resistance R_i) with respect to the transit angle $\theta = \omega T_L$, for (a) the square-law SCLC solid-state diode ($N = 0$, $\mu = \text{const}$), and (b) the planar SCLC vacuum diode.

The impedance of the semiconductor region can be calculated if the a.c. boundary condition is specified. Now we find that an a.c. boundary condition was not apparently used for SCLC injection. An explanation can be found immediately:

it may be seen from equations (4.56) and (4.58) that K_{21} and K_{22} vanish automatically for $E_0(0) = 0$ ($v_0(0) = 0$) This is an interesting property and will be discussed at the beginning of the next chapter.

Let us consider, for example, the case when the current is completely injection-limited (pure thermionic emission is a standard example). By definition, there are no variations of the injected particle current, i.e.

$$\underline{\mathbf{j}}_1(0) = 0 \tag{4.96}$$

and from the a.c. continuity equation

$$\underline{\mathbf{J}}_1 = i\omega\varepsilon\,\underline{\mathbf{E}}_1(0). \tag{4.97}$$

For zero space charge ($\zeta = 0$) we have from equations (4.94) – (4.97)

$$\underline{\mathbf{V}}_1 = \frac{1}{i\omega C_0}\underline{\mathbf{J}}_1 \tag{4.98}$$

$$\underline{\mathbf{E}}_1(L) = \underline{\mathbf{E}}_1(0) = \frac{\underline{\mathbf{J}}_1}{i\omega\varepsilon} \tag{4.99}$$

i.e. the electric field is uniform, the particle current is negligible and the impedance $\underline{\mathbf{V}}_1(\underline{\mathbf{J}}_1)$ is simply due to the 'geometrical' capacitance (per unit area)

$$C_0 = \frac{\varepsilon}{L} \tag{4.100}$$

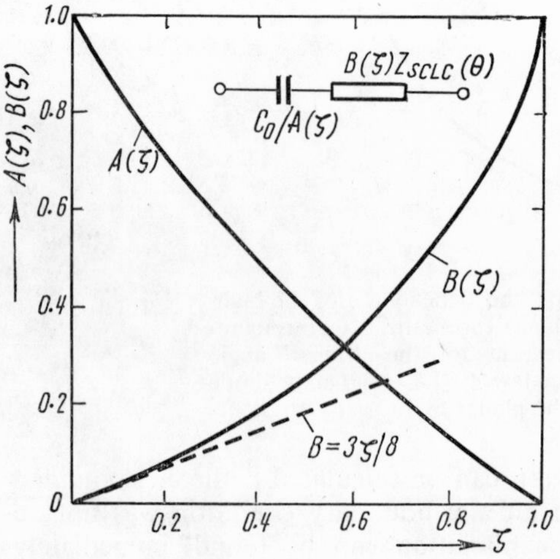

Figure 4.3. The coefficients entering in the expansion (4.101) of the emission-limited current diode ($N = 0$, $\mu = \text{const}$).

of the semiconductor region. In fact, the particle current is not exactly zero, because a small steady-state current, J_0, flows and is modulated by the a.c. voltage. The general expression for the a.c. impedance of the emission-limited-current diode with an arbitrary degree of space charge is given by

$$Z = \frac{1}{i\omega C_0} A(\zeta) + B(\zeta) Z_{\mathrm{SCLC}}(\theta) \tag{4.101}$$

i.e. a combination of the SCLC impedance, equation (4.90) and the impedance of the geometrical capacitance. $A(\zeta)$ and $B(\zeta)$ are plotted in Figure 4.3.

If another injection mechanism is considered, the impedance of the diode space is given, for example, by equation (4.18) where K_{ij} will be taken from equations (4.76) – (4.79) and $\sigma_1(0)$ from equation (4.16), etc.

4.8 An Explicit and General Form of the Transport Coefficients

The transport coefficients for semiconductors were derived in Section 4.5 under quite general conditions. Their form is, however, too complicated and not suitable for analytic calculations. Because the velocity-field dependence, $v = v(E)$ was not yet specified, these coefficients contain integrals and their form should depend upon the particular approximation used for $v = v(E)$.

The transport coefficients derived in the preceding Section have an explicit and simple form but they are not sufficiently general, because we assumed $N = 0$ and $\mu = $ const.

We suggest below an analytical and explicit form of the transport coefficients, K_{ij}. These coefficients will be calculated by taking into account the fixed space-charge ($N =$ const. $\neq 0$). The velocity field-dependence will be approximated by a straight line (i.e. the differential mobility $\mu_d = \mathrm{d}v/\mathrm{d}E$ is assumed constant). Of course, if the steady-state electric field in the semiconductor region is highly non-uniform, such an approximation would be inacceptable. This difficulty may be avoided by dividing the semiconductor into two or more regions each of them characterized by $\mu_d =$ const. Thus, this new approach may be considered sufficiently general.

The steady-state behaviour of a semiconductor region where $N \neq 0$ and $\mu_d =$ constant was studied in Section 3.10. The a.c. transport coefficients will be expressed as a function of the parameters introduced in that steady-state theory.

The transport coefficients, K_{ij} are given here without derivation. The relation between the d.c. velocity and the carrier transit-time, T_x, is given by equation (3.92).

This expression should be introduced in equations (4.55)—(4.58) (where $s = i\omega$). The integrals can be performed without difficulty. The result is

$$K_{11} = \frac{v_0(0)\,T_L^2}{\varepsilon}\left\{\Delta\,Q(\theta_d + i\theta) + \frac{\Delta - 1}{i\theta}\left[R(i\theta + \theta_d) - R(\theta_d)\right]\right. \tag{4.102}$$

$$K_{12} = v_0(0)\,T_L\,R(i\theta + \theta_d) \tag{4.103}$$

$$K_{21} = \frac{T_L}{\varepsilon}\,\frac{v_0(0)}{v_0(L)}\,[\exp-(\theta_d + i\theta)]\,\{\Delta R(-i\theta - \theta_d) - (\Delta - 1)\,R(-i\theta)\} \tag{4.104}$$

$$K_{22} = \frac{v_0(0)}{v_0(L)}\left[\exp(-i\theta + \theta_d)\right] \tag{4.105}$$

where

$$R(z) = \frac{1 - \exp(-z)}{z}, \quad Q(z) = \frac{\exp(-z) + z - 1}{z^2} \tag{4.106}$$

(see also equations 4.81 and 4.82) and θ_d and Δ were defined in Sections 3.9 and 3.10, respectively.

The above coefficients are of the form

$$K_{ij} = K_{ij}(i\theta) \tag{4.107}$$

where $\theta = \omega T_L$ is a normalized frequency. We wish to know how these coefficients depend upon the semiconductor parameters and upon the steady-state bias. This dependence can be easily found by noting the fact that the cathode velocity $v_0(0)$ and the anode velocity $v_0(L)$ are given by equations (3.98) and (3.96), respectively. It may be shown that the K_{ij} coefficients are given by relations of the form

$$K_{11} = \frac{L}{eN\,\mu_d}\,F_{11}(i\theta;\,\theta_d,\,\Delta) \tag{4.108}$$

$$K_{12} = L\,F_{12}(i\theta;\,\theta_d,\,\Delta) \tag{4.109}$$

$$K_{21} = \frac{1}{e\,N\mu_d}\,F_{21}(i\theta;\,\theta_d,\,\Delta) \tag{4.110}$$

$$K_{22} = F_{22}(i\theta;\,\theta_d,\,\Delta). \tag{4.111}$$

The F_{ij} functions (dimensionless quantities) are real but transcendental functions of three variables $i\theta$, θ_d and Δ. We recall the fact that θ_d is the normalized transit-time (3.93)

$$\theta_d = T_L/\tau_d, \quad \tau_d = \varepsilon/e\,N\mu_d \tag{4.112}$$

and Δ is the injection level, (3.72)

$$\Delta = p_0(0)/N. \tag{4.113}$$

The K_{ij} coefficients also depend upon the physical parameters of the semiconductor region. Only two parameters should be specified: the length, L, and the bulk differential conductivity*).

If we wish to know the frequency dependence, equation (4.107), we need the above two parameters and also the dimensionless quantities, equations (4.112) and (4.113) which depend upon the d.c. biasing of the semiconductor region. Now, with reference to the $\theta_d - \Delta$ plots shown in Figures 3.24—3.27, we stress the fact that the normalized current $\overline{J}_0$ (or $\overline{J}$) and voltage $\overline{V}_0{}'$ (or $\overline{V}'$) determine completely θ_d and Δ. It is interesting to note that the injection mechanism does not enter *directly* into these calculations. The K_{ij} coefficients can be calculated exclusively as a function of bulk properties, if the injected current and the voltage drop are specified. We do not forget, of course, that the injected current depends upon the voltage drop, and the injection mechanism should be specified in order to calculate this dependence. But the fact we wish to outline here is that two identical semiconductor regions with the same injected current and the same voltage drop, have exactly the same transport coefficients, even if the injection mechanism is different (i.e. SCL injection or pure thermionic emission).

In the limits of the above approximations ($N =$ const., $\mu_d =$ const., negligible diffusion), the problem of bias dependence is completely and most elegantly solved. For any semiconductor region, the $K_{ij} = K_{ij}(\theta)$ may be calculated by use of the graphs from Figures 3.24—3.27. Of course, the F_{ij} functions from equations (4.108)—(4.111) can be also represented. However, this will require a too large space. We have to represent the real and imaginary parts of 4 complex functions. Four physical situations should be considered (majority and minority carrier injection in positive and negative mobility materials, see Table 3.1). For each case in part we represent $\mathcal{Re}F_{ij}$ or $\mathcal{Im}F_{ij}$ as a function of θ, $|\Delta|$ and $|\theta_d|$ being the parameters. If we select a minimum of five values for $|\Delta|$ and represent families of characteristics with θ_d as a parameter, the total number of diagrams will be $4 \times 2 \times 4 \times 5 = 160$. This is still a minimum which may be unsatisfactory for detailed calculations.

On Table 4.1 there are several special cases which may be studied by using equations (4.102) — (4.113).

Table 4.1

	$p_0(0)$	N	μ_d	Δ	θ_d
1. Bulk neutrality	N	finite	finite	1	finite
2. No fixed space charge	finite	0	finite	$\pm\infty$	0
3. Negligible injected charge ($J_0 \to 0$)	$\to 0$	finite	finite	$\to 0$	finite
4. Saturated velocity	finite	finite	0	finite	0
5. Space-charge-limited injection	$+\infty$	finite	finite	$\pm\infty$	finite

*) This conductivity has no physical sense for minority carrier injection or for injection in insulator with deep traps ($N < 0$).

The transport coefficients derived in this Section differ from Yoshimura's coefficients [17], [74] in two respects.

First, the present results are valid for $v/E \neq$ constant, in particular for negative mobility materials. Yoshimura's coefficients are restricted to the low-field constant-mobility case.

Secondly, the transport coefficients derived here are more conveniently expressed as a function of the d.c. bias conditions. We recall the fact that Yoshimura has actually studied only two particular cases, namely the third and the fifth from those indicated in Table 4.1.

Very often, the semiconductor region is assumed to be quasi-neutral: the electric field is almost uniform and the d.c. carrier velocity, v_0 should be also uniform. Because $\Delta = 1$ (Table 4.1), the transport coefficients, equations (4.102) – (4.105) become

$$K_{11} = \frac{L^2}{\varepsilon v_0} Q(\theta_d + i\theta) = \frac{L^2}{\varepsilon v_0} \frac{[\exp - (\theta_d + i\theta)] + (\theta_d + i\theta) - 1}{(\theta_d + i\theta)^2} \tag{4.114}$$

$$K_{12} = L\,R(i\theta + \theta_d) = L\frac{1 - [\exp - (\theta_d + i\,\theta)]}{\theta_d + i\theta} \tag{4.115}$$

$$K_{21} = \frac{L}{\varepsilon v_0} [\exp - (\theta_d + i\theta)]\; R(-\theta_d - i\theta) = \frac{L}{\varepsilon v_0} \frac{1 - [\exp - (\theta_d + i\theta)]}{\theta_d + i\theta} \tag{4.116}$$

$$K_{22} = \exp - (\theta_d + i\theta). \tag{4.117}$$

Here, θ_d has the sign of the differential mobility (N should be positive, majority carrier injection).

Another special case is the so-called high-field drift region which was studied extensively by Wright [86] – [90]. The drift velocity is assumed to be saturated. Therefore $\mu_d = 0$ and $\theta_d = 0$ whereas N and Δ are finite. We note, however, that the transport coefficients (4.102) – (4.105) do no longer depend upon Δ. We have

$$K_{11} = \frac{L^2}{\varepsilon v_l} Q(i\theta) = \frac{L^2}{\varepsilon v_l} \frac{[\exp - i\theta] + i\theta - 1}{(i\theta)^2} \tag{4.118}$$

$$K_{12} = LR(i\theta) = L\frac{1 - [\exp - i\theta]}{i\theta} \tag{4.119}$$

$$K_{21} = \frac{L}{\varepsilon v_l} R(i\theta) = \frac{L}{\varepsilon v_l} \frac{1 - [\exp - i\theta]}{i\theta} \tag{4.120}$$

$$K_{22} = \exp - i\theta. \tag{4.121}$$

We recall the fact that the C_{ij} coefficients defined by equations (4.19) and (4.20) may be obtained by using equations (4.21) — (4.24). For a high-field (saturated velocity) region we obtain ($T_L = L/v_l$)

$$C_{11} = \frac{L^2}{\varepsilon v_l} \frac{1}{i\theta} = \frac{1}{i\omega C_0} \quad (4.122)$$

$$C_{12} = -\frac{L}{i\omega\varepsilon} R(i\theta) = \frac{T_L}{C_0} \frac{(\exp - i\theta) - 1}{(i\theta)^2} \quad (4.123)$$

$$C_{21} = 0 \quad (4.124)$$

$$C_{22} = \exp - i\theta. \quad (4.125)$$

4.9 Other Transport Coefficients

The main result of this chapter is represented by the transport coefficients for semiconductor regions. These were derived in a general form in Sections 4.5 and 4.6.

Similar transport coefficients can be calculated for an insulator region with trapping centres. We have extensively studied the case of shallow traps. The results can be found in a number of papers [70], [82], [91] — [93] and will not be repeated here. A typical example of an insulator diode with SCL current is analysed in Section 5.10.

Wadhwa and Sisodia [75] calculated a matrix of nine coefficients which describe the small signal a.c. behaviour by taking into account the mobility relaxation time. Their results will be briefly discussed in Section 5.11.

Problems

4.1 Show that the current flowing through a solid-state space free of impurities and defects and with a constant carrier mobility can be expressed as

$$\frac{J_0}{J_{SCLC}} = \frac{16\left(\frac{2}{\zeta} - 1\right)^3}{\left[3\left(\frac{2}{\zeta} - 1\right)^2 + 1\right]^2} \quad (P.4.1)$$

where J_{SCLC} is given by equation (4.85) and ζ is the space-charge factor (4.74). This dependence is plotted in Figure 4.1.

4.2 Verify that in the conditions of Problem 4.1, the carrier transit-time is given by

$$\frac{T_L(\zeta)}{T_{L,0}} = 1 + \frac{1}{3\left(\frac{2}{\zeta} - 1\right)^2}, \quad T_{L,0} = \frac{L^2}{\mu V_0}. \tag{P.4.2}$$

4.3 Derive directly, from the basic equations, the transport coefficients (4.118) – (4.121) of the high field drift region (saturated velocity).

4.4 How can K_{ij} be expressed as a function of the control characteristic j_c and the neutral characteristic j_n (see Sections 3.6, 3.7 and 4.5)?

5

Small-signal Behaviour of Space-charge-limited Current Devices

5.1 General Properties

The existence of space-charge-limited (SCL) currents in solids is well established. SCL currents have been measured in CdS and other wide-gap materials [7], [67], [68] as well as in semiconductors (Ge, Si, GaAs) [15], [55], [56], [94] – [102].

The injection under SCLC conditions requires an ohmic contact, a so-called 'infinite reservoir of mobile carriers'. Assume hole injection into a symmetrical structure. Due to the accumulation of charge carriers inside the semiconductor bulk, a maximum of potential should exist at a certain point $x = x_0$ (Figure 5.1). As the applied voltage increases, the plane $x = x_0$ moves towards the injecting contact (positively biased). If the voltage is sufficiently large, the above mentioned plane, called the virtual cathode, almost coincides with the cathode proper ($x = 0$), i.e.

$$x_0 \to 0, \text{ where } E_0(x_0) = 0 \tag{5.1}$$

Equation (5.1) may be a suitable boundary condition for the *steady-state* electric field. This condition is sufficient to calculate the steady-state current, J_0, if the diffusion current is neglected (Section 3.11). We recall the fact that, between the

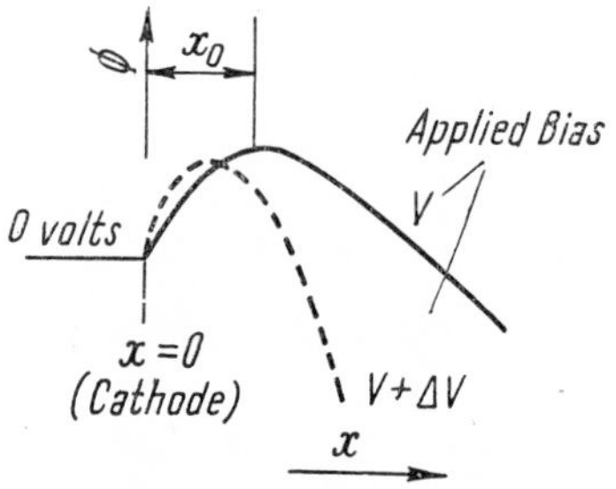

Figure 5.1. The potential distribution in the vicinity of the cathode for SCL hole injection in a semiconductor region. The dashed curve indicates how this distribution changes when the voltage bias increases.

virtual cathode and the cathode proper, the electric field opposes to the injection of holes into the bulk. There is the diffusion current which transports a part of the injected holes beyond the potential minimum. At $x = x_0$ the current flow should be entirely a diffusion current, because the drift velocity is zero. The elementary theory which neglects diffusion avoids this difficulty by assuming that

$$p_0(0) \to \infty \tag{5.2}$$

('infinite reservoir' at $x = 0$). Thus the drift current $ep_0(0)\, v_0(0))$ may be finite. However, an inconsistency still remains, because an infinite gradient of carrier density at $x = 0$ would imply a very large gradient of carrier concentration and a non-negligible diffusion current in the immediate vicinity of the cathode ($x \gtrsim 0$). The neglect of diffusion in the elementary *steady-state* theory is justified by the experimental data and by the results of more exact theories.

A *dynamic* boundary condition of the form (5.1) may be suggested, namely

$$E(x,\ t)\,|_{x=0} = 0. \tag{5.3}$$

However, such a condition is neither evident, nor justified by other data.

General Formulae. Almost the entire chapter is based upon the model introduced in Section 3.5, i.e.: a one-dimensional geometry, uniform and constant fixed space-charge and negligible diffusion are assumed.

The steady-state characteristic of an SCLC diode may be calculated by use of equations (3.34) and (3.35) [103], [104] where the cathode field, E_c, should be replaced by zero. Thus one obtains

$$V_0 = \varepsilon \int_0^{E_0(L)} \frac{E_0\, v_0(E_0)\, \mathrm{d}E_0}{J_0 - e N v_0(E_0)} \tag{5.4}$$

$$L = \varepsilon \int_0^{E_0(L)} \frac{v_0(E_0)\, \mathrm{d}E_0}{J_0 - e N v_0(E_0)}\,. \tag{5.5}$$

The transport coefficients, K_{ij}, are given by equations (4.55) — (4.58) where $s = i\omega$ and $v_0(0) = 0$ (SCLC condition):

$$K_{11} = \frac{1}{\varepsilon} \int_0^{T_L} \mathrm{d}y\, [J_0 - e N v_0(y)]\, [\exp - i\omega y] \int_0^{y} \frac{v_0(\eta) \exp i\omega\eta}{J_0 - eNv_0(\eta)}\, \mathrm{d}\eta \tag{5.6}$$

$$K_{21} = \frac{1}{\varepsilon} \frac{J_0 - e N v_0(T_L)}{v_0(T_L)} \int^{T_L} \frac{v_0(\eta) \exp i\omega(\eta - T_L)}{J_0 - e N v_0(\eta)}\, \mathrm{d}\eta \tag{5.7}$$

$$K_{12} = K_{22} = 0. \tag{5.8}$$

Here v_0 should be given as a function of T_x. This function is given implicitly by equation (4.47) rewritten here for $E_c = 0$

$$T_x = \varepsilon \int_0^{E_0(x)} \frac{\mathrm{d}E_0}{J_0 - e N v_0(E_0)}\,. \tag{5.9}$$

The device impedance results from equations (4.14) and (5.7) [105]

$$Z_{SCLC} = \frac{\underline{\mathbf{V}}_1}{\underline{\mathbf{J}}_1} = K_{11}. \tag{5.10}$$

Note the fact that because K_{12} and K_{22} vanish in equations (4.14) and (4.15), the device behaviour apparently does not depend upon the a.c. boundary condition. This does not mean, however, that $\underline{\mathbf{E}}_1(0)$ may have any value. The fact can be shown (Problem 5.1) that $\underline{\mathbf{E}}_1(0)$ must be zero

$$\underline{\mathbf{E}}_1(0) = 0. \tag{5.11}$$

The same result is obtained formally by introducing $E(0, t)$ given by equation (4.13) into equation (5.3).

We point out that C_{12} and C_{22} in equations (4.19) and (4.20) are also zero, i.e. an *injected* particle current $\underline{\mathbf{j}}_1(0)$ does not influence the diode behaviour (see also the discussion on the 'injected' noise in chapter 9). We describe this relative independence of the diode behaviour of the boundary conditions saying that the potential maximum at $x = 0$ completely isolates the semiconductor bulk ($x > 0$) from the processes taking place at $x \leqslant 0$. This isolation is somewhat natural, because of the infinite density of charge carriers at $x = 0$ (condition (5.2)) [106].

The incremental (low-frequency or d.c.) resistance of an SCLC diode is always positive. With reference to equation (3.59), the contact resistance is zero and the incremental resistance is equal to the bulk resistance, which is always positive.

However, the high-frequency dynamic resistance of certain SCLC diode may be negative. This possibility will be investigated thoroughly in the subsequent Sections.

5.2 Small-signal Behaviour of SCLC Silicon Diodes at High Bias Fields

For a start, we shall assume the fact that the injected charge density is everywhere much higher than the background ion density

$$J_0 = e\,p_0(x)\,v_0(x) \gg e\,N\,v_0(x). \tag{5.12}$$

The steady-state equations (5.4) and (5.5) may be written

$$J_0 = C_0 \int_0^{E_0(L)} v_0(E_0)\,\mathrm{d}E_0\,, \quad C_0 = \frac{\varepsilon}{L} \tag{5.13}$$

$$V_0 = \frac{L \int_0^{E_0(L)} E_0\,v_0(E_0)\,\mathrm{d}E_0}{\int_0^{E_0(L)} v_0(E_0)\,\mathrm{d}E_0}\,. \tag{5.14}$$

The diode impedance, equation (5.10), becomes

$$Z = \frac{1}{\varepsilon}\int_0^{T_L} dy\,[\exp - i\omega y]\int_0^{y} v_0(\eta)\,[\exp i\omega\eta\; d\eta \tag{5.15}$$

where v_0 should be given as a function of T_x. A relation between E_0 and T_x is necessary, and this is given by equations (5.9) and (5.12)

$$J_0\,T_x = \varepsilon E_0(x). \tag{5.16}$$

The above formulae can be used for an arbitrary velocity-field dependence. Similar results were first derived by Dascalu [82] and Kroemer [83].

We shall discuss first the case of a semiconductor with positive mobility and saturated velocity at higher field intensities, for example a p^+ip^+ (or n^+in^+) silicon device. The central region is assumed intrinsic, i.e. $N = 0$. In practice, the bias current J_0 should be sufficiently high to inject a mobile charge density which is much greater than the thermal equilibrium density inside the central region (equation (5.12)) and thus N is negligible.

Consider as an example a velocity-field dependence of the form

$$v = v_l\,[1 - \exp\,(-\,E/E_l)] \tag{5.17}$$

which is appropiate for electrons in silicon [98]. Here, v_l is the limit velocity ($E \to \infty$) and v_l/E_l is the low-field mobility ($E \to 0$). From equations (5.13) and (5.14) we obtain [107]

$$\frac{J_0}{\varepsilon E_l v_l/L} = \exp\,(-\,a) + a - 1,\quad a = \frac{E_0(L)}{E_l} \tag{5.18}$$

$$\frac{V_0}{E_l L} = \frac{(a+1)\exp\,(-\,a) + a^2/2 - 1}{\exp\,(-\,a) + a - 1} \tag{5.19}$$

The $J - V$ characteristic is plotted in Figure 5.2. For the low-current region ($a \to 0$)

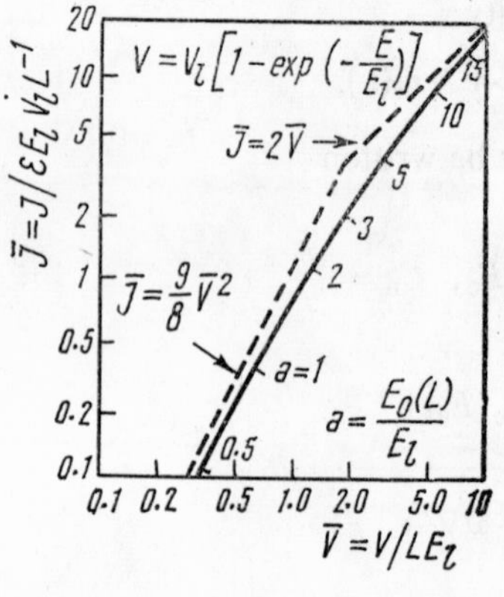

Figure 5.2. The normalized steady-state $J - V$ characteristic of a SCLC diode with zero fixed space charge ($N = 0$) and field-dependent mobility ($v - E$ relationship given by equation. (5.17)).

we reobtain the well known square law, equation (4.85), (calculated for $\mu = \text{const}$). The high-field ($a \to \infty$) asymptote is

$$J_0 = 2\varepsilon v_l V_0/L^2 \tag{5.20}$$

and may be calculated directly by assuming saturated velocity throughout the entire conduction space.

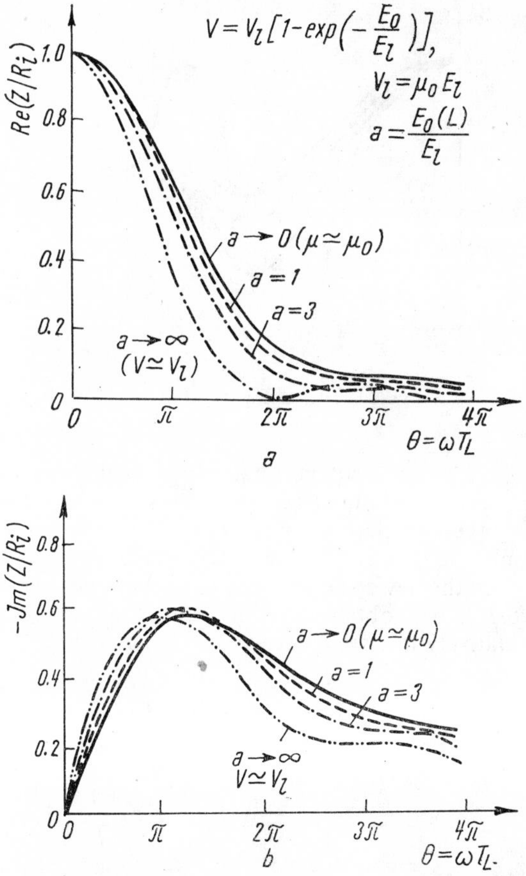

Figure 5.3. The real (a) and imaginary (b) parts of the normalized impedance, plotted *vs.* the transit-angle. The parameter a indicates the steady-state operating point on the characteristic shown in Figure 5.2.

The small-signal impedance, equation (5.15), becomes [103], [106], [107]:

$$Z = R_i\left(1 - a + \frac{a^2}{2} - \exp - a\right)^{-1}\left\{\frac{a^2}{i\theta} + \right.$$
$$\left. + \frac{1}{1 - i\theta/a}\left[1 - \exp(-a) + \left(\frac{a}{\theta}\right)^2\left(1 - \exp - a\right)\right]\right\} \tag{5.21}$$

where R_i is the low-frequency (incremental) resistance $R_i = Z|_{\theta\to 0}$ and θ is the transit angle.

The real and imaginary parts of the normalized impedance Z/R_i are plotted in Figure 5.3 with a as a parameter. The series resistance is always non-negative. The reactance is negative, suggesting a more suitable equivalent circuit consisting of a conductance shunted by a parallel positive capacitance. The frequency dependence of the parallel conductance (note the oscillatory behaviour) is indicated in Figure 5.4.

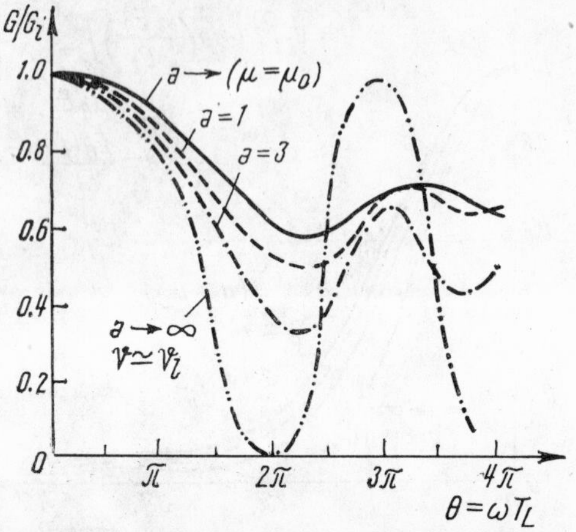

Figure 5.4. The frequency dependence of the parallel conductance of the SCLC diode with field-dependent mobility (see Figure 5.2).

The low-frequency equivalent circuit consists of R_i shunted by a (low-frequency) dynamic capacitance, C. This low-frequency capacitance may be calculated by expanding the impedance, equation (5.21), in power-series with respect to $i\theta$ and retaining only the first and the second term ($\theta \to 0$). As an example we shall discuss the case when the device is biased in the square law region. By introducing $a \to 0$ in equation (5.21), one obtains the impedance, equation (4.90) [10], [14], [17], [77], [78]. At relatively low frequencies

$$Z|_{\theta\to 0} \simeq R_i\left(1 - \frac{i\theta}{4}\right), \quad \theta = \omega T_L \tag{5.22}$$

or

$$Y \simeq \frac{1}{R_i}\left(1 + \frac{i\theta}{4}\right). \tag{5.23}$$

As the reader may easily convince himself, the following equality takes place (Problem 1.4) [10], [77]:

$$R_i C_0 = T_L/3, \quad C_0 = \varepsilon/L \tag{5.24}$$

and thus, one obtains

$$Y \simeq \frac{1}{R_i} + i\omega\left(\frac{3}{4}C_0\right) \tag{5.25}$$

which shows that the dynamic capacitance is three-quarters of the geometrical capacitance C_0.

Due to the mobility-field dependence ($a \neq 0$), the dynamic capacitance shows a slight dependence upon the steady-state bias, as shown in Figure 5.5 [107], falling gradually from $\frac{3}{4}\,C_0$ ($\mu = \text{const}$) to $\frac{2}{3}\,C_0$ ($v = \text{const}$).

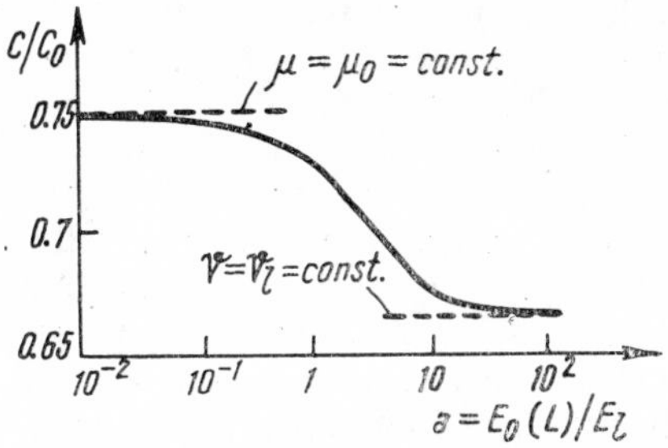

Figure 5.5. The bias-dependence of the low frequency dynamic capacitance of the SCLC diode with field-dependent mobility.

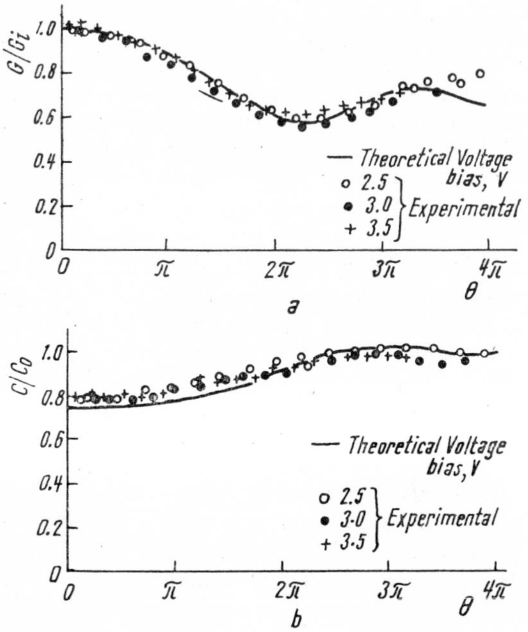

Figure 5.6. Experimental frequency characteristics measured for a punch-through $p^+\nu p^+$ silicon diode biased in the square-law region of the $J-V$ characteristic [109] ($T = 300$ K, $L = 108$ μm, 38,500 Ω cm resistivity of the ν layer, 1.33 mm² area of the injecting contact):

Figure (a) shows the parallel conductance and Figure (b) — the parallel capacitance.

The fact that the dynamic capacitance is not equal to C_0 is, by itself, not surprising because the SCLC diode is not a parallel-plate condenser. The electric charge whose storage determines a capacitive effect, is, actually, not on the electrodes, being non-uniformly distributed in the semiconductor bulk. More interesting is the fact that the dynamic capacitance cannot be calculated in a quasi-stationary manner, i.e. by taking into account the variation of the total charge injected by the steady-state bias current. We shall discuss this aspect in the nexy Section.

The a.c. small signal theory was verified rather well by measurements in the square-law region. The value of $\frac{3}{4}$ C_0 for the low-frequency capacitance was reported in many papers [10], [15], [55], [99], [108], [109]. The high-frequency characteristics up to $\theta = 3\pi \ldots 4\pi$ were measured by Shao and Wright [10] for a single-crystal CdS diode, and by Dascalu [106], [109] for silicon diodes. The results of the latter author are reproduced, in part, in Figure 5.6. The solid line is the theoretical result derived from equation (5.21) for $a \to 0$. The experimental results are normalized by use of the results of direct or indirect measurements. It may be seen that the experimental dots fit the theory quite well*).

The *transient* response to a voltage step of small amplitude was studied by Baron *et al.* [110] for two limit cases, namely $\mu =$ const. ($a \to 0$) and $v =$ saturated ($a \to \infty$). The reader may reobtain Baron's results [110] by starting from the general results obtained in Section 4.5. Here we reproduce from [110] the time-dependence of the small-signal component of the current (Figure 5.7). It may be shown that the steady-state (constant) value is reached within a time of the order of the carrier transit-time.

Baron *et al.* [110] indicated that the transient small-signal current has three components. These components may be identified in equation (4.35). The last term in this equation is the displacement current. The second term in the right-hand-side is due to the velocity variation (or modulation) of the carriers injected by the d.c. bias. The first term corresponds to the variation of density ($\mathrm{d}E_1/\mathrm{d}x = ep_1/\varepsilon$) of the charge carriers moving at the velocity determined by the d.c. electric field. Note the fact that if the velocity is saturated ($a \to \infty$) the component due to velocity variation will be zero.

The mobile charge injected at $t = 0$ by the voltage step of amplitude V_1 is finite and equal to $\varepsilon V_1/L$ (because the cathode field $E_1(0)$ should be zero)†). This charge forms initially a sheet of infinite small thickness. At $t > 0$, the charge injected and $t = 0$ and after $t = 0$ moves towards the anode. The time of arrival is the d.c. transit-time, T_L, (which has the value (4/3) $L^2/\mu V_0$ for $\mu =$ const. and L/v_l for $v =$ const.) and this moment is indicated by a change in the variation of $J_1(t)$, because a new component occurs (modulation of the carrier density at the anode). However, there is an important difference between the two variations indicated in Figure 5.7. In the case $v =$ saturated there is *a discontinuity* in the current variation. The explanation will be fully understood in Chapter 12 (large signal transient response). We note here that the amount of charge injected at $t = 0$ will be dispersed by electrostatic forces in the case $\mu =$ const. This is not *mathematically* possible for $v =$ const.

---

*) Measurements were performed for six values of the d.c. bias (only three characteristics are plotted in Figure 5.6) and clearly show that the *shape* of the frequency characteristic does not depend upon the operating point if this point is chosen in the square-law region.

†) At the instant of the step application ($t = 0$) surge of current should occur to provide this charge. But the actual form of this current impulse will depend upon the external circuit and the diode capacitance C_0. This fact is very important from experimental point of view but is completely ignored in the present theoretical analysis.

Nevertheless, in practice, the carrier velocity will not be rigorously saturated at low-field intensities and in the immediate vicinity of the cathode where the field should be low under SCLC conditions.

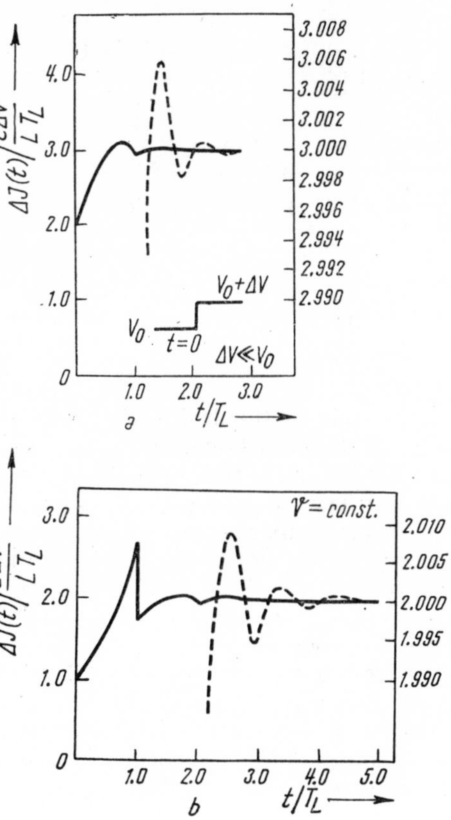

Figure 5.7. Response of unipolar SCL current to a small voltage step superimposed at $t = 0$ on the steady-state bias. The current was calculated by Baron *et al.* [110] for the following cases:
(a) constant carrier mobility, (b) saturated carrier velocity. In both cases the fixed space charge was neglected.

5.3 Dynamic and Stationary Capacitances of SCLC Diodes

In this Section, we shall show that the dynamic low-frequency capacitance should be lower than the stationary capacitance and this is due to the effect of the finite time required for charge transport. In other words, the reactive behaviour of the SCLC diode cannot be obtained from quasi-stationary calculations, even at very low frequencies.

The stationary capacitance is

$$C_{st} = \partial Q / \partial V. \tag{5.26}$$

where Q is the total electric charge between electrodes (both C_{st} and Q are calculated per unit of electrode area). By the Gauss theorem

$$\varepsilon[E(L) - E(0)] = Q \tag{5.27}$$

and, because $E(0) = 0$, C_{st} becomes

$$C_{st} = \varepsilon[\partial E(L)/\partial V]. \tag{5.28}$$

We shall now write the current continuity equation (4.1) at $x = L$

$$J(t) = j(L, t) + \varepsilon \frac{\partial E(L, t)}{\partial t} \tag{5.29}$$

and, by use of equation (5.28), we obtain:

$$J(t) = j(L, t) + C_{st} \frac{\mathrm{d}V}{\mathrm{d}t}. \tag{5.30}$$

Let us imagine a detailed equivalent circuit by splitting the device admittance into two components corresponding respectively, to the particle current at the anode (the first term in the right-hand-side of equation 5.30) and to the displacement current at the anode (the second term in equation (5.30)). The second part of the admittance should be exactly the admittance of the stationary capacitance C_{st}, as shown by equation (5.30). The device capacitance is equal to C_{st} only if the particle current at the anode is in phase with the applied voltage (i.e. the corresponding partial admittance is a pure resistance). However, at very high frequencies $j\,(L, t)$ should be delayed with respect to $V(t)$, due to the transport processes.

We wish to outline the fact that the reactive behaviour due to $j(L, t)$ should be taken into account also at low frequencies and the device capacitance is never equal to C_{st}. The demonstration is based upon the fact that a space charge exists in the semiconductor bulk. The electric field is non-uniform and consequently the displacement current is also non-uniform (at a given time instant, t, it depends upon the distance, x). The particle current should be also non-uniform at an arbitrary t, because the total current $J(t)$ is constant ($\partial J/\partial x = 0$) The non-uniformity of j shows that transit-time effects are important and the particle current has also a 'reactive behaviour'. Therefore, if the stationary capacitance, C_{st}, due to the displacement current is taken into account, an additional reactance due to the particle current should be also considered.

At relatively low frequencies, the particle current reactance should be *inductive*, due to the *delay* of the particle current with respect to the applied voltage so that we expect a dynamic capacitance lower than C_{st}. With reference to the diode studied in the preceding section, we note that as the bias current increases (*a* increases in Figure 5.2) the stationary capacitance also increases from $\left(\frac{3}{2}\right) C_0$ to $2\,C_0$, while the dynamic capacitance (represented in Figure 5.5) *decreases* from $\left(\frac{3}{4}\right) C_0$ to $\left(\frac{2}{3}\right)$ C_0.

The dynamic low-frequency capacitance equals C_{st} only if the semiconductor is neutral, i.e. the electric field is uniform and equal to the bias field ($C_{st} = C_0$). In this case the displacement current does not depend upon x and j must also be constant: transit-time effects do not occur. We expect the difference $C_{st} - C$ to increase as the non-uniformity of the electric field increases.

The difference between the dynamic capacitance and the stationary capacitance of the square-law SCLC diode was first explained by Drăgănescu and Dascalu [111], who also suggested an equivalent circuit (Figure 5.8a) with a *negative capacitance*

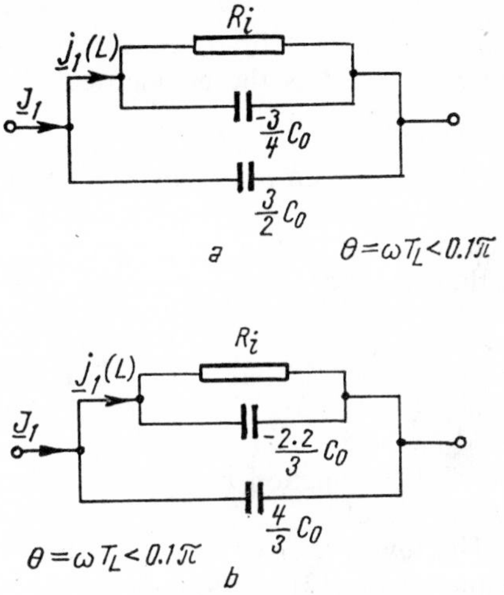

Figure 5.8. Small-signal low-frequency equivalent circuit indicating the inductive effect of carrier inertia by introducing a *negative capacitance* for the reactive effect due to the particle current at the anode. The calculations were made (a) for the SCLC solid-state diode ($N = 0$, $\mu = \text{const}$) [111] and, (b) for the planar vacuum diode [112], [113].

representing the inductive effect of carrier inertia. The situation is similar for the SCLC planar vacuum diode [112], [113] as shown by the detailed equivalent circuit of Figure 5.8b.

It is well known that the dynamic capacitance of the $p - n$ junction is exactly half of the stationary capacitance, equation (5.26). This fact was explained also as an effect of carrier inertia (the delay required by diffusion of minority carriers) [114], [115].

More recently, Cherry [116], has indicated again the generality of these results ($C \neq C_{st}$) for electronic devices requiring charge transport (see also Section 5.12).

5.4 SCLC Negative-mobility Diode, in the Limit of Zero Doping (Kroemer [83])

We have shown (Section 5.2) the fact that the small-signal impedance of the silicon SCLC diode biased at high fields does not exhibit a negative real part. This conclusion does not depend upon the particular $v - E$ relationship used, equation (5.17), but is general for SCLC semiconductor diodes with positive differential mobility ($dv/dE \geq 0$). The SCLC impedance may have a negative real part (negative resistance) if the mobility dv/dE is negative on a certain region of the $v - E$ characteristic.

These properties of the pure SCLC flow (*negligible fixed space-charge*) will be discussed below, and other interesting results, including the effect of the diffusion current will be also presented following the results obtained by Kroemer [83].

After a partial integration, the impedance, equation (5.15) can be written

$$Z = -\frac{1}{i\omega\varepsilon}[\exp - i\omega T_L]\int_0^{T_L} v_0 [\exp i\omega\eta]\, d\eta + \frac{1}{i\omega C_0} \tag{5.31}$$

(where $C_0 = \varepsilon/L$ is the geometrical capacitance). The series resistance is

$$R(\omega) = \mathcal{R}e\, Z = \frac{1}{\omega\varepsilon}\int_0^{T_L} v_0 \sin\omega\,(T_L - \eta)\, d\eta \tag{5.32}$$

and the reactance is given by

$$X(\omega) = \frac{1}{\omega\varepsilon}\int_0^{T_L} v_0 \cos\omega\,(T_L - \eta)\, d\eta - \frac{1}{\omega C_0}. \tag{5.33}$$

Here, v_0 is a function of $T_x = \eta$, which follows directly from $v = v\,(E)$ by use of equation (5.16).

The low-frequency (incremental) resistance, R_i, will be discussed first. By use of equations (5.13), (5.14) and (5.16) we obtain successively

$$\begin{aligned} R_\iota = \mathcal{R}e\, Z\Big|_{\omega\to 0} &= \frac{1}{\varepsilon}\int_0^{T_L} v_0(\eta)\,(T_L - \eta)\, d\eta = \\ &= \frac{L\,T_L}{\varepsilon} - \frac{1}{\varepsilon}\int_0^{T_L} v_0(\eta)\,\eta\, d\eta = \frac{L\,T_L}{\varepsilon} - \frac{V_0}{J_0}. \end{aligned} \tag{5.34}$$

The total transit-time T_L is given by equation (5.16) for $x = L$ and is introduced in equation (5.34) to give

$$R_i = \frac{L}{J_0}\left[E_0(L) - \frac{V_0}{L}\right]. \tag{5.35}$$

Because the electric field E_0 starts from zero at $x = 0$ and increases monotonously until $x = L$ (Section 3.4) it is clear that $E_0\,(L)$ should exceed the 'average' (or bias) field V_0/L and the incremental resistance R_i should be always positive, i.e. no negative resistance is possible on a d.c. characteristic. This fact has been pointed out by Shockely [14], and may be explained intuitively. With an increase in the applied voltage the carrier velocity will decrease ($dv/dE < 0$) but the injected space charge will increase and compensate for the decrease of carrier velocity, such that the current will continue to increase as the voltage bias increases.

Kroemer has shown later that this positive external conductance 'is strictly a low-frequency phenomenon, based on the charge readjustment with varying voltage. If the voltage varies fast enough so that the charge readjustment cannot follow it, the decrease in drift velocity with increased internal field should override the charge increase, and a negative external conductance should result...' [117].

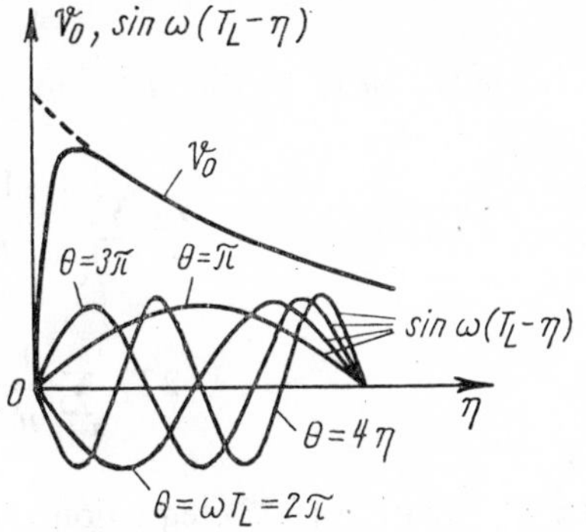

Figure 5.9. Figure used for the evaluation of the high-frequency resistance (5.32) (see also [83]).

The possibility of a high-frequency negative conductance (or resistance) may be demonstrated starting from equation (5.32). The integral in equation (5.32) becomes negative if $v_0 = v_0(\eta)$ has a descendent portion and the frequency $\frac{\omega}{2\pi}$ is properly chosen, as may be seen with reference to Figure 5.9 [83]. This figure clearly shows that no negative resistance is possible for $\theta = \omega T_L < \pi$. However, if $\theta = 2\pi$, the device resistance, equation (5.32), will become negative if the device is biased deep into the negative mobility region, so that $v_0 = v_0(T_x)$ looks like shown by the solid line. The resistance will be positive for $\theta = \omega T_L = 3\pi$ and may be again negative around $\theta = 4\pi$. It can be seen also that no negative resistance occurs if dv/dE is always positive (Figure 5.9). As Kroemer indicated, it is not possible 'to obtain a negative resistance from purely SCL flow with transit-time shift alone' [83] i.e. a negative mobility region is necessary. However, it will be later shown that if a fixed space charge does exist ($N \neq 0$) or, if the current is not space charge limited, an h.f. negative resistance may be obtained as the result of transit-time effects in *positive mobility* semiconductors. Finally, Figure 5.9 shows that the positive mobility region for small T_x (near the cathode) reduces considerably the magnitude of the negative resistance which can be obtained at $\theta = 2\pi$, 4π etc. This region cannot be avoided because the electric field should increase from zero at $x = 0$ (SCL injection) to the threshold value where dv/dE becomes negative. Under 'anomalous cathode boundary-conditions' (non-ohmic cathode) the electric field at $x = 0$ may be positive and the entire semiconductor can be biased in the negative mobility region (see the dashed curve on Figure 5.9), with a corresponding increase in the h.f. negative resistance. All these will be discussed in much more detail in the following two chapters.

Kroemer [83] indicated a sufficient (but not necessary) criterion for the existence of an h.f. negative resistance. This criterion is: an h.f. negative conductance must exist if the incremental resistance $R_i = R(0)$ is larger than the static resistance V_0/J_0. The proof is simple. From equations (5.13), (5.14), (5.16), (5.32) and (5.34) we obtain

$$\frac{V_0}{J_0} - R(0) = \frac{2}{\varepsilon} \int_0^{T_L} v_0(\eta) \left(\eta - \frac{T_L}{2}\right) \mathrm{d}\eta. \tag{5.36}$$

In the range $0 < \eta < T_L$ the factor $\eta - T_L/2$ can be expanded into a Fourier series

$$\begin{aligned} \eta - \frac{T_L}{2} &= 2T_L \sum_{n=1}^{\infty} \frac{(-1)^{n-1}}{n} \sin \frac{2\pi n}{T_L} \left(\eta - \frac{T_L}{2}\right) = \\ &= 2T_L \sum_{n=1}^{\infty} \frac{1}{n} \sin \frac{2\pi n}{T_L} (T_L - \eta) \,. \end{aligned} \tag{5.37}$$

Because of equation (5.32), equation (5.36) can now be written

$$\frac{V_0}{J_0} - R(0) = 8\pi \sum_{n=1}^{\infty} R\left(n \frac{2\pi}{T_L}\right) \tag{5.38}$$

and $V_0/J_0 - R(0)$ can be negative only if at least one of the 'R' coefficients are negative (i.e. the series resistance, equation (5.32), should be negative at the transit time frequency $f_T = 1/T_L$ or a multiple of it). Thus, $V_0/J_0 < R(0)$ implies at least one frequency range with $R(\omega) < 0$. At low bias fields, the slope of the static $J - V$ characteristic will be steeper than J_0/V_0 (i.e. $R(0) < V_0/J_0$). Therefore, an inflection point on the static characteristic*) indicates that the h.f. resistance must be negative at frequencies comparable to, or higher than, the transit time frequency $f_T = 1/T_L$.

Figure 5.10 indicates the frequency and bias dependence of the parallel conductance, $G(\omega)$, calculated numerically [83] by use of the velocity-field dependence indicated in Figure 5.11. The susceptance is always positive (capacitive) as shown in Problem 5.5, and is not represented here. The frequency characteristics (Figure 5.10) show that if the device is biased deep into the negative-mobility region (Figure 5.11), $G(\omega)$ will become negative, at least for one frequency range. If the bias field is further increased, the depth of the negative conductance dip will decrease (not shown in Figure 5.10) because there is, of course, an optimum of electric field and negative-mobility distribution inside the device (Figure 5.11).

---

*) Such an inflection point can be seen on the experimental $J_0 - V_0$ characteristics measured by Yamashita [101] for $n^+ - n$GaAs $- n^+$ diodes characterized by a low $N_D L$ product (N_D small and negligible at high bias fields).

An interesting feature of the frequency characteristics represented in Figure 5.10 is the oscillation of $G(\omega)$ with increasing ω. An analytical calculation reveals some properties which do not depend upon the particular $v = v(E)$ dependence used. Following Kroemer [83], we shall calculate the v.h.f. ($\omega T_L \gg 1$) conductance and

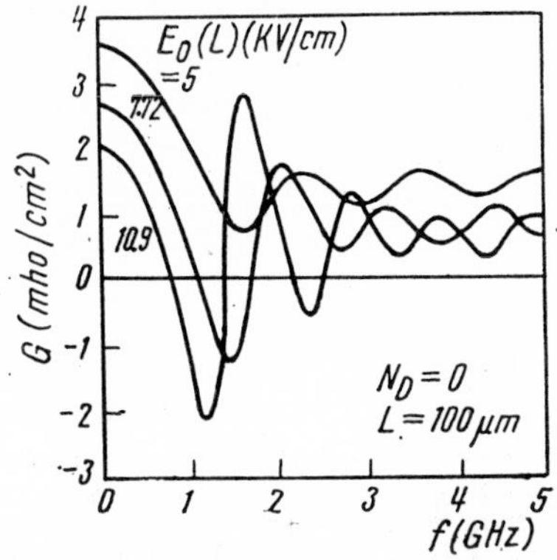

Figure 5.10. Conductance of a 100 μm GaAs diode, from 0 to 5 GHz and for several bias fields V_0/L, as calculated by Kroemer [83].

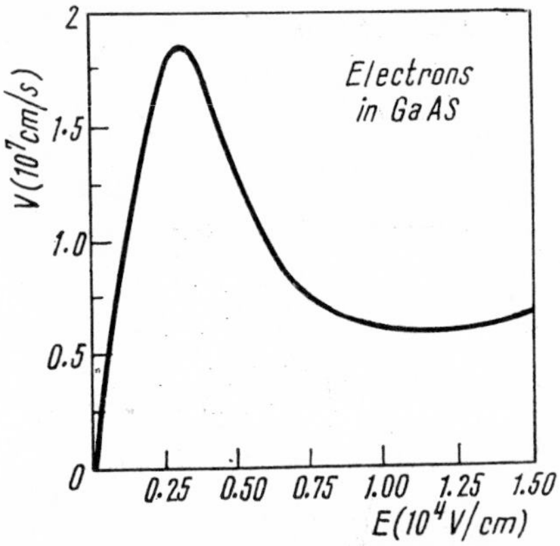

Figure 5.11. The velocity-field dependence used by Kroemer [83] for electrons in GaAs.

capacitance. The integrals occurring in equations (5.32) and (5.33) may be evaluated by making successive partial integrations. The first step yields

$$R(\omega) = \frac{1}{\omega^3 \varepsilon}\left[v_0(L) - \int_0^{T_L} \frac{dv_0}{d\eta} \cos \omega (t - \eta)\, d\eta \right] \tag{5.39}$$

$$X(\omega) = -\frac{1}{\omega C_0} + \frac{1}{\omega^2 \varepsilon}\int_0^{T_L} \frac{dv_0}{d\eta} \sin \omega (t - \eta)\, d\eta \tag{5.40}$$

where the SCL boundary condition $v_0(0) = 0$ was used. As $\omega \to \infty$ the resistance and the reactance vanish asymptotically, the former like $1/\omega^2$, the lattter as $1/\omega$. Note that if the integrals in equations (5.39) and (5.40) are developed by further partial integrations they are proportional to $1/\omega$ and thus we have

$$R(\omega)\,|_{\,\omega\to\infty} \simeq \frac{v_0(L)}{\varepsilon\omega^2} \tag{5.41}$$

$$X(\omega)\,|_{\,\omega\to\infty} \simeq -\frac{1}{\omega C_0} \tag{5.42}$$

if only the terms in $1/\omega$ and $1/\omega^2$ are taken into account. The parallel conductance $G(\omega)$ becomes

$$G(\omega)\,\Big|_{\omega\to\infty} = \frac{R(\omega)}{R^2(\omega)+X^2(\omega)}\Bigg|_{\omega\to\infty} \to \omega^2 C_0^2 R(\omega) \to \frac{\varepsilon v_0(L)}{L^2} \tag{5.43}$$

and the susceptance is

$$B(\omega)\,\Big|_{\omega\to\infty} = -\frac{X(\omega)}{R^2(\omega)+X^2(\omega)}\Bigg|_{\omega\to\infty} \to -\frac{1}{X(\omega)} \to \omega C_0. \tag{5.44}$$

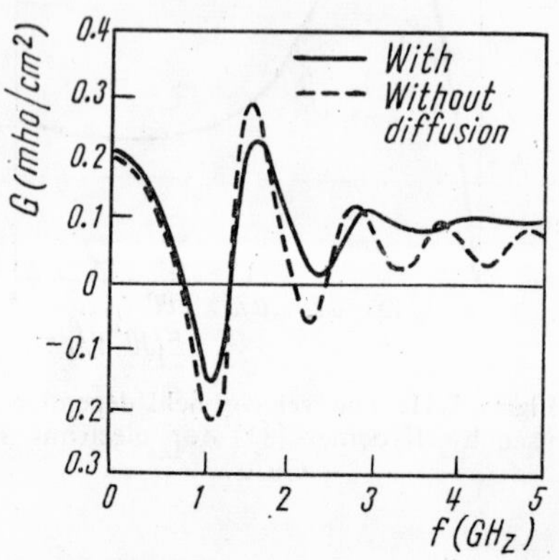

Figure 5.12. Conductance versus frequency calculated by Kroemer [83] including diffusion effects (solid curve) and without diffusion (dashed curve), for a bias field of 4.81 KV/cm (see Figures 5.10 and 5.11).

The asymptotic behaviour ($\omega \to \infty$) of the SCLC diodes is very simple: a constant and positive conductance, equation (5.43), is shunted by the geometrical capacitance $C_0 = \varepsilon/L$. However, there exist prolonged oscillations of conductance before the asymptotic value, equation (5.43), is attained practically. Kroemer pointed out that such a behaviour is not confirmed by experiment and indicated that if diffusion is taken into account in theoretical calculations, these oscillations will appear strongly damped [83].

Figure 5.12 shows theoretical results with and without diffusion. Below the transit-time frequency, the agreement between these results is good and can be even better if some computation details (the position of the virtual cathode) are properly adjusted. The differences occur at higher frequencies: the second range of negative conductance disappears, the conductance oscillations are strongly attenuated and the v.h.f. conductance increases slightly.

The problem of incorporating diffusion in *analytical* calculations is particularly difficult. It was shown in Section 3.11 that the steady-state calculations, taking diffusion into account, are exceedingly complicated or impossible to develop analytically. This fact complicates the small-signal problem*). Sometimes, diffusion

---

*) Nigrin [118] found an exact small-signal solution for SCL injection in a metal-insulator-semi-insulator-metal structure (the space charge varactor) by taking diffusion into account and neglecting trapping and thermal carrier-generation. However, the solution was calculated for zero steady-state current, a situation which is peculiar to the device studied by Nigrin.

is neglected in the steady-state and the small signal solution uses the steady-state solution by also taking into account the a.c. diffusion current [119]. This approximation was often made in calculating the a.c. impedance of a semiconductor sample which is neutral in the steady-state [76], [120] — [122].

In the above cited paper [83], Kroemer also performed analytical calculations by including a d.c. and a.c. diffusion current. The results are qualitative in nature and demonstrate that in the presence of diffusion (no matter which the details of this diffusion process are) the frequency dependence of the parallel conductance above the transit-time frequency will show less pronounced oscillations than in the absence of diffusion*⁾. The former behaviour is confirmed by experiment.

5.5 Small-signal Impedance of the SCLC Resistor

In this Chapter, the fixed charge density has been considered negligible so far ($N = 0$). We shall assume that N is finite but, for the beginning, the carrier mobility will be assumed constant

$$v = \mu E, \ \mu = \text{const.} \tag{5.45}$$

The a.c. theory of Section 4.8 may be applied directly. The small signal impedance is again given by K_{11} (since $K_{12} = 0$ in equation 4, 18, because of $v_0(0) = 0$). Note the fact that $\Delta = p_0(0)/N$ tends to infinity due to the boundary condition (5.2), and Δ may be pulled out from the parentheses. Then, we can write successively

$$\begin{aligned} \frac{v_0(0)\, T_L^2}{\varepsilon}\,\Delta &= \frac{e v_0(0)\, p_0(0)\, T_L^2}{eN\varepsilon} = \frac{J_0}{eN\varepsilon}\,\theta_d^2\,\tau_d^2 = \\ &= \frac{J_0\,\tau_d}{e\,NL}\;\frac{L}{eN\mu}\,\theta_d^2 = J_0 R_0\,\theta_d^2 \end{aligned} \tag{5.46}$$

where we used the notations introduced in Section 3.10. Here

$$R_0 = L/eN\mu \tag{5.47}$$

---

*⁾ We reproduce from Kroemer's paper [83] the following physical interpretation: 'In the absence of diffusion, there would exist an internal feedback loop that would cause fresh carriers to be reinjected whenever the field ahead of the virtual cathode rises periodically as a result of earlier carrier density waves leaving the crystal at the anode. The feedback is positive for integer multiples of the transit-time frequency, negative for half-integer multiples, thus causing the strong conductance oscillations. The fact that in the presence of diffusion the entire current in the virtual cathode plane must be diffusion current breaks up the reinjection feedback loop. It is this breaking-up, rather than any spatial attenuation of the carrier density waves by diffusion, that causes the qualitatively different and much more rapid disappearance of the conductance oscillations'.
The effect of carrier diffusion is further discussed in Section 5.6 and the corresponding Appendix.

is the resistance of a semiconductor bar of length L, unitary conductive section, and conductivity $eN\mu$. Because here μ_d (introduced at the beginning of Section 3.10) is the low-field mobility, θ_d may be replaced by θ_r given by equation (3.69)

$$\theta_r = T_L/\tau_r\,, \ \tau_r \to \tau_d \mid_{\mu_d=\mu} = \varepsilon/eN\mu. \tag{5.48}$$

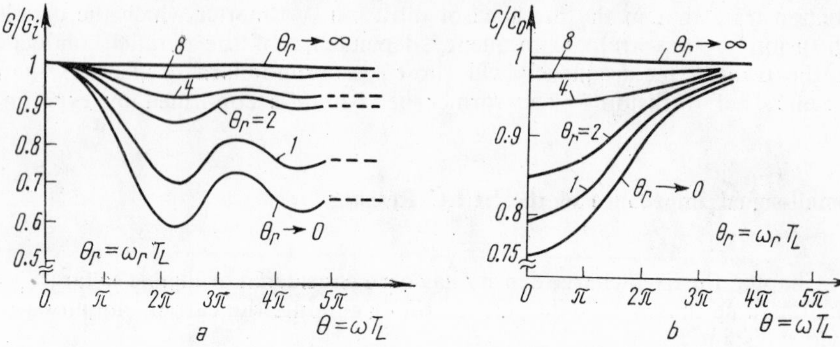

Figure 5.13. Normalized frequency characteristics for the SCLC resistor: (a) the parallel conductance and (b) the parallel capacitance. The carrier mobility is assumed constant. The value of θ_r indicates the steady-state bias, as shown in the text.

Then, the diode impedance, equation (4.102), becomes

$$Z = R_0 \bar{J}_0\, \theta_r^2 \left[Q\,(\theta_r + i\theta) + \frac{R\,(i\theta + \theta_r) - R(\theta_r)}{i\theta} \right] \tag{5.49}$$

and may be written in normalized form

$$\bar{Z} = \frac{Z}{Z\mid_{\theta\to 0}} = \bar{Z}\,(\theta; \theta_r)\,. \tag{5.50}$$

In this Section we shall consider a $p^+\pi p^+$ (or $n^+\nu n^+$) structure with SCL current and $\mu \simeq$ const. This device has a symmetrical characteristic and is sometimes called the SCLC resistor. Here N is positive and θ_r takes positive values from zero to infinity. The normalized steady-state characteristic is plotted in Figure 3.21 ($\Delta \to \infty$). Figure 5.13 shows the frequency dependence of the parallel conductance and parallel capacitance for several operating points (several values of θ_r). For $\theta_r \to 0$, i.e. in the square-law region $J \propto V^2$ of the steady-state characteristic, we re-obtain the results known from Section 5.2 for $a \to 0$. For $\theta_r \to \infty$, i.e. in the linear (ohmic) region, the conductance and capacitance are frequency independent*) [54].

*) Chisholm and Yeh [123] performed an approximative calculation by assuming a d.c. electric field identical with that for $N = 0$ and by solving the differential equation of the a.c. electric field as shown in the previous chapter. The results differ substantially from the 'exact' ones given above (except for very small $|\theta_r|$). This comparison indicates the importance of knowing the accurate steady-state distribution when performing small-signal calculations.

Figure 5.14 represents the bias dependence of the low-frequency dynamic capacitance. As predicted in Section 5.3, this capacitance is equal to the geometrical capacitance $C_0 = \varepsilon/L$ only if the semiconductor is neutral.

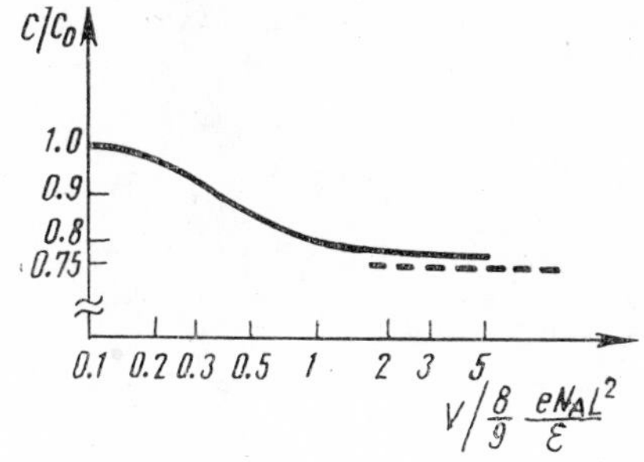

Figure 5.14. The bias dependence of the low frequency dynamic capacitance of the SCLC resistor.

5.6 The Punch-through Diode

The conduction mechanism in a $p^+ \nu p^+$ (or $n^+ \pi n^+$) device, called the punch-through diode was discussed in Section 1.2. The steady-state characteristic (μ = const) can be obtained from the general theory of Section 3.9, by introducing $\Delta \to \infty$ (SCL emission). Figure 5.15 shows the normalized $J - V$ characteristic. Several values of the parameter $|\theta_r|$ are indicated on this curve. θ_r is given again by equation (5.48) but here τ_r is negative ($N < 0$) and it is not a true relaxation time ($-N$ is donor concentration and μ the hole mobility).

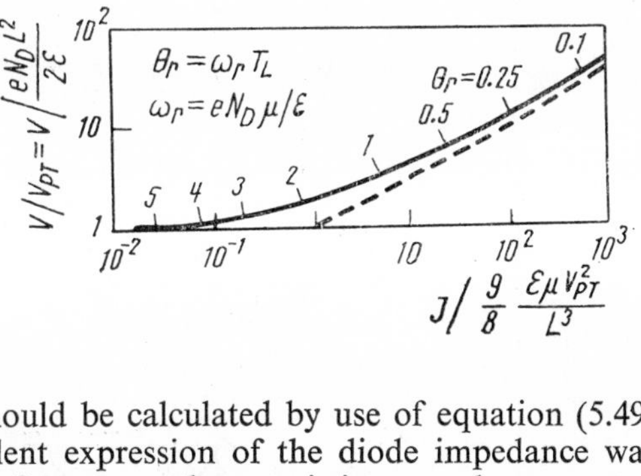

Figure 5.15. The normalized steady-state characteristic of the punch-through diode (negligible diffusion and constant mobility), first calculated by Shockley and Prim [13].

The small-signal characteristics should be calculated by use of equation (5.49) where θ_r is now negative. An equivalent expression of the diode impedance was first derived by Yoshimura [17]. The frequency characteristics are also presented and discussed by Birdsall and Bridges in their monograph [74].

In Figure 5.16 we also present the frequency characteristics of the SCLC punch-through diode [57]. The reactive behaviour is always capacitive.

Figure 5.16 shows the fact that a $p^+ \nu p^+$ (or $n^+ \pi n^+$) structure biased in the punch-through region (steep increase of the current with bias voltage, Figure 5.15) should

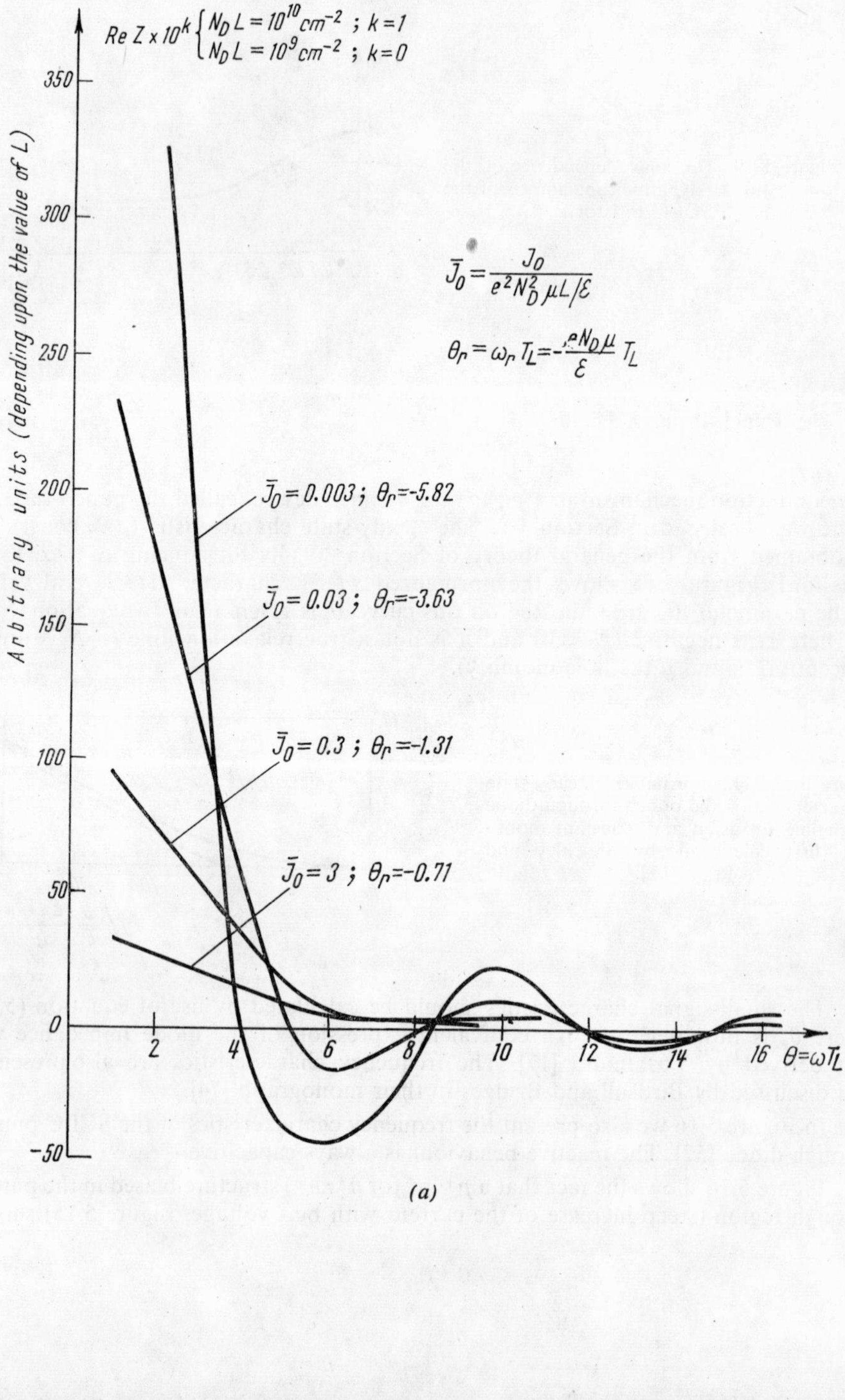
Re Z x 10^k { N_D L = 10^10 cm^-2 ; k=1
N_D L = 10^9 cm^-2 ; k=0
Arbitrary units (depending upon the value of L)
350
300
250
200
150
100
50
0
-50
J̄_0 = J_0 / (e^2 N_D^2 μL/ε)
θ_r = ω_r T_L = -(eN_D μ/ε) T_L
J̄_0 = 0.003 ; θ_r = -5.82
J̄_0 = 0.03 ; θ_r = -3.63
J̄_0 = 0.3 ; θ_r = -1.31
J̄_0 = 3 ; θ_r = -0.71
2
4
6
8
10
12
14
16
θ=ωT_L

(a)

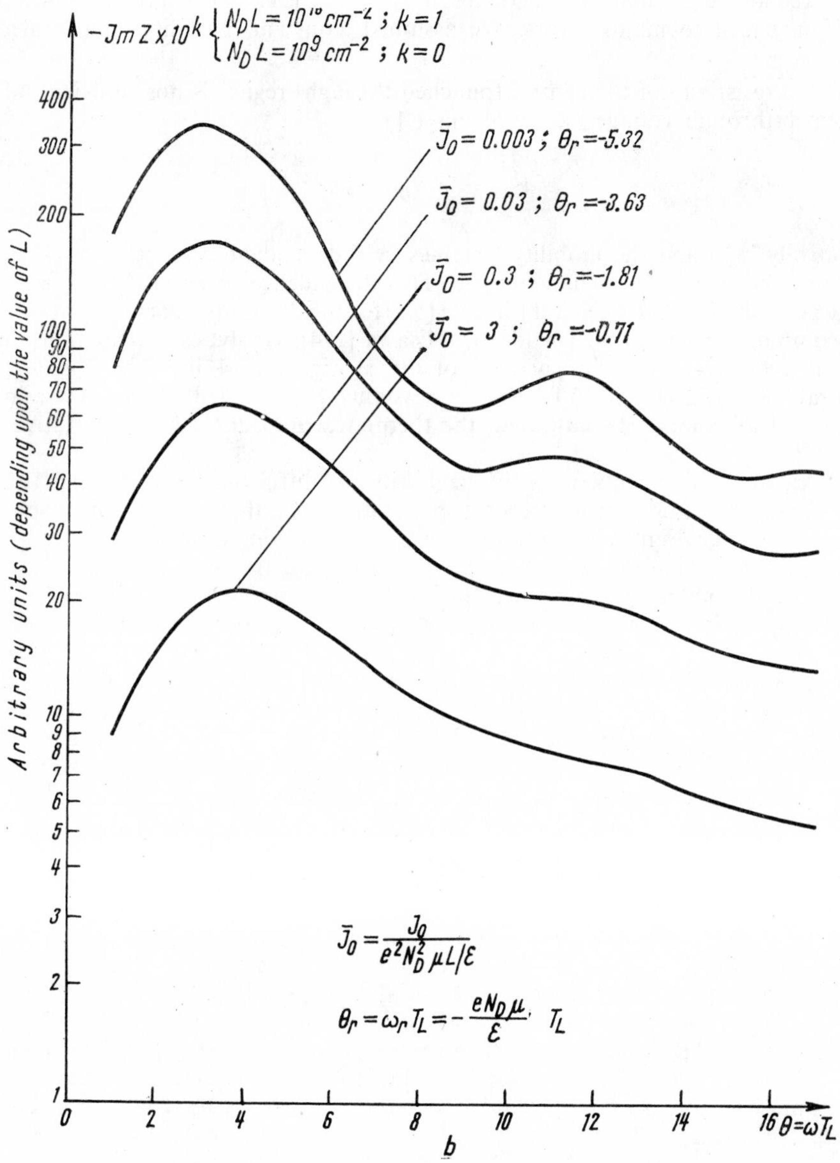

Figure 5.16. Frequency characteristics calculated for the SCLC punch-through diode.

exhibit a negative resistance in a frequency range around the transit-time frequency. Such a negative resistance was first calculated by Shockley [14].

In the above calculations, there are, however, at least two major simplifications which may lead to major errors. We assumed constant mobility and neglected diffusion.

If the resistivity of the central (punched-through) region is not sufficiently low, the punch-through voltage (see Problem 1.1).

$$V_{PT} = e\, N_D\, L^2/2\varepsilon, \tag{5.51}$$

is relatively high and the mobility becomes field dependent even in the low-current region of the steady-state characteristic. This dependence, $\mu = \mu\,(E)$, can be taken into account by using the general formula (5.6) for the diode impedance. Some numerical computations were also made by Mişawa [124]. Analytical calculations show that the decrease of the d.c. carrier mobility at higher field intensities is an unfavourable situation for the h.f. negative resistance [17], [125]. If the carrier velocity is as sumed as completely saturated, the theoretical impedance has no negative real part.

A second major problem is the neglection of diffusion. Note the fact that in the low-current region, just after the punch-through, the carrier density on the top of the barrier (Figure 3.29) is relatively low*) and a condition like (5.2) can hardly be justified. This indicates that the diffusion current is important in the vicinity of the potential maximum. Theoretical [61], [126] and experimental [15], [98] results indicated the effect of the diffusion current†), in the ν region (especially at low bias currents).

It was not until recently that experiments have evidenced negative resistence effects in punch-through diodes [18] — [20], [127]. It is assumed that the carrier velocity is nearly saturated in these devices. Because the elementary theory [17] ($v \simeq v_{sat}$) gives a very small negative resistance, and also because the experimental results indicate a maximum of the absolute value of the negative resistance at frequencies lower than expected (for a transit angle of $3\pi/2$ instead of almost 2π) there exists a tendency to improve the small-signal theory of this device. One possibility is to take an accurate velocity-field dependence into account, a dependence exhibiting a gradual transition from constant mobility to (almost) saturated carrier velocity [128]. The fact was also pointed out that an additional delay of the a.c. current with respect to the a.c. voltage may result from the slow diffusion process of charge carriers before they reach the potential maximum [18].

A major problem of the analytical theory is including the effect of the diffusion current. This problem was also discussed in Section 5.4. An alternative is presented below. We will neglect the small a.c. diffusion current at and beyond the potential maximum but more suitable boundary conditions will be used. The d.c. carrier

---

*) If the depletion layer of the 'collector' junction just punches through the central region, this density will be the minority carrier density in the (residual) neutral base.

†) The longer the extrinsic Debye length (the larger the resistivity) of the central layer as compared to the device length, the larger the diffusion effect [13], [15], [98], [106].

lensity at the potential maximum (which is located approximately at the emitter unction, $x = 0$) is finite

$$p_0(0) = \text{finite} \tag{5.52}$$

nd is determined by taking into account both drift and diffusion. On the other and, by definition, at the potential maximum

$$E_0(0) = 0 \quad \text{and} \quad v_0(0) = 0. \tag{5.53}$$

The a.c. boundary condition is no longer $\mathbf{E}_1(0) = 0$. We shall calculate the a.c. particle current at $x = 0$ as follows (see the preceding chapter for similar calculations):

$$\mathbf{j}_1(0) = ep_0(0)\,\mathbf{v}_1(0) + e\,\mathbf{p}_1(0)\,v_0(0) =$$

$$= ep_0(0)\,\mu\,\mathbf{E}_1(0) = \sigma_1(0)\,\mathbf{E}_1(0) \tag{5.54}$$

where

$$\sigma_1(0) = e\,\mu p_0(0) = \text{finite} \tag{5.55}$$

s a sort of a.c. conductivity of the $x = 0$ plane (virtual cathode). Therefore, the .c. boundary condition determined by equations (5.54) and (5.55) is exactly of the orm of equation (4.16). We stress the fact that such a boundary condition was irst suggested and justified by Shockley [16] and then used by Wright [87], [88], nd others [128], [129], The problem of calculating the a.c. response is also discussed n the Appendix to this Section.

At the time of writing this book, the analytical linear theory including diffusion s not entirely satisfactory and we believe that this is an interesting and important ield of theoretical research. This should be, in fact, a refinement of the theory, which is much like the problem of including the effect of initial velocities in calculating the a.c. response of the electron vacuum devices [130].

.7 Transferred-electron Small-signal Amplifier

'he transferred-electron (or negative mobility) amplifier is an SCLC diode constructed rom a negative mobility semiconductor (like GaAs or InP) with ohmic contacts Section 1.4). This device was studied widely and its small signal properties are well stablished [29], [131].

Let us first assume the fact that the bulk space-charge may be neglected. The xternal characteristic $J - (V/L)$ is identical to the neutral characteristic (Section .7).

$$J = j_c(E_c) = eNv(E_c) = eNv(V_0/L) = eNv_l$$

because the electric field is uniform and equal to the bias field. A small-signal calculation which uses the boundary condition $\underline{E}_1(0) = 0$ yields the following impedance

$$Z = \frac{L^2}{\varepsilon v_l} \frac{\exp(-s) + s - 1}{s^2}, \quad s = \frac{T_L}{\tau_d}(1 + i\omega\tau_d) \tag{5.56}$$

where τ_d is the differential relaxation time, equation (3.94)

$$\tau_d = \frac{\varepsilon}{eN(\mathrm{d}v/\mathrm{d}E)_0}. \tag{5.57}$$

Here N is positive (majority carrier injection) but the differential mobility $(\mathrm{d}v/\mathrm{d}E)$, calculated for $E_0 = V_0/L$ may be either positive or negative, depending upon the d.c. bias (Figure 3.13). The impedance, equation (5.56) was first calculated by McCumber and Chynoweth [76]. It depends upon frequency through the transit-angle $\theta = \omega T_L$. The frequency characteristics are plotted in Figure 5.17. At low frequencies $(\theta \to 0)$ one obtains

$$Z|_{\theta\to 0} = R_i = \frac{L^2}{\varepsilon v_l} \frac{\exp(-\theta_d) + \theta_d - 1}{\theta_d^2}, \quad \theta_d = \frac{T_L}{\tau_d} \tag{5.58}$$

Note the fact that R_i is positive, even when the device is biased in the negative mobility region $(\theta_d < 0)$. This result is quite surprising, because the $J - V$ characteristic, equation (5.55), *does* exhibit a negative slope if the differential mobility is negative. In other words, the a.c. result, equation (5.58), is in contradiction to the steady-state model which is the basis of the a.c. calculation. This inconsistency will be removed by using the proper d.c. boundary condition. We have already shown in Section 3.8 that an SCL-emission diode like the n^+nn^+ (or p^+pp^+) device studied here has a static (low frequency) resistance which is always positive.

McCumber and Chynoweth's results [76] indicate the existence of a high-frequency negative resistance (Figure 5.17) for negative θ_d (negative differential mobility)*). It may also be shown [76] that the a.c. impedance as a function of the complex variable p (which replaces $i\omega$ in equation (5.56)) has no singularities in the finite ω plane. Consequently, the device is always stable when operated under constant current (or a.c. open circuit) conditions. On the other hand, $Z = Z(p)$ does have a denumerable number of zeros. It may be shown that the condition for a.c. short circuit stability reduces to [76]

$$\theta_d = \frac{T_L}{\tau_d} > -2.09 \tag{5.59}$$

---

*) These calculations were repeated by also taking diffusion into account [76]. Suematsu and Nishimura [132] indicated the fact that the conductance oscillations are rapidly attenuated in the presence of diffusion and that the conductance becomes practically constant above the transit-time frequency.

and may be also written

$$\frac{d}{dE} \ln v(E) > -\frac{2.09\,\varepsilon}{eNL} \,. \tag{5.60}$$

The device is short circuit stable only if the doping-length product NL is sufficiently low*). If the doping-length product is higher than the critical value, the device bi-

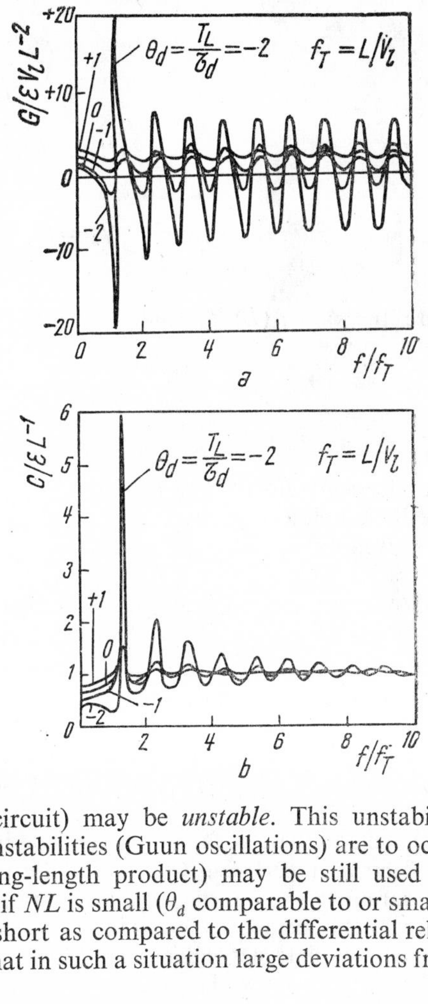

Figure 5.17. The frequency dependence of the parallel conductance (a) and capacitance (b) of a negative mobility sample as calculated by McCumber and Chynoweth [76].

ased with a constant voltage (a.c. shortcircuit) may be *unstable*. This unstability criterion must be met if the large signal instabilities (Guun oscillations) are to occur [76]. The stable device (subcritical doping-length product) may be still used for *small-signal amplification* [131]. However, if NL is small (θ_d comparable to or smaller than unity) the transit-time is relatively short as compared to the differential relaxation time. It was shown in Section 3.9 that in such a situation large deviations from neutrality do occur.

---

*) Initially, McCumber and Chynoweth [76] have indicated a critical doping-length product of 2.7×10^{11} cm^{-2} for GaAs at room temperature.

Consequently, a correct theory of small-signal amplification should take into account the deviations from the d.c. neutrality. Calculations neglecting diffusion may be performed by using the formulae presented in Section 5.1. We shall write equations (5.4) and (5.5), respectively

$$\overline{\mathcal{V}}_0 = \int_0^a \frac{\overline{E}\, \overline{v}\,(\overline{E})\, \mathrm{d}\overline{E}}{\overline{\mathcal{J}}_0 - \nu\, \overline{v}\,(\overline{E})} \tag{5.61}$$

$$1 = \int_0^a \frac{\overline{v}\,(\overline{E})\, \mathrm{d}\overline{E}}{\overline{\mathcal{J}}_0 - \nu\, \overline{v}\,(\overline{E})} \tag{5.62}$$

$$\overline{\mathcal{V}}_0 = V/E_M L \tag{5.63}$$

$$\overline{\mathcal{J}}_0 = \frac{J}{\varepsilon E_M\, v_M/L}. \tag{5.64}$$

Here the $v = v\,(E)$ dependence is normalized as follows

$$\overline{v} = \overline{v}\,(\overline{E}),\ \overline{v} = \frac{v}{v_M},\ \overline{E} = \frac{E}{E_M} \tag{5.65}$$

(the electron velocity and electric field are normalized to the peak velocity and the threshold field, respectively, see Figure 3.23)*). The dimensionless parameter, ν, is given by

$$\nu = \frac{\mathrm{e}NL}{\varepsilon E_M} \propto NL \tag{5.66}$$

and, for a certain semiconductor, it is proportional to the doping-length product. Finally, a is defined as

$$a = E_0(L)/E_M. \tag{5.67}$$

We conclude that, as far as the semiconductor is specified (for example one studies the electron injection in GaAs), the normalized $J - V$ characteristic depends upon the parameter ν, i.e. one obtains a family of normalized characteristics. We shall illustrate this below, by use of the results communicated by Mahrous *et al.* [133]. The velocity-field dependence used in calculations is shown in Figure 5.18. Figures 5.19 and 5.20 show the electric field and the potential distribution, respectively.

---

*) A similar normalization may be used for other kinds of $v = v(E)$ dependence, and $v(1)$ may not be equal to unity.

Figure 5.21 represents normalized $J - V$ characteristics for several values of the parameter υ. Here, for electrons in GaAs, we have:

$$= \frac{eN_D L}{\varepsilon E_M} = 4.65 \times 10^{-11}\, N_D L \; (N_D L \text{ in cm}^{-2}). \tag{5.68}$$

Figures 5.19 and 5.20 indicate the fact that for bias fields V_0/L below the threshold field, E_M, the electric field is approximately uniform. The non-uniformities are more pronounced for small ν (compare $\nu = 2$ with $\nu = 15$), i.e. for low doping-length products, as we have already shown.

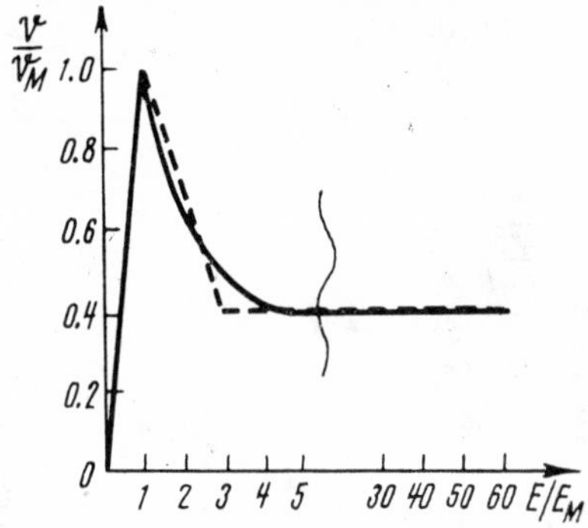

Figure 5.18. Velocity-field dependence used by Mahrous *et al.* [133] for electrons in GaAs.

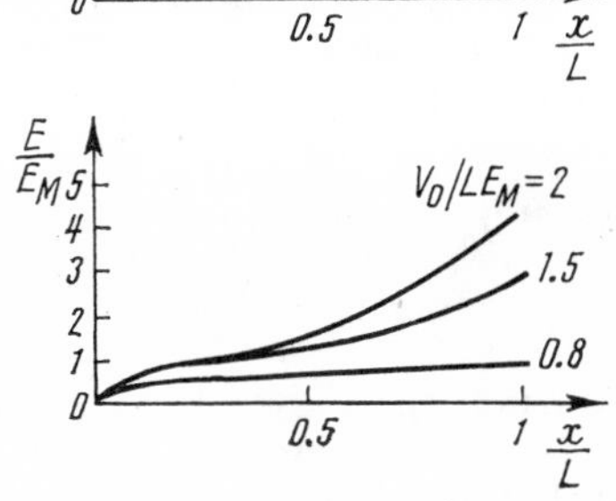

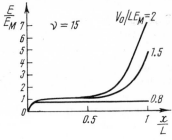

Figure 5.19. Static field distribution for $\nu = 15$ and $\nu = 2$ [133].

As the bias field rises above the threshold field ($V_0/L = 1.5\ E_M$ and $2E_M$ in Figures 5.19 and 5.20), the electric field becomes very non-uniform in the vicinity of the anode (this non-uniformity is less important for small ν). Note the fact that if ν is relatively large, a large increase of the bias-field leaves the field in the cathode region almost unmodified: all supplementary voltage drop occurs on the anode region. Because the carrier flow is largely determined by the field in the cathode region, the current has to change very little with the applied voltage. This is confirmed by Figure 5.21: the $J - V$ characteristic for large ν, exhibits a current saturation above the threshold field. This feature is well confirmed by experimental measurements.

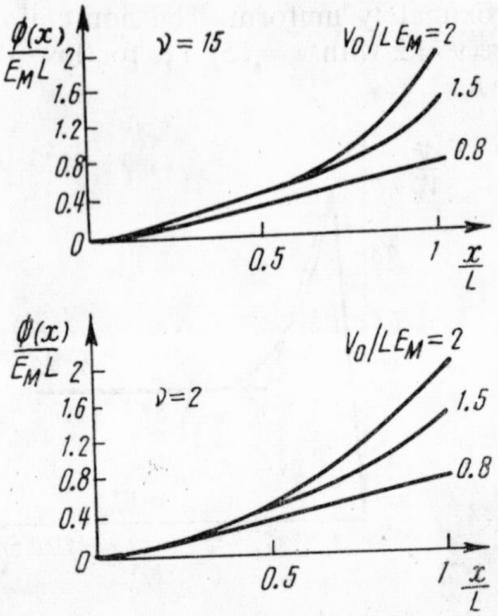

Figure 5.20. Static potential distribution [133].

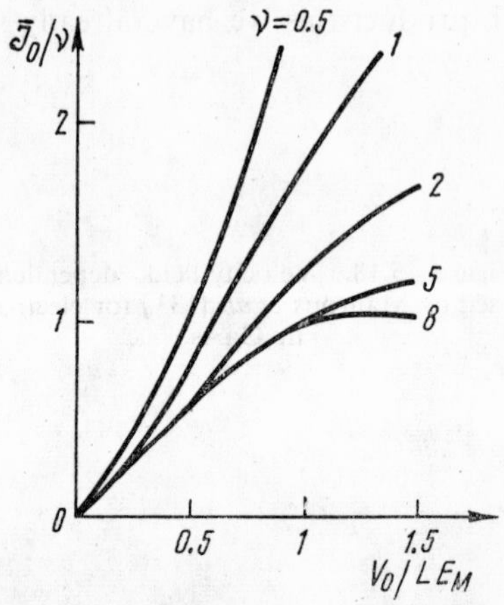

Figure 2.21. Normalized steady-state characteristics [133].

If ν is taken very small we shall approach the limit case studied by Kroemer [83] (Section 5.4), i.e. zero doping. The $J - V$ characteristic will exhibit an inflection point above the threshold field [83].

We shall discuss now the small-signal behaviour and the stability of the steady-state régime. The a.c. impedance is given by equations (5.6) and (5.10). For a given $\overline{v} = \overline{v}\,(\overline{E})$ relationship, the normalized impedance should be of the form

$$\overline{Z} = Z/R_i = \overline{Z}\,(\theta;\, \nu,\, \overline{\mathcal{J}}_0) \tag{5.69}$$

and we suggest that the reader finds its value. The normalized frequency characteristics will depend upon the parameter ν and upon the bias current, $\overline{\mathcal{J}}_0$, as well (see Figure 5.21). We illustrate this dependence by reproducing the results obtained by Holmstrom [134] for a model identical with that of Mahrous *et al.* [133], except for the details of the $v - E$ dependence for electrons in GaAs. Figure 5.22 shows these results for several values of ν and $\overline{\mathcal{J}}_0$. The impedance is normalized to the incremental resistance at low current intensities. The frequency is normalized to the reciprocal of the dielectric relaxation time at low bias fields.

The first two diagrams in Figure 5.22 are calculated for $\nu = 3$. In the first case (Figure 5.22a) we have $\mathcal{J}_0/\nu = J/eN_D v_M = J_0/j_n(E_M) = 1.1$ and no negative resistance occurs. If the bias current increases such as $\mathcal{J}_0/\nu = 1.2$ (Figure 5.22c), the series resistance will become negative. A negative resistance also occurs for $\mathcal{J}_0/\nu = 1.1$ provided that ν has a larger value ($\nu = 4$), i.e. a larger doping-length

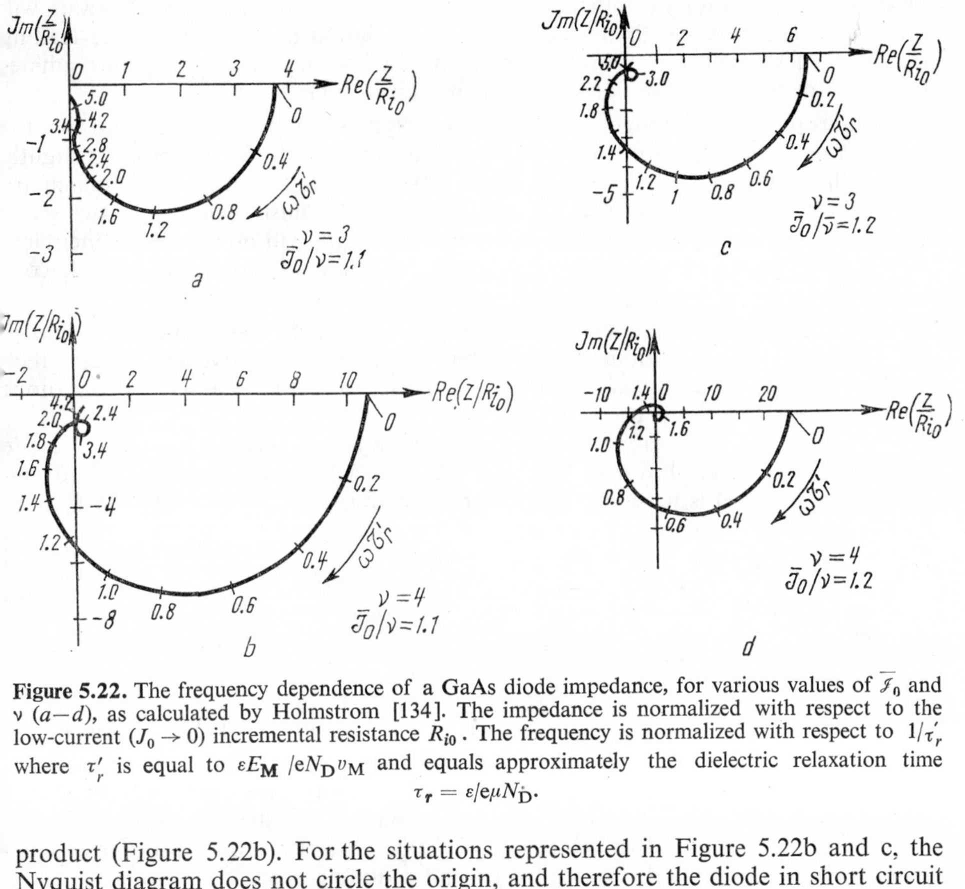

Figure 5.22. The frequency dependence of a GaAs diode impedance, for various values of $\mathcal{J}_0$ and ν (*a–d*), as calculated by Holmstrom [134]. The impedance is normalized with respect to the low-current ($J_0 \to 0$) incremental resistance R_{i0}. The frequency is normalized with respect to $1/\tau_r'$ where τ_r' is equal to $\varepsilon E_M / eN_D v_M$ and equals approximately the dielectric relaxation time $\tau_r = \varepsilon / e\mu N_D$.

product (Figure 5.22b). For the situations represented in Figure 5.22b and c, the Nyquist diagram does not circle the origin, and therefore the diode in short circuit does not oscillate*). In the case depicted in Figure 5.22d, the a.c. short-circuited diode is unstable. The diode can also oscillate if a small resistance is connected

*) If the poles of the admittance (i.e. the zeros of the impedance) lie in the lower half of the complex frequency plane, then the diode will be stable under short circuit or constant-voltage conditions. By applying the Nyquist theorem and taking into account the fact that the impedance has no poles (at finite distance), one obtains the rule indicated in the text. The complete Nyquist diagram will be obtained by adding the branch of negative frequencies which is symmetrical with respect to the horizontal axis, because $Z(-\omega)$ is the complex conjugate of $Z(\omega)$, as the reader may easily verify (see also [79]).

in series. If the load impedance is complex, the oscillations can or cannot occur, depending upon the fact whether or not the Nyquist diagram of the series circuit (load impedance plus diode impedance) circles the origin.

If the diode biased by a constant voltage is a.c. unstable, any accidental fluctuation may lead to oscillations (see Chapter 13). It will be shown that, in a medium of negative mobility, an internal perturbation may grow and if the doping-length product is high enough (of the order of 10^{12} cm^{-2}), stable high-field domains will travel successively through the device, their formation and disappearance leading to spikes of current (Gunn oscillations). The oscillation frequency is approximately equal to the reciprocal of the domain transit time.

Mahrous and Hartnagel [135] used the Nyquist analysis to demonstrate the fact that domain formation can be inhibited in diodes with the NL product slightly higher than 10^{12} cm^{-2}, if a suitable load impedance is applied. If the imaginary part of the external impedance is capacitive, the Nyquist diagram of the series circuit will be transferred downwards so that the origin will no longer be encircled. It was also demonstrated *experimentally* [135] that domain formation can be controlled by reactive loading.

We assume now a subcritically-doped diode, i.e. a diode which is stable in a.c. short circuit. The diode may exhibit, however, a negative resistance which can be used for *small-signal amplification* in a reflection-type amplifier (the diode connected to a matched transmission line).

We note the fact that self sustained oscillations are possible even for a diode impedance like those shown in Figures 5.22b and 5.22c, but a suitable load impedance (inductive) is necessary. However, in contrast to the transit-time (Gunn) oscillations, the oscillation frequency will be determined by the external circuit (reactive tuning) and not by the diode itself*).

5.8 Non-uniformly Doped SCLC Diodes

The semiconductor between the electrodes of the SCLC diode may be more or less non-uniformly doped, due to the technological processes used to fabricate the structure. In this section we discuss large non-uniformities which are introduced with the purpose of ameliorating the operation of the structure†).

Charlton *et al.* [136] indicated that a supercritically doped transferred-electron device with a cathode doping notch and a doping profile increasing from the cathode to the anode may act as a stabilised amplifier at high bias voltages. We consider here a device (Figure 5.23) provided with a thin doping notch near the cathode and the remainder of the semiconductor *uniformly* doped. We shall show that the electric field inside such a structure is less non-uniform than in a usual,

---

*) The frequency of transit-time oscillations may be slightly modified by changing the d.c. bias (which changes the velocity of the domain).

†) Kroemer has shown that the incremental resistance of an SCLC semiconductor diode with *non-uniform* doping should be always positive [62].

uniformly-doped device. It is interesting to note that Thim [137] has demonstrated recently the fact that the effect of the thermal noise in negative-mobility amplifiers can be reduced if an almost uniform steady-state field is established inside the device.

The uniformly-doped negative mobility amplifier has a highly non-uniform anode field (Figure 5.19). The existence of such a large non-uniformity can be

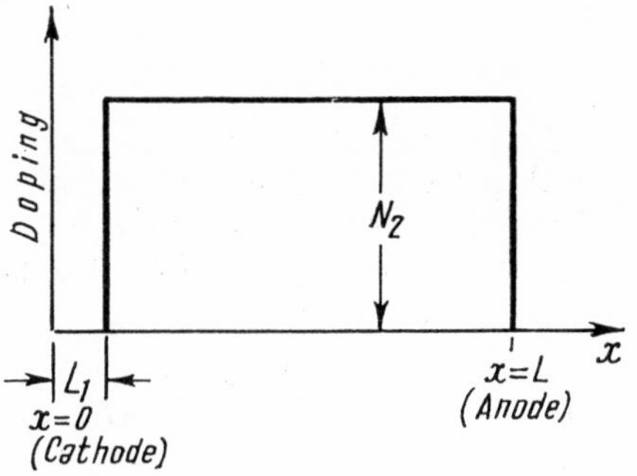

Figure 5.23. Doping profile of a negative-mobility semiconductor sample.

demonstrated by using the theory developed in Chapter 3. With reference to equation (3.35) and (3.44) we may write

$$x = \frac{\varepsilon}{eN} \int_0^{E(x)} \frac{j_n(E)\,\mathrm{d}E}{J_0 - j_n(E)} \tag{5.70}$$

(the cathode field was taken equal to zero). Let us consider a bias current J_0 which is sufficiently high to provide an anode field above the threshold field. The entire semiconductor should be accumulated and the electric field increases monotonously

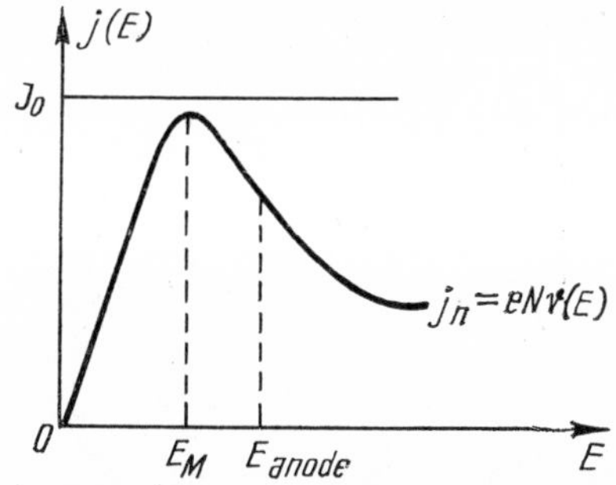

Figure 5.24. $j - E$ diagram used for estimating the field non-uniformity.

from $x = 0$ to $x = L$ (Chapter 3). The carrier density has a minimum inside the semiconductor. If the doping-length product is sufficiently large, this minimum value will be close to the thermal equilibrium value (N). This situation is described by Figure 5.24. The reader may see easily that the quantity under the integral sign is positive, first increases with $E(x)$ and then increases very rapidly as $E(x)$ is comparable to the threshold field; it then decreases with increasing E. According to equation (5.70), the field gradient should increase rapidly in the vicinity of electrodes and may change very little inside the semiconductor bulk (Figure 5.19).

In a semiconductor SCLC structure like that depicted in Figure 5.23, the interface field E_{12} $(x = L_1)$ may be expressed as a function of the bias current J_0. By applying equation (3.44) to the first region (doping notch with $N_1 \to 0$) one obtains

$$J_0 = \frac{\varepsilon}{L_1} \int_0^{E_{12}} v(E)\, \mathrm{d}E\,. \tag{5.71}$$

$J_0 = J_0(E_{12})$ may be considered as the control characteristic of the second region, whose neutral characteristic is $j_{n2} = eN_2 v(E)$ (Figure 5.25). We can write the distance x' from the interface $1 - 2$ (Figure 5.23) as

$$x' = \frac{\varepsilon}{eN_2} \int_{E_{12}}^{E(x)} \frac{j_{n2}(E)\, \mathrm{d}E}{J_0 - j_{n2}(E)}\,. \tag{5.72}$$

For a given J_0, the shape of the field distribution may be found by use of equation (5.72) and Figure 3.15. If the bias current J_0 is smaller than the crossover current J_X (Figure 3.15), the anode region will be depleted entirely. The important point is that the field will increase to a value in excess of the threshold in a small fraction of the sample length (the cathode region is relatively thin). Over a large portion of the sample, the electric field can be almost uniform. Figure 5.26 shows the results of computer calculations (including diffusion) made by Charlton

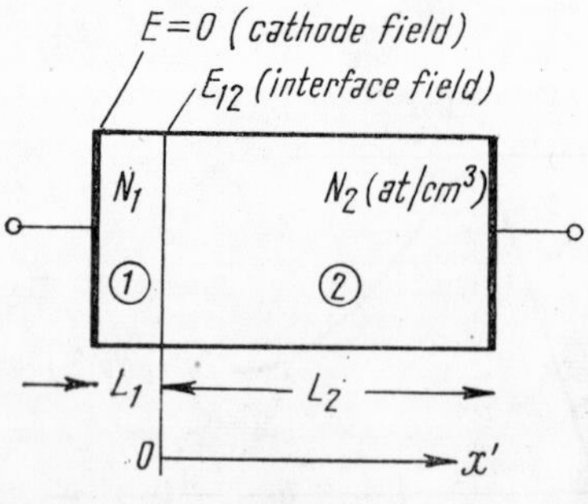

Figure 5.25. Another $j - E$ diagram (see the text).

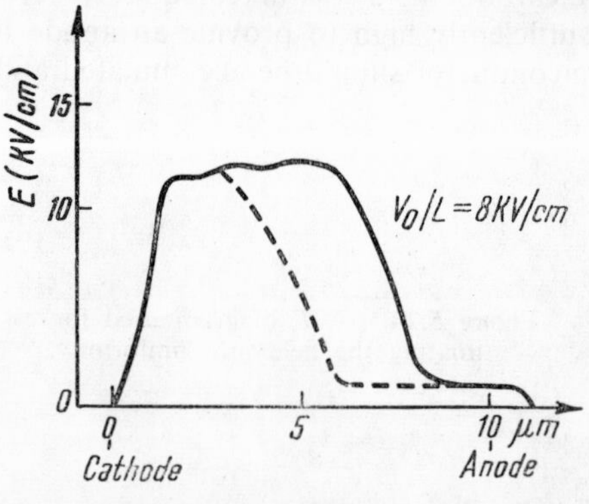

Figure 5.26. The static electric field distribution for a GaAs doping profile as indicated in the text, and bias fields as shown, according to Charlton *et al.* [136].

et al. [136] for a GaAs sample with $L_1 = 1 \mu$m, $L_2 = 10 \mu$m, and a variation of doping from 10^{15} to 2×10^{15} cm^{-3} towards the anode ($N_2 \neq$ const.). The computer simulation has shown that if the bias field is sufficiently high, the device will be stable. The anomalous boundary condition at the 1 — 2 interface has prevented the formation of domains. A similar situation was discussed by Kroemer [52] and we refer readers to his outstanding paper. Clearly, if diffusion is neglected, the small-signal stability may be also studied by using the a.c. transport coefficients derived in the preceding chapter.

The theory developed before can also be applied to another non-uniformly doped SCLC structure. This is the transit-time oscillator with velocity limited injection, suggested by Wright [90]. The device consists of a sandwich structure: an injecting contact, a highly doped cathode region, a lowly doped anode region and the collecting contact. Let us assume SCL *minority* carrier injection. In fact the device is a non-uniform punch-through structure. Sometimes it is named a double drift-region structure [128]. If the injected bias current is relatively small, from equation (5.70) we obtain

$$L_1 = \frac{\varepsilon}{eN_1} \int_0^{E_{12}} \frac{J_{n1}(E)\,\mathrm{d}E}{J_0 - j_{n1}(E)\,\mathrm{d}E} \simeq \frac{\varepsilon}{e\,|N_1|} E_{12} \tag{5.73}$$

If the doping-length product $|N_1|L_1$ of the cathode region is chosen properly, the interface field E_{12} will have such a value that the anode region will be a high-field region and the carrier velocity will be practically saturated*). Because the mobile charge injected in region 1 is very small, the electric field is determined by the fixed space charge and its variations are negligible. The a.c. particle current in region 1 is due simply to the carrier-density modulation (see Chapter 7) and the a.c. particle current at the distance x from the injecting plane exhibits simply a delay corresponding to the transit angle, θ_x, to that plane. If $\underline{\mathbf{j}}_1(0) = \underline{\mathbf{J}}_1$ is the particle current at the cathode, its value at the interface should be

$$\underline{\mathbf{j}}_1(L_1) = \underline{\mathbf{j}}_1(0) \exp\left(-\,i\theta_1\right) \tag{5.74}$$

where $\theta_1 = \omega T_{L1}$, and T_{L1} is the transit-time across the cathode region. Equation (5.74) is the a.c. boundary condition for the second drift region. Wright [90] has shown that the cathode region provides the necessary phase lag to get an external negative resistance (a simplified expression of the a.c. impedance [90] is given in Problem 5.8).

We note that the a. c. behaviour of a one-dimensional and non-uniformly doped unipolar device may be easily studied by using the transport coefficients of Chapter 4, provided that the doping profile may be approximated by segments of uniform impurity concentration†). An alternative method meant for taking non-uniform doping into account was suggested by Shoji [138].

---

*) From Poisson equation it is evident that the electric field should increase monotonously from the injecting to the collecting contact.

†) Diffusion may be important, however, especially at the boundaries between the regions of uniform doping, where the actual impurity gradient is larger.

5.9 SCLC Solid-state Diodes with Convergent Geometries

The most important geometries after the plane-parallel structure are the cylindrical and spherical electrode diodes*). Assume negligible diffusion, constant mobility and a semiconductor free of fixed space charge. The steady-state characteristic is of the form [139], [140]

$$I_0 = \varepsilon \mu V_0^2 F$$

where F is a factor depending upon geometrical parameters of the device.

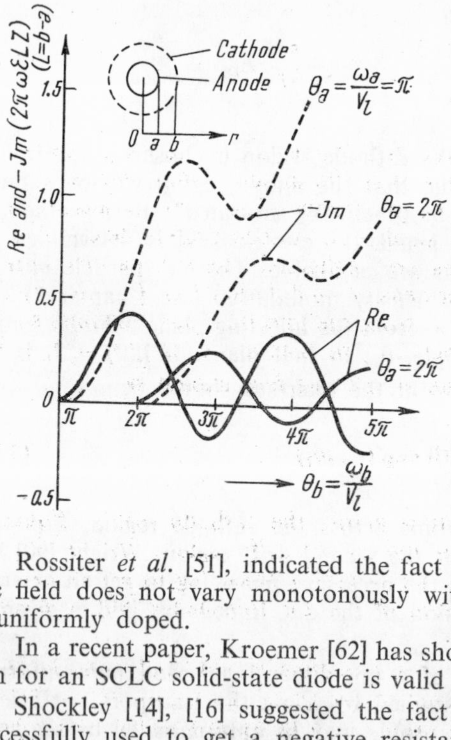

Figure 5.27. Frequency characteristics calculated by Lundström and Wierich [141] for a circular (see inset) velocity-saturated p^+pp^+ diode with outer contact injecting. The series resistance (solid curves) and the absolute value of the capacitive reactance both normalized to $[2\pi \varepsilon \omega (b - a)]^{-1}$ and plotted versus $\theta_b = \omega_b/v_l$, for two values (π and 2π) of $\theta_a = \omega_a/v_l$. A negative resistance is obtained when $\theta_b - \theta_a$ is approximately a multiple of 2π.

Rossiter *et al.* [51], indicated the fact that in a spherical geometry, the electric field does not vary monotonously with distance even if the semiconductor is uniformly doped.

In a recent paper, Kroemer [62] has shown that the positive conductance theorem for an SCLC solid-state diode is valid irrespective of the device geometry.

Shockley [14], [16] suggested the fact that convergent geometries might be successfully used to get a negative resistance device. Later on, Lundstrom and Wierich [141] calculated the impedance of a 'circular' semiconductor diode and have indeed found a transit-time negative resistance.

The frequency characteristics plotted in Figure 5.27 were calculated for a 'circular' (cylindrical) structure (inset in Figure 5.27) with outer contact injecting

*) There exist, however, more practical structures, e.g. the planar structure (two contacts on the same side of a semiconductor layer) [29].

(ohmic). Negligible diffusion and saturated drift velocity*) are assumed throughout the calculations. We recall the fact that in similar conditions the resistance of a plane-parallel sample has no negative values (Section 5.2).

5.10 Transit-time Negative-resistance of an SCLC Diode with Shallow Traps

There exist other possibilities of obtaining an h.f. negative resistance. We have suggested elsewhere that, in certain conditions, an SCLC insulator diode with traps should exhibit such a negative resistance [82], [92], [142], [143].

We write down the basic equations of our model:

$$J = e\mu pE + E\frac{\partial E}{\partial t} \tag{5.75}$$

(constant mobility, negligible diffusion),

$$\frac{\partial p_t}{\partial t} = \frac{p}{\tau_f} - \frac{p_t}{\tau_t} \tag{5.76}$$

(single-level trapping; if the level is shallow τ_f and τ_t are constants (See Section 3.13)†),

$$\frac{\partial E}{\partial x} = \frac{e(p + p_t)}{\varepsilon}\,. \tag{5.77}$$

These equations will be solved with the standard SCLC boundary condition $E(0, t) = 0$. We have studied the a.c. small-signal behaviour††) by applying the perturbation method of Chapter 3. We are interested here in the special case of *slow* shallow trapping, which is defined by

$$\tau \gg T_L \tag{5.78}$$

where

$$\tau = \tau_t\tau_f/(\tau_t + \tau_f) \tag{5.79}$$

---

*) Lundström and Wierich [141] have also improved their calculations by taking into account the low-field region in the vicinity of the cathode. The d.c. electric field at the inner contact (anode) may become very high and may lead to carried multiplication.

†) Equation (5.76) is a consequence of equation (3.124). We have to take into account the fact that $p_D = p_t$, $\frac{\partial n_D}{\partial t} = -\frac{\partial p_D}{\partial t} = \frac{\partial p}{\partial t} - \frac{\partial n}{\partial t}$ and here $n = 0$. The time constants τ_t and τ_f are given by equations (3.125) and (3.126).

††) These results were generalized for shallow trapping levels having an arbitrary distribution in energy but uniform distribution in the bulk [68], [91].

is a relaxation time which is characteristic for the inertia occurring in the re-establishment of the quasi-thermal equilibrium between the free and trapped carriers (equilibrium which is disturbed by the signal). If the condition (5.78) is satisfied, the probability of trapping during the transit of an injected carrier is very small. The 'efective' carrier mobility is practically equal to the carrier mobility in a trap-free crystal, and the 'effective' transit-time is equal to the transit-time in the absence of trapping*). The ratio of the free carrier density to the total (free and trapped) carrier-density is denoted by δ (Section 3.13). δ may be close to unity ($\delta \lesssim 1$), or, conversely, very small as compared to unity ($\delta \ll 1$). We have already shown that

$$J_0 = \frac{9}{8}\,\varepsilon\mu\delta\, V_0^2/L^3,\ \ \delta = (1 + \tau_t/\tau_f)^{-1}. \tag{5.80}$$

The normalized frequency characteristics of the SCLC (insulator) diode with slow shallow traps are plotted in Figure 5.28, with δ as a parameter. Note the fact that the series resistance and reactance are normalized to the h. f. resistance†)

$$R_i' = \frac{2\delta}{1+\delta} R_i < R_i \tag{5.81}$$

(R_i is the low-frequency or incremental resistance). The interesting point is that, if the trapping is severe (δ small), the device exhibits a negative dynamic resistance at frequencies comparable to the carrier transit-time or a multiple of it.

We shall derive below the impedance for the limit case $\delta \to 0$ (or $\delta \ll 1$). The usual perturbation technique will be used again. The free hole density and the trapped hole density in equation (5.76) should be replaced, respectively, by

$$p(x, t) = p_0(x) + \underline{\mathbf{p}}_1(x) \exp(i\omega t),\ \ |\underline{\mathbf{p}}_1| \ll p_0 \tag{5.82}$$

$$p_t(x, t) = p_{t0}(x) + \underline{\mathbf{p}}_{t1} \exp(i\omega t),\ \ |\underline{\mathbf{p}}_{t1}| \ll p_{t0} \tag{5.83}$$

The corresponding d.c. relation is

$$p_0/\tau_f = p_{t0}/\tau_t \tag{5.84}$$

and thus we may write

$$\delta = \frac{p_0(x)}{p_0(x) + p_{t0}(x)} = \frac{1}{1 + \tau_t/\tau_f} = \text{const.} \tag{5.85}$$

---

*) If $\tau \ll T_L$, an injected carrier will be trapped many times before it will reach the anode (fast trapping). Consequently, the 'effective' mobility decreases and the carrier transit-time increases. Note the fact that in the steady-state no distinction occurs between 'slow' and 'fast' traps.

†) At frequencies comparable to $1/\tau$, the diode resistance is accompanied by a capacitive effect accounted for by the trapping-releasing processes). If $\omega\tau \gg 1$ but $\omega T_L = \theta \ll 1$, the diode acts again as a simple a.c. resistance which is denoted here by R'_i [82].

Because δ will be assumed very small we must have

$$\delta \ll 1 \rightarrow \tau_t \gg \tau_f, \quad \tau \simeq \tau_f, \quad p_0 \ll p_{t0}. \tag{5.86}$$

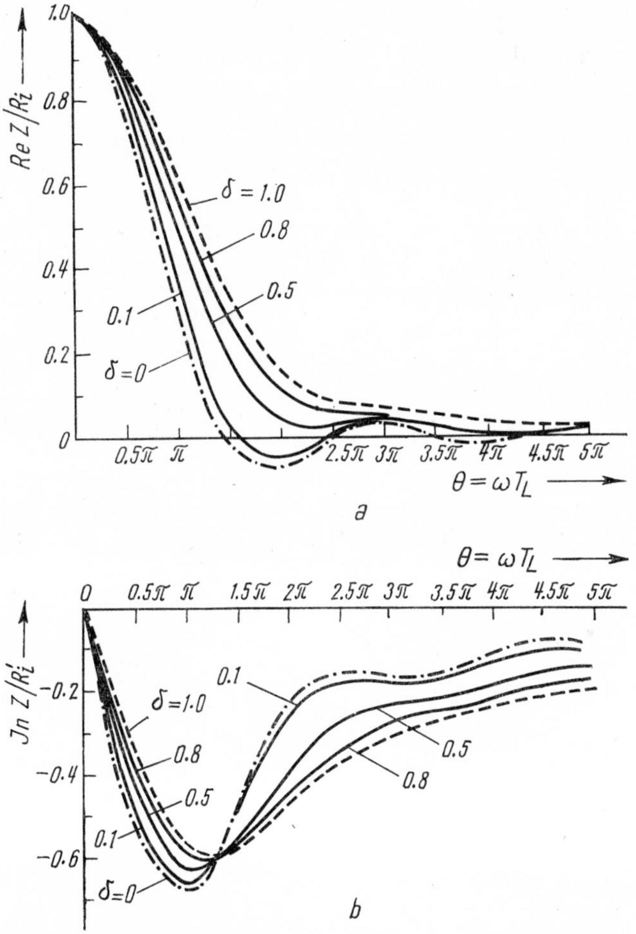

Figure 5.28. Real (a) and imaginary (b) part of the normalized impedance represented versus transit angle, for several values of the trapping parameter. The impedance is normalized to the h.f. resistance, equation (5.81) [92].

The a.c. part of equation (5.76) is

$$i\omega \underline{\mathbf{p}}_{t1} = \frac{\underline{\mathbf{p}}_1}{\tau_f} - \frac{\underline{\mathbf{p}}_{t1}}{\tau_t}. \tag{5.87}$$

By assuming the fact that the frequency is considerably higher than the relaxation frequency ($\omega \gg 1/\tau$) and by taking into account $\tau \simeq \tau_s$ from equation (5.86), one obtains

$$|\underline{\mathbf{p}}_{t1}| \ll |\underline{\mathbf{p}}_1| \tag{5.88}$$

and the a.c. part of the Poisson equation (5.77) will become

$$\frac{d\underline{\mathbf{E}}_1}{dx} \simeq \frac{e\,\underline{\mathbf{p}}_1}{\varepsilon}. \tag{5.89}$$

This is because at relatively high frequencies ($\omega\tau \gg 1$) the electrons injected *by the signal* remain free (the trapping cannot follow the rapid variation of the a.c. signal).

We write down again the a.c. part of equation (5.75). This is

$$\underline{\mathbf{J}}_1 = e\underline{\mathbf{p}}_1 v_0 + ep_0 \underline{\mathbf{v}}_1 + i\omega\, \varepsilon \underline{\mathbf{E}}_1 \tag{5.90}$$

and, for the problem studied here, it becomes

$$\underline{\mathbf{J}}_1 = \mu\varepsilon E_0 \frac{d\underline{\mathbf{E}}_1}{dx} + \mu\varepsilon\delta \frac{dE_0}{dx} \underline{\mathbf{E}}_1 + i\omega\varepsilon\underline{\mathbf{E}}_1 \tag{5.91}$$

where

$$E_0(x) = \eta \sqrt{x} \quad , \quad \eta = \sqrt{\frac{2J_0}{\mu\varepsilon\delta}}. \tag{5.92}$$

Because δ is very small, we obtain

$$\underline{\mathbf{J}}_1 \simeq \mu\varepsilon E_0 \frac{d\underline{\mathbf{E}}_1}{dx} + i\omega\varepsilon\underline{\mathbf{E}}_1. \tag{5.93}$$

The solution of this equation with the usual boundary condition $\underline{\mathbf{E}}_1(0) = 0$, takes the form

$$\underline{\mathbf{E}}_1(x) = \frac{\underline{\mathbf{J}}_1}{i\omega\varepsilon} [1 - \exp bx^{1/2}], \; b = -\frac{2i\omega}{\mu\eta}. \tag{5.94}$$

A second integration yields the a.c. potential, the total a.c. voltage drop and then the impedance. The final result is [143]

$$Z = \frac{\underline{\mathbf{V}}_1}{\underline{\mathbf{J}}_1} = \frac{2L}{i\omega\varepsilon} \frac{(1 + i\theta)\exp(-i\theta) - 1 - \theta^2/2}{(-i\theta)^2} \tag{5.95}$$

where

$$\theta = \omega T_L \quad , \quad T_L = \int_0^L \frac{\mathrm{d}x}{\mu E_0(x)} = \frac{2\sqrt{d}}{\mu\eta}. \tag{5.96}$$

The impedance, equation (5.95), may be also written

$$Z = R_i' \left[\frac{3}{\theta^3} (\sin\theta - \theta\cos\theta) - i\frac{3}{\theta}\left(1 + \frac{\theta^2}{2} - \cos\theta - \theta\sin\theta\right)\right] \tag{5.97}$$

(where R_i' is given by equation (5.81), $\delta \to 0$). Clearly, the series resistance becomes negative at frequencies situated around the transit-time frequency ($\theta = 2\pi$) or around a multiple of this frequency (see also Figure 5.28).

The effect of trapping in SCLC solid-state devices is discussed by the author in much more detail in a number of papers [68], [82], [91], [92], [142] – [144].

5.11 Effect of Mobility Relaxation Time

We have already shown in Chapter 2 that the motion of a 'free' carrier in solids may be described by an equation which takes collisions into account. By assuming one-dimensional motion in an 'external' electric field, one obtains from equation (2.22), for a hole (charge $+e$)

$$m\frac{\mathrm{d}v}{\mathrm{d}t} + m\nu_c v = eE \tag{5.98}$$

where m is the electron mass and ν_c the collision frequency*). The above equation contains two limit cases. Without collisions ($\nu_c = 0$) we reobtain the motion equation of a charged particle in vacuum. The opposite case is the collision-dominated motion (very frequent collisions as compared to the time interval when the electric field acts upon the charged carrier). The first term in equation (5.98) may be neglected and one obtains

$$v = \mu E \quad , \quad \mu = e/m\nu_c \tag{5.99}$$

Let us first discuss the steady-state regime. Assume, for simplicity, a zero fixed space-charge. For $\nu_c \to 0$ the $J - V$ characteristic should be identical with the Child law ($J \propto V^{3/2}$) of the planar vacuum diode. If $\nu_c \to \infty$ the square-law ($J \propto V^2$) of the 'ideal' SCLC solid-state diode will be obtained. We expect a smooth transition from one case to another, as ν_c takes positive, finite values†).

*) We underline the fact that $\mathrm{d}v/\mathrm{d}t$ is the total derivative and does *not* vanish for time-independent processes.

†) The transition from inertia-limited flow (in vacuum) to mobility-limited flow (in the gas medium) for electron flow in electron valves, was studied by Ingold [145]. Our problem is, up to a certain point, identical with the situation discussed by this author.

Wadhwa and Sisodia [75] studied the SCLC diode with deep traps (or the punch-through structure, the equations and results being identical) by starting from equation (5.98). Their main objective was to derive the a.c. small-signal properties. Wadhwa and Sisodia obtained a matrix of *nine* coefficients which should be identical with the Llewellyn-Peterson coefficients if the collision frequency and the fixed-space charge density are both taken equal to zero. If these coefficients are used in conjunction with the SCLC boundary conditions, the diode impedance is obtained. Its analytical expression is, however, exceedingly complicated and will not be reproduced here. Some computed frequency characteristics indicate a negative resistance effect in the transit-time frequency region. We recall the fact that even if the mobility relaxation time is negligible ($\nu_c \to \infty$), the device will still exhibit a transit-time negative resistance: we re-obtain the case studied by Shockley [14] and Yoshimura [17] (Section 5.6).

Wadhwa and Sisodia expect their theory to be useful for certain materials and especially at low temperatures, where the collision frequency is low and should be taken as finite [75].

5.12 SCLC Solid-state Triodes

In this Section we briefly discuss SCLC solid-state triodes (or transistors) which can be described by a one-dimensional model (with plane-parallel electrodes). Such a model is sketched in Figure 5.29. The mobile carriers are injected under SCLC conditions*) into the emitter-base (gate) region: the magnitude of the emitter current is controlled by the base-emitter voltage. The particle current reaching the base is then re-injected (with no or little modification) into the base-collector space. The collector current should be approximately equal to the emitter current. In general, we expect the collector-emitter voltage to have no influence upon the

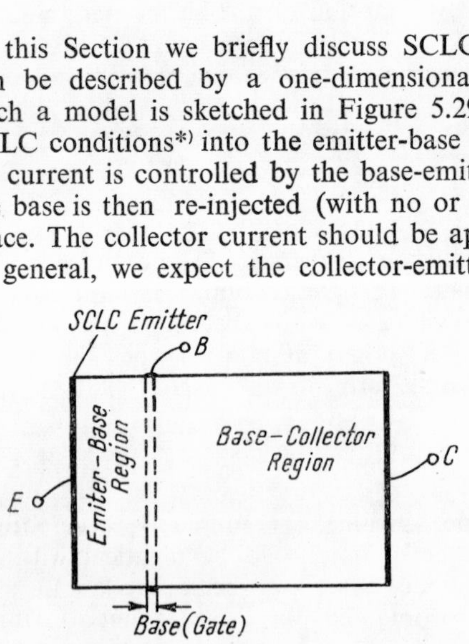

Figure 5.29. Structure of an SCLC solid-state triode (transistor) with plane-parallel electrodes.

collector current and thus the output d.c. characteristics (emitter grounded) should be of the pentode type.

*) The injection mechanism can be, however, different in another structures, i.e. tunnel emission or Schottky-type emission [146].

Practical versions of the above model were constructed experimentally, but the difficulty of making a suitable base (which should be almost 'permeable' for charge carriers but sufficiently thick to have a negligible 'distributed' resistance) is a major drawback of such a structure. We quote here the SCL heterojunction transistor suggested by Wright [147] and realised experimentally later [148]. Another possibility suggested in the past was the (SCLC) hot-carrier triodes with thin-film metal base [146].

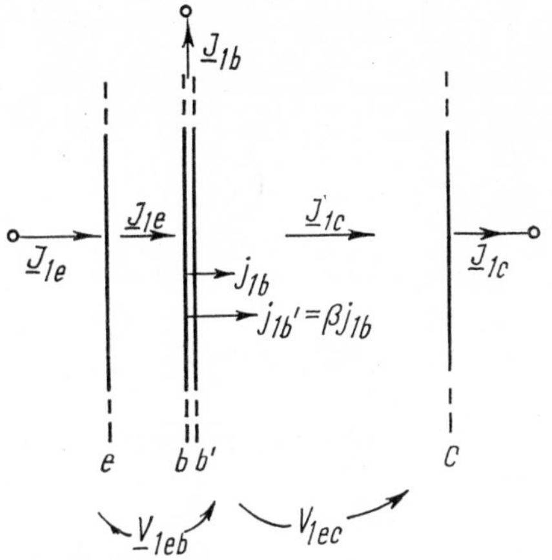

Figure 5.30. A.c. components of voltages and currents for an SCLC solid-state triode. The notations j and J are used for currents instead of current densities.

We shall discuss, in principle, the possibility of applying the theory developed in Chapter 4, in the study of the a.c. small-signal behaviour of an SCLC solid-state transistor. The particle current leaving the base region will be given by (Figure 5.30)

$$\underline{\mathbf{j}}_{1b}' = \beta \underline{\mathbf{j}}_{1b} \tag{5.100}$$

where β is a transport factor (complex number). If the particle current crosses the base by diffusion, as it does in the heterojunction transistor, the factor β may be calculated as shown by Shockley [14] and Wright [87].

The emitter-base and base-collector regions, respectively (Figure 5.30) may have different properties and the a.c. transport coefficients will have different expressions. For example, the emitter-base space may be a dielectric with traps (like in the CdS—Si heterojunction transistor constructed by Brojdo *et al.* [148]) and the base-collector space, a semiconductor with field-dependent carrier mobility. Here, we shall prefer the C_{ij} coefficients defined by equations (4.19) and (4.20) because we wish to require the continuity of the particle-current at interfaces. With the notation of Figure 5.30 we write

$$\underline{\mathbf{V}}_{1eb} = e_{11}\, \underline{\mathbf{J}}_{1e} + e_{12}\, \underline{\mathbf{j}}_{1e} \tag{5.101}$$

$$\underline{\mathbf{j}}_{1b} = e_{21}\, \underline{\mathbf{J}}_{1e} + e_{22}\, \underline{\mathbf{j}}_{1e} \tag{5.102}$$

for the base-emitter region, and

$$\underline{\mathbf{V}}_{1bc} = c_{11}\,\underline{\mathbf{J}}_{1c} + c_{12}\,\underline{\mathbf{j}}_{1b}' \tag{5.103}$$

$$\underline{\mathbf{j}}_{1c} = c_{21}\,\underline{\mathbf{J}}_{1c} + c_{22}\,\underline{\mathbf{j}}_{1b}' \tag{5.104}$$

for the base-collector region (the a.c. voltage drop on the base region was neglected). From Figure 5.30 one obtains

$$\underline{\mathbf{J}}_{1e} = \underline{\mathbf{J}}_{1b} + \underline{\mathbf{J}}_{1c} \tag{5.105}$$

$$\underline{\mathbf{V}}_{1ec} = \underline{\mathbf{V}}_{1eb} + \underline{\mathbf{V}}_{1bc}\,. \tag{5.106}$$

Here $\underline{\mathbf{J}}_{1e}$, $\underline{\mathbf{J}}_{1b}$, $\underline{\mathbf{J}}_{1c}$ are the external (circuit) currents flowing in the emitter, base and collector terminals, respectively.

Under SCL injection conditions (neglecting diffusion), e_{12} and e_{22} should be zero, as shown at the beginning of this chapter. Therefore

$$\underline{\mathbf{V}}_{1eb} = e_{11}\,\underline{\mathbf{J}}_{1e} \tag{5.107}$$

$$\underline{\mathbf{j}}_{1b} = e_{21}\,\underline{\mathbf{J}}_{1e} = \frac{e_{21}}{e_{11}}\,\underline{\mathbf{V}}_{1eb} \tag{5.108}$$

and then

$$\underline{\mathbf{j}}_{1b}' = \beta\,\underline{\mathbf{j}}_{1b} = \beta\,\frac{e_{21}}{e_{11}}\,\underline{\mathbf{V}}_{1eb}\,. \tag{5.109}$$

Equation (5.103) may be written

$$\underline{\mathbf{V}}_{1bc} = c_{11}\,\underline{\mathbf{J}}_{1c} + \beta\,\frac{e_{21}}{e_{11}}\,c_{12}\,\underline{\mathbf{V}}_{1eb}\,. \tag{5.110}$$

The above relations are sufficient to determine the a.c. response of the SCLC triode. For example, let us define the following parameters (grounded emitter)

$$\underline{\mathbf{J}}_{1b} = Y_{11}\,\underline{\mathbf{V}}_{1eb} + Y_{12}\,\underline{\mathbf{V}}_{1ec} \tag{5.111}$$

$$\underline{\mathbf{J}}_{1c} = Y_{21}\,\underline{\mathbf{V}}_{1eb} + Y_{22}\,\underline{\mathbf{V}}_{1ec}\,. \tag{5.112}$$

By use of equations (5.105) – (5.107), (5.110) – (5.112) one obtains

$$Y_{11} = \frac{1}{e_{11}} + \frac{1}{c_{11}}\left(1 + \beta\frac{e_{21}}{e_{11}}c_{12}\right) \tag{5.113}$$

$$Y_{12} = -\frac{1}{c_{11}} \tag{5.114}$$

$$Y_{21} = -\frac{1}{c_{11}}\left[1 + \beta\frac{e_{21}}{e_{11}}c_{12}\right] \tag{5.115}$$

$$Y_{22} = \frac{1}{c_{11}} \tag{5.116}$$

The frequency characteristics of the SCLC 'dielectric' triode were calculated by Brojdo [149] by taking into account the effect of mobility field dependence. The effect of shallow traps in the emitter base region was discussed by Dascalu [150]. Brian [151] has also calculated the effect of recombination and carrier transport by diffusion in the base layer, for an SCLC heterojunction transistor. He also found experimentally that, at moderately high frequencies, the processes occurring during charge transport in the base region are much more important than the transit-time effects in the emitter-base region, which were outlined by Brojdo [149]. A recent paper by Ladd and Feucht [152] suggests the fact that the transit-time in the base-collector region may be longer than the transit-time in the emitter-base region. These authors also discussed the effect of parasitic resistances and capacitances and concluded that the heterojunction transistor with SCLC emitter should be roughly equivalent to the homojunction bipolar transistor with respect to its high-frequency performances.

Equations (5.113) – (5.116) are sufficiently general to include all results previously obtained and to derive new ones. As an example we shall consider some general properties of the equivalent circuit. We do prefer a 'π-type' circuit (Figure 5.31). Thus, one obtains

$$Y_a = Y_{11} + Y_{12} = \frac{1}{e_{11}} - Y_m \tag{5.117}$$

$$Y_b = -Y_{12} = \frac{1}{c_{11}} \tag{5.118}$$

$$Y_c = Y_{12} + Y_{22} = 0 \tag{5.119}$$

$$Y_m = Y_{21} - Y_{12} = -\beta\frac{e_{21}\,c_{12}}{e_{11}\,c_{11}}\,. \tag{5.120}$$

We shall denote by

$$Y = \frac{1}{e_{11}} = Y_{\text{conv}} + Y_d \tag{5.121}$$

the SCLC admittance of the emitter-base region. This admittance may be written as the sum of a 'convection' admittance Y_{conv} defined as (see Figure 5.30)

$$Y_{\text{conv}} = \underline{\mathbf{j}}_{1b}/\underline{\mathbf{V}}_{1eb} = e_{21}/e_{11} \tag{5.122}$$

and a 'displacement' admittance $Y_d = Y - Y_{\text{conv}}$. These partial admittances correspond, respectively, to the particle (convection) current and to the displacement current calculated at the collecting plane of the emitter-base region.

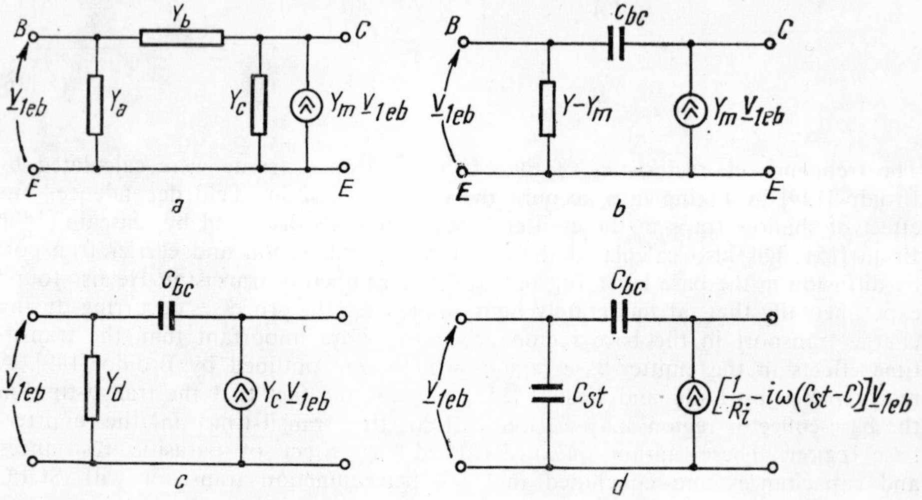

Figure 5.31. π-type configuration of the a.c. small-signal circuit of a transistor (triode).

We shall assume now the fact that the injected current in the base-collector region is relatively small (i.e. the mobile injected charge is negligible). By taking into account equations (4.14), (4.15), (4.19) – (4.24), (4.46), (4.57), (4.58) and (4.62) one obtains

$$C_{11} = \frac{1}{i\omega\, C_{bc}} \tag{5.123}$$

$$-(C_{12}/C_{11}) = \frac{T_2}{L_2}\int_0^1 v_{02}\exp(-i\theta_2)\,\mathrm{d}\xi = F(\theta_2). \tag{5.124}$$

Here C_{bc} is the geometrical capacitance (ε_2/L_2) of the base-collector space, $\theta_2 = \omega T_{L2}$ and T_{L2} the transit-angle and the transit-time of carriers in the base-collector region, respectively. The d.c. velocity in this region v_{02} should be replaced as a function of the normalized transit-time $\xi = T_{x2}/T_{L2}$. Clearly, if the transit angle θ_2 is very small, the ratio introduced by equation (5.124) may be approximated by unity, i.e. $F(0) = 1$.

Therefore, the equivalent circuit of Figure 5.31a can be replaced by that given in Figure 5.31b where, according to equations (5.120), (5.122) and (5.124)

$$Y_m = Y_{\text{conv}}\, \beta\, F(\theta_2). \tag{5.125}$$

Here, Y_c, β and $F(\theta_2)$ indicate the effect upon the transconductance of the carrier transport in the emitter-base, base and base-collector regions (Figure 5.29), respectively. We shall simplify these results by restricting the discussion to frequencies which are sufficiently low to assume $F(\theta_2) \simeq 1$. On the other hand we shall consider a base transport factor which is equal to unity ($\beta \simeq 1$). Thus, we obtain the equivalent circuit represented in Figure 5.31c. At very low frequencies we must have (Section 5.3)

$$Y_d \simeq i\omega C_{st} \tag{5.126}$$

where C_{st} is the stationary capacitance, equation (5.26), and

$$Y_m = \frac{1}{R_i} - i\omega\,(C_{st} - C) \tag{5.127}$$

where C is the dynamic, low frequency capacitance ($C < C_{st}$). Finally, let us consider a zero fixed space-charge and a constant mobility in the emitter-base region. As shown in Section 5.3 we have $C = \left(\frac{3}{4}\right) C_{eb}$ and $C_{st} = \left(\frac{3}{2}\right) C_{eb}$, where C_{eb} is the geometrical capacitance (ε_1/L_1) of the emitter-base space. The low-frequency equivalent circuit is represented in Figure 5.31d and is more rigorous than the a.c. equivalent circuit suggested initially by Wright [147]. We stress the fact that the dynamic base-emitter capacitance is $\left(\frac{3}{2}\right) C_{eb}$ [150] and not $\left(\frac{3}{4}\right) C_{eb}$ [147], [152]. A similar situation for the vacuum triode is known for a long time. For example, the dynamic gate-cathode capacitance is equal to the 'stationary' capacitance of the gate-cathode space C_{st}. Moreover, the high-frequency input resistance was calculated by Drăgănescu [112], [113] by using a two-term series expansion of the displacement admittance of the triode.

Problems

5.1. By using the expression of the transport coefficients (4.65) and (4.66), show that if the steady-state electric field vanishes at $x = 0$, the a.c. field at the same plane will be also zero.

5.2. Assume that the velocity-field dependence may be approximated by a relationship of the form $v \propto E^M$, where $0 \leqslant M \leqslant 1$. Calculate the small-signal impedance starting from equation (5.15) and verify the result by introducing $M = 0$ (saturated velocity) and $M = 1$ (constant mobility).

Note: This problem was solved in reference [82]. See [81] for the low-frequency equivalent circuit.

5.3. Consider a rectifying square-law insulator (dielectric) diode which exhibits high-frequency dielectric losses inside the conduction space. Show that the 'electronic' admittance cannot be obtained by simply substracting the admittance measured with forward bias (SCL injection) and reverse bias (no charge injection), respectively.

Hint: In the former case the a.c. electric field is not uniform.

5.4. Show that the small-signal transient response of an SCLC diode to a voltage step can be obtained by the inverse Laplace transform of the electronic admittance.

Note: The total admittance is given by the electronic admittance in parallel with the geometrical capacitance C_0 of the same diode (one-dimensional model).

5.5. Demonstrate the fact that the reactance, equation (5.33) of the SCL impedance, equation (5.31), is always capacitive.

5.6. Show that if the cathode velocity is not zero, the parallel conductance (5.43) at frequencies much higher than the transit-time frequency will be given by

$$G(\omega)\big|_{\omega\to\infty} \to \frac{\varepsilon\,[v_0(L) - v_0(0)\cos\theta]}{L^2}\,. \qquad \text{(P.5.1)}$$

Correlate this result with the fact that the parallel conductance of the SCLC saturated velocity diode ($a \to \infty$ in Figure 5.4) seems to have an oscillatory behaviour up to very high frequencies.

Note: Equation (P.5.1) is derived by assuming $\mathbf{E}_1(0) = 0$ and $E_0(0) \neq 0$. Such a situation can occur for a unipolar solid-state diode with trapping at the cathode interface (the control characteristic has a region which is almost vertical, as shown in Section 5.6).

5.7. Calculate the low-frequency dynamic capacitance of an SCLC solid-state diode by assuming: negligible diffusion, constant mobility, no trapping.

Hint: Use a series development of the diode impedance $Z(\theta)$ for small θ, starting from the general expression (5.6) and then introducing $\mu = $ const. The result is

$$\frac{C}{C_0} = \frac{(e^{-\theta_r} + \theta_r - 1)\left[3\,(e^{-\theta_r} - 1) + \dfrac{\theta_r^2}{2}\,e^{-\theta_r} + 2\theta_r\,e^{-\theta_r} + \theta_r\right]}{[\theta_r(1 + e^{-\theta_r}) + 2\,(e^{-\theta_r} - 1)]^2} \qquad \text{(P.5.2)}$$

where C_0 is the geometrical capacitance ε/L and θ_r is defined by equation (5.48).

5.8. Show that the impedance of the double drift-region diode described in Section 5.8 is given approximately by [90]

$$Z = \frac{L_2^2}{\varepsilon\, v_{\text{sat}}} \left\{ \frac{2}{i\theta_2} + \frac{2 \exp(-i\theta_1)}{\theta_2^2} [1 - \exp(-i\theta_2)] \right\}. \tag{P.5.3}$$

Hint: Use equation (5.74) and assume saturated velocity in the second region (length L_2).

5.9. Demonstrate the fact that the v.h.f. resistance of the SCLC dielectric diode with *fast* shallow traps (Section 5.10) is positive [92].

6

Small-signal Negative Resistance of Injection-controlled Devices

6.1 Injection-controlled Devices

We have shown theoretically in the previous chapter that certain SCLC diodes should exhibit a high-frequency negative resistance, which can be used to construct an h.f. amplifier or oscillator. Only the small-signal behaviour was considered. This theory can be used for small-signal amplification and for small-signal stability of large-signal oscillators and amplifiers.

The above negative resistance is a transit-time effect or a mixture of transit-time and negative-mobility effects. A better understanding of these phenomena will be achieved by discussing the importance of the injection mechanism.

The space-charge-limited current requires an infinite injection (ohmic cathode) and, at a given applied voltage, its magnitude is entirely determined by the semiconductor bulk. It was shown (see also the next chapter) that in SCLC diodes, the drift transit-time itself hardly provides a sufficient delay to determine a negative resistance effect. On the other hand, the boundary condition of the zero electric field at the cathode is unfavourable for negative-mobility devices, because a part of the semiconductor (the cathode region) is biased in the low-field positive mobility region of the velocity-field characteristic and the overall a.c. power generation is not efficient.

The concept of injection-controlled-current (ICC) device opens wide possibilities for negative resistance solid-state diodes. As a large variety of injection mechanisms can be imagined, the ICC devices could also be extremely diverse.

We classify here these devices into two broad categories: devices with non-ohmic cathode and devices with delayed injection. This classification is somewhat artificial and ambiguous and is used here merely for theoretical calculations i.e. from the point of view of a.c. boundary conditions and transport coefficients used.

6.1.1 Non-ohmic-cathode Devices

By definition we assume an injected particle current (at $x = 0$) which is an instantaneous function of the cathode field

$$j_c = j_c(E_c)\,. \tag{6.1}$$

The modified thermionic emission (Schottky-emission), the field emission and other types of injection mechanisms discussed in Section 3.3 fall into this category. The corresponding a.c. boundary condition (4.16) is rewritten here for convenience

$$\underline{\mathbf{j}}_1 (0) = \sigma_1 (0)\, \underline{\mathbf{E}}_1 (0), \quad \sigma_1(0) = \frac{\mathrm{d}j_c}{\mathrm{d}E_c}. \tag{6.2}$$

The catode a.c. conductivity $\sigma_1(0)$ is the slope of the control characteristic (Section 3.6) and is a real number, positive, zero or negative.

We assume, for simplicity, that the entire drift space, between cathode ($x = 0$) and anode ($x = L$) is homogeneous and may be described by a set of four transport coefficients, K_{ij}, defined by equations (4.14) and (4.15). Therefore, the device impedance is given by equation (4.18), transcribed below

$$Z = K_{11} + \frac{K_{12}}{\sigma_1(0) + i\omega\,\varepsilon}. \tag{6.3}$$

The effect of a non-ohmic boundary condition is twofold. First, the distribution of the steady-state electric field depends upon the injection mechanism and may be controlled conveniently. For example, the entire semiconductor may be biased in the negative mobility region of the $J - V$ characteristic. This 'steady-state' effect is implicitly contained in K_{11} and K_{12}. Secondly, the a.c. boundary condition, $\sigma_1 (0) =$ given, finite, appears explicitly in the impedance equation (6.3). We shall see in the next chapter that this new boundary condition modifies the injection and propagation of the space-charge wave.

6.1.2 Delayed-injection Devices

By definition we assume the fact that the a.c. particle current $\underline{\mathbf{j}}_1$ injected in the principal drift region, at $x = 0$ can be related to the total current flowing in this region as follows:

$$\underline{\mathbf{j}}_1 (0) = \beta\, \underline{\mathbf{J}}_1 \tag{6.4}$$

where β is a complex quantity. Therefore, the impedance of this drift region, characterized by K_{ij}, becomes

$$Z = K_{11} + \frac{K_{12}\,(1 - \beta)}{i\omega\,\varepsilon}. \tag{6.5}$$

Usually, $\underline{\mathbf{j}}_1(0)$ is delayed with respect to $\underline{\mathbf{J}}_1$, and this aids to the negative resistance effect. This additional delay is sometimes due to the charge transport (by drift and/or diffusion) in a 'control region' situated at $x < 0$, between the actual source of charge carriers and the emitting plane $x = 0$. If the charge carriers are generated by the avalanche multiplication process inside the control region, the required delay is provided by the inertia of the avalanche process itself [28], [29], [153]. Finally, note the fact that the a.c. voltage drop on this control region is negligible with respect to that on the principal drift region, then the total impedance (neglecting parasitics, of course), will be given by equation (6.5).

6.2 Small-signal Theory

The study of negative-resistance injection-controlled current (ICC) devices will demonstrate the usefulness of the general theory put forward in Chapter 4. We argue this as follows.

First, the transport coefficients K_{ij} or C_{ij} are suitable for incorporating various boundary conditions, as these encountered in ICC devices. Sometimes, ICC devices consist of several cascaded drift regions and the theory allows the study of such a situation. We shall see in Chapter 9 that the injection noise may be also evaluated by using the transport coefficients *).

Secondly, the theory of Chapter 4 neglects diffusion, which is a more reasonable assumption for the drift region of ICC devices than for SCLC devices. Hence, the general theory of transport coefficients may prove more useful for the former category of devices.

Thirdly, the explicit analytical form of the transport coefficients, which has been derived in Section 4.8, founds its principal application for ICC devices. This is because of the fact that the steady-state electric field is less nonuniform than in a SCLC device and very often the velocity-field dependence may be assumed linear for a certain range of bias-field intensities. We recall the fact that the coefficients of Section 4.8 are derived by assuming a constant differential mobility. This simplification is acceptable and represents a generalization of the majority of analytical theories, which assume a constant d.c. velocity by postulating either d.c. neutrality or velocity saturation.

6.3 Impedance of a Diode with Non-ohmic Cathode

In this Section we assume the following idealized device: it consists of a homogeneous and one-dimensional drift region: the charge carriers are injected at $x = 0$ and collected at $x = L$. The particle current at $x = 0$ obeys a law of the form of (6.1), where the time, t, does not appear explicitly. This solid-state conduction

---

*) This injection noise is conventionally zero for SCL injection, because the potential maximum (minimum) 'isolates' the conduction space from all electrical processes taking place at $x < 0$, as shown in the introductory Section of Chapter 5.

space is free of traps and the electronic impurities present here are completely ionize
Diffusion current is neglected. The bias conditions are so chosen that the veloci
field dependency may be approximated by a straight line (Section 3.10).

The device impedance is given by equation (6.3) where K_{11}, K_{12}, $\sigma_1(0)$ are giv
by equations (4.102), (4.102) and (6.2), respectively. After algebraic manipulatior
one obtains

$$Z = \frac{1}{i\omega C_0} + \frac{v_0(0)\, T_L^2}{\varepsilon} \{\theta_d \Delta (1 + i\theta\Gamma)\, [1 - (i\theta + \theta_d) - \exp - (i\theta + \theta_d)] - - (\theta_d + i\theta)\,[1 - \exp - (\theta_d + i\theta)]\}\, \{i\theta\, (\theta_d + i\theta)^2\, (1 + i\theta\,\Gamma)\}^{-1} . \quad (6.$$

Here, all parameters have already been defined, except Γ. We have introduc

$$\Gamma = \frac{\varepsilon}{\sigma_1(0)} \frac{1}{T_L} = \frac{\tau_c}{T_L} \quad (6.$$

where

$$\tau_c = \frac{\varepsilon}{\sigma_1(0)} \quad (6.$$

is a sort of (differential) relaxation time of the cathode plane ($x = 0$). By taki
into account the transformation (5.46), the impedance, equation (6.6), may
written more conveniently as

$$Z = \frac{1}{i\omega C_0} + R_d\, \Phi\, (i\theta;\, \theta_d,\, \Delta,\, \Gamma) \quad (6.$$

where

$$R_d = \frac{L}{e N \mu_d} \quad (6.1$$

is a sort of 'differential' resistance of the neutral semiconductor biased in the regic
$dv/dE = \mu_d =$ const. (for $N < 0$ the neutrality is not possible and equation (6.1
is merely a definition). The dimensionless function Φ is given by

$$\Phi = \bar{J}_0 \left(\frac{\theta_d^2}{\Delta}\right) \{\theta_d \Delta (1 + i\theta\Gamma)\, [1 - (i\theta + \theta_d) - \exp - (i\theta + \theta_d)] - - (\theta_d + i\theta)\, [1 - \exp - (\theta_d + i\theta)]\} \times \{i\theta\, (\theta_d + i\theta)^2\, (1 + i\theta\,\Gamma)\}^{-1} \quad (6.1$$

where the normalized current, $\bar{J}_0$, is defined by equation (3.92) and depends on
upon Δ and θ_d.

The impedance, equation (6.9), may be also written

$$Z = R_d \Psi(i\theta;\ \theta_d,\ \Delta,\ \Gamma) \tag{6.12}$$

where

$$\Psi = \frac{\theta_d}{i\theta} + \Phi. \tag{6.13}$$

Note the fact that $R_d C_0$ is exactly equal to the differential relaxation time τ_d defined by equation (3.94).

Clearly, the incremental resistance is

$$R_i = R_d \Psi(0;\ \theta_d, \Delta, \Gamma) \tag{6.14}$$

and the normalized impedance should be of the form

$$\overline{Z} = Z/R_i = \overline{Z}(i\theta;\ \theta_d,\ \Delta,\ \Gamma). \tag{6.15}$$

Of course, this is not the single possible choice of parameters. Note the fact that in equation (6.6), Γ is always multiplied by $i\theta$, and thus the cathode relaxation time τ_c introduced by equation (6.8) occurs only multiplied by $i\omega$, i.e.

$$i\theta\,\Gamma = i\omega\, T_L(\tau_c/T_L) = i\omega\,\tau_c = i\,\theta_c. \tag{6.16}$$

However, we have prefered to have only one frequency dependent parameter, the transit angle $\theta = \omega\, T_L$, instead of two (θ and θ_c).

The normalized frequency characteristics derived from $\overline{Z} = \overline{Z}(\theta)$ depend upon *three* adimensional parameters. θ_d and Δ may be determined directly from Figures 3.24–3.27 if the bias conditions and the device parameters are known. Γ introduces the a.c. boundary condition, i.e. the effect of the injecting (non-ohmic contact). We shall stress the fact that θ_d and Δ should depend implicitly upon the d.c. boundary condition (which allows one to calculate J_0 for a given V_0) as shown in Section 3.9.

The reader will note the fact that we have here *four* time constants, namely the signal period

$$T = 2\,\pi/\omega \tag{6.17}$$

the d.c. transit time T_L, the bulk differential relaxation time, equation (3.94)*)

$$\tau_d = \frac{\varepsilon}{e\,N\,\mu_d} \tag{6.18}$$

*) We recall the fact that this is merely a notation for $N < 0$, because μ_d refers to holes and N to ionized donors.

and τ_c, the cathode differential relaxation time, equation (6.8). These time constants occur through the adimensional parameters

$$\theta = \frac{T_L}{T}\ , \quad \theta_d = \frac{T_L}{\tau_d}\ , \quad \theta\Gamma = \frac{\tau_c}{T}. \tag{6.19}$$

The injection level $\Delta = p_0(0)/N$ defined in Section 3.9 is an independent parameter; it takes somehow into account the d.c. boundary condition.

The variety of devices and operation conditions which can be described by equation (6.15) may be better imagined if one considers the particular cases which are contained in this expression *).

6.4 Particular Cases

The most important particular cases which are described by the results of the preceeding section are summarized in Table 6.1.

Table 6.1

	N	$v = v(E)$	$\sigma_1(0)$	J_0	Δ	θ_d	Γ	
1	finite	$\mu =$ const.	$\to\infty$	finite	$\to\pm\infty$	finite	0	SCL current
2	positive	$\frac{dv}{dE} =$ const.	real	finite	1	finite	finite	d.c. neutrality
3	finite, negative	$\mu =$ const.	real	$\to 0$	$\to 0$	finite	finite	negligible mobile space charge
4	$N = 0$	$\frac{dv}{dE} =$ const.	real	finite	$\to\infty$	$\to 0$	finite	wide-gap intrinsic semiconductor or insulator without traps
5	finite	$v =$ const., saturated	real	finite	finite	$\to 0$	finite	saturated-velocity
6	finite	$\frac{dv}{dE} =$ const.	$\to 0$	constant	finite	finite	$\to\infty$	constant-current injection (emitter-current-limited injection)

*) The above expression is, however, at three levels of generality below the general results derived in Chapter 4, namely the following simplifying assumptions were introduced: (a) a single drift-region device; (b) a boundary condition of the form (6.2), where $\sigma_1(0)$ is a real quantity and (c) a straight-line approximation for $v = v(E)$.

The impedance (6.9) may be also written as

$$Z = \frac{1}{i\omega C_0} + R_d\theta_d^2\bar{J}_0(\theta_d, \Delta)\left[\Phi_1(i\theta;\, \theta_d) + \frac{1}{\Delta}\Phi_2(i\theta;\, \theta_d, \Gamma)\right] \tag{6.20}$$

where Φ_1 and Φ_2 follow from equations (6.9) and (6.11)

$$\Phi_1 = \frac{\theta_d[1 - (i\theta + \theta_d) - \exp - (i\theta + \theta_d)]}{i\theta(\theta_d + i\theta)^2} \tag{6.21}$$

$$\Phi_2 = \frac{[\exp - (i\theta + \theta_d)] - 1}{i\theta(\theta_d + i\theta)(1 + i\theta\Gamma)}. \tag{6.22}$$

The SCLC impedance (No. 1 in Table 6.1) may be expressed as a function of Φ_1, by introducing $\Delta \to \pm\infty$ in equations (3.9.2) and (6.20)*)

$$Z_{SCLC} = \frac{1}{i\omega C_0} + R_d \frac{\theta_d^2}{\exp(-\theta_d) + \theta_d - 1}\Phi_1(i\theta, \theta_d) \tag{6.23}$$

whereas the zero-current impedance ($\Delta \to 0$) may be written as a function of Φ_2

$$Z|_{J_0 \to 0} = \frac{1}{i\omega C_0} + R_d \frac{\theta_d^2}{\exp(-\theta_d) - 1}\Phi_2(i\theta;\, \theta_d, \Gamma). \tag{6.24}$$

In the last expression we shall assume constant carrier mobility and minority carrier injection ($\theta_d < 0$). Then, if the transit-time is large ($|\theta_d|$ large), the second term in equation (6.24) is small or even negligible and $Z \simeq 1/i\omega C_0$, i.e. the zero-current impedance (6.24) reduces to the impedance of the geometrical capacitance C_0. This situation was discussed by Yoshimura [17] ($\Gamma \to \infty$). Clearly, for an arbitrary Δ, the diode impedance should be a combination of the SCLC impedance, equation (6.23), and the zero-current impedance, equation (6.24). For example, the total series resistance is

$$R = \mathcal{R}e Z = A_1 \mathcal{R}e \Phi_1(i\theta;\, \theta_d) + A_2 \mathcal{R}e \Phi_2(i\theta;\, \theta_d, \Gamma) \tag{6.25}$$

where A_1 and A_2 are real quantities, which depend upon the d.c. bias, but not upon frequency. Therefore, the frequency characteristics are a linear combination of those calculated for two particular cases. As an example we consider the following

---

*) This impedance should coïncide with that calculated by Yoshimura [17]. Because of the d.c. boundary condition $E_0(0) = 0$, the unique possible approximation for $v = v(E)$ is $v = \mu E$, $\mu = \text{const.}$

case: minority carrier injection ($N < 0$), constant mobility ($\mu = $ const.), emitte current-limited (ECL) injection ($\Gamma \to \infty$). Yoshimura [17] found that both t SCLC impedance and the zero-current impedance should exhibit a negative resis ance around the transit-time frequency, if $|\theta_d|$ is sufficiently large (above 3, f example). It is now clear that for the ECL impedance at any injection level Δ, t series resistance becomes negative provided that $|\theta_d| > 3$.

Another particular situation which we discuss here is the impedance of a u polar semiconductor diode which is quasi-neutral in the steady-state. We sh introduce $\Delta = 1$ (Table 6.1) in the general expression (6.20). Because $\bar{J}_0$ defin by (3.9.2) becomes $\bar{J}_0 = 1/\theta_d$ for $\Delta = 1$, we obtain

$$Z = \frac{1}{i\omega C_0} + R_d\theta_d[\Phi_1(i\theta;\,\theta_d) + \Phi_2(i\theta;\,\theta_d,\,\Gamma)]. \quad (6.2$$

Some algebraic manipulations which are omitted here, lead to

$$Z = \frac{L^2}{\varepsilon v_0}\,\frac{(1 + i\theta\Gamma)(\theta_d + i\theta) + (\theta_d\Gamma - 1)[1 - \exp -(\theta_d + i\theta)]}{(\theta_d + i\theta)^2(1 + i\theta\Gamma)} \quad (6.2$$

which is exactly the result obtained by Hariu *et al.* [153]. Because N is positiv θ_d has the sign of the differential mobility. v_0 is the d.c. velocity, which is unifor throughout the entire sample.

We shall discuss the effect of the a.c. boundary condition. Assume, first, $\sigma_1(0) \to \infty$, i.e. $\Gamma \to 0$. In this case equation (6.27) reduces to the impedance, equati (5.56), first calculated by McCumber and Chynoweth [76]. If $\theta_d < 0$, this i pedance has a negative real part at certain transit angles (Section 5.7). We note t fact that $\sigma_1(0) \to \infty$ does not necessarily mean that the injected contact is ohm $\sigma_1(0)$ can be also very large for a cathode-semiconductor interface with trappi states: the control characteristic of such a contact may exhibit a very steep slop as shown in Figure 3.10.

The second important case is the ECL injection, defined by zero a.c. co ductivity of the cathode, $\sigma_1(0) = 0$, such as $\Gamma \to \infty$. Therefore, we obtain fro equation (6.27) the impedance calculated by Atalla and Moll [121] (in the spec case when the a.c. diffusion current is zero, see Problem 6.1). We note the fa that for $\theta_d < 0$, the impedance has a negative real part at any frequency, except zer

If Γ is finite, positive (such as for Schottky-effect emission cathode, for e ample) and $\theta_d < 0$, the diode resistance can be also negative, as shown in Figu 6.1, reproduced after Hariu *et al.* [153]. A similar effect was also discussed by Y *et al.* [154]. In the second paper [154] it was shown that the steady-state electr field can be uniform indeed in a unipolar semiconductor diode with a non-ohm injecting electrode. Detailed computer calculations were made for a GaAs devi with Schottky-barrier-emission cathode.

In this chapter we shall discuss in detail two examples of unipolar diodes wi non-ohmic injecting cathode. First of these is the transit-time silicon diode wi saturated drift velocity. The cathode field is sufficiently high to provide velocity saturation in the entire semiconductor bulk. The injection may take place by thermionic emission (modified by the image-force effect) over the potential barrier [155], by tunnel emission [48], etc. (Section 3.3).

The negative resistance obtained from such a device is a pure transit-time effect. The injection mechanism is, however, essential. (We recall the fact that the impedance of the saturated velocity diode with SCLC boundary condition does

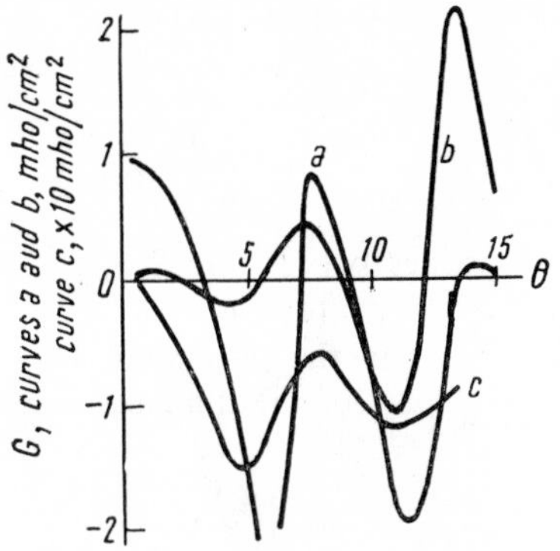

Figure 6.1. Conductance of injection-limited negative-mobility solid-state diodes calculated by Hariu *et al.* [153]. With our notations $\Delta = 1$, an $\mathrm{d}1/\tau_\mathrm{d}$, $1/\tau_\mathrm{c}$ are equal, respectively, to: $-10^9\,\mathrm{s}^{-1}$ and $5\times10^9\,\mathrm{s}^{-1}$ (curve *a*), $-5\times10^9\,\mathrm{s}^{-1}$ and $5\times10^9\,\mathrm{s}^{-1}$ (curve *b*), $-10^9\,\mathrm{s}^{-1}$ and $10^2\,\mathrm{s}^{-1}$ (curve *c*). The transit angle is $\theta = \omega T_\mathrm{L}$.

not have a negative real part). The device can be used as a local oscillator at frequencies lower and comparable to the transit-time frequency [18], [20], [156]–[159].

The second example is a transferred-electron device which exhibits a negative conductance at frequencies much higher than the transit-time frequency and can be used, in principle, for broad-band amplification [153]. This negative conductance is entirely due to the a.c. negative mobility.

6.5 Saturated-velocity Transit-time Diode

The device discussed in this section is characterized by (a) non-ohmic injecting cathode, and (b) saturated carrier velocity in the entire semiconductor bulk.

The device impedance may be obtained by direct integration of the basic equations, by introducing $\theta_d \to 0$ (Δ and Γ are finite) into equation (6.20), or by using the matrix given at the end of Section 4.8. The final result is

$$Z = \frac{L^2}{\varepsilon v_l}\left[\frac{1}{i\theta} + \frac{\exp(-i\theta) - 1}{(i\theta)^2(1 + i\theta\Gamma)}\right], \quad \theta = \theta T_L = \frac{\omega L}{v_l}. \tag{6.28}$$

We stress the fact that the impedance depends upon the injection mechanism and upon the d.c. bias, only through the dimensionless parameter Γ, which is defined by equation (6.7) as the ratio of the cathode relaxation time to the carrier transit time. For a given device only the former can be modified by changing the d.c. bias.

The fact that the impedance $Z = Z(i\theta)$ does not depend explicitly upon th injection mechanism determined the following situation: the above result was ca culated by several authors studying devices with different injection mechanism The impedance, equation (6.28), was obtained by Wright [88] for punch-throug injection, by Claassen and Harth [48] for tunnel emission, by Haus *et al.* [16 for thermoionic emission over the potential barrier of a metal-semiconducto structure biased just below the punch-through, and by Weller [155] for Schott emission.

The incremental resistance is

$$R_i = Z|_{\theta\to 0} = \frac{L^2}{2\varepsilon v_l}(2\Gamma + 1) \qquad (6.2$$

and takes values between $L^2/2\varepsilon v_l$ for SCLC injection ($\Gamma = 0$) and ∞ for ECL i jection ($\Gamma \to \infty$). The normalized frequency characteristics $\bar{Z} = Z/R_i = \bar{Z}($ depend only upon the parameter Γ.

The real part of the impedance, equation (6.28), is

$$R = \mathcal{R}eZ = \frac{L^2}{\varepsilon v_l}\,\frac{1 - \cos\theta + \theta\Gamma\sin\theta}{\theta^2(1 + \theta^2\Gamma^2)} \qquad (6.3$$

and can be negative at certain transit angles. For Γ = finite and positive, the resis ance oscillates from positive to negative values as shown in Figure 6.2 [57]. Th first frequency domain where $R < 0$, is $\theta_1 < \theta < 2\pi$, when $\theta_1 = \theta_1(\Gamma)$. If Γ b comes very small ($\sigma_1(0) \to \infty$, SCLC injection), then $\theta_1 \to 2\pi$ and the negati frequency disappears, as expected (Section 5.2). If the a.c. cathode conductivi is low ($\sigma_1(0)$ = small), then Γ will be large and $\theta_1 \to \pi$. Also, the maximum a solute value of the negative resistance should occur for $\theta = \theta_m \simeq 3\pi/2$ [155]. Ho ever, for $\Gamma \gg 1$, the series resistance calculated for $\theta = 3\pi/2$ decreases rapid with increasing Γ.

The 'optimization' of this negative resistance was discussed by Claassen an Harth [48], by Haus *et al.* [160] and by Weller [155].

We consider below the metal-silicon-metal diode with Schottky-barrier c thode and calculate the parameter Γ and its bias dependence. The Schottky-emissio current, equation (3.21), is repeated here for convenience

$$J_0 = J_{sat} = J_R \exp(E_c/E_R)^{1/2} \qquad (6.3$$

(J_R is the pure thermionic-emission current, E_C is the cathode field and E_R is 'critical' value of the electric field for the image-force effect, see Section 3.3). B use of equations (5.1), (6.2), (6.7) and (6.31), one obtains

$$\Gamma = \frac{2\varepsilon v_l E_R}{LJ_R}\,\frac{(\ln J_0/J_R)^2}{J_0/J_R} \qquad (6.3$$

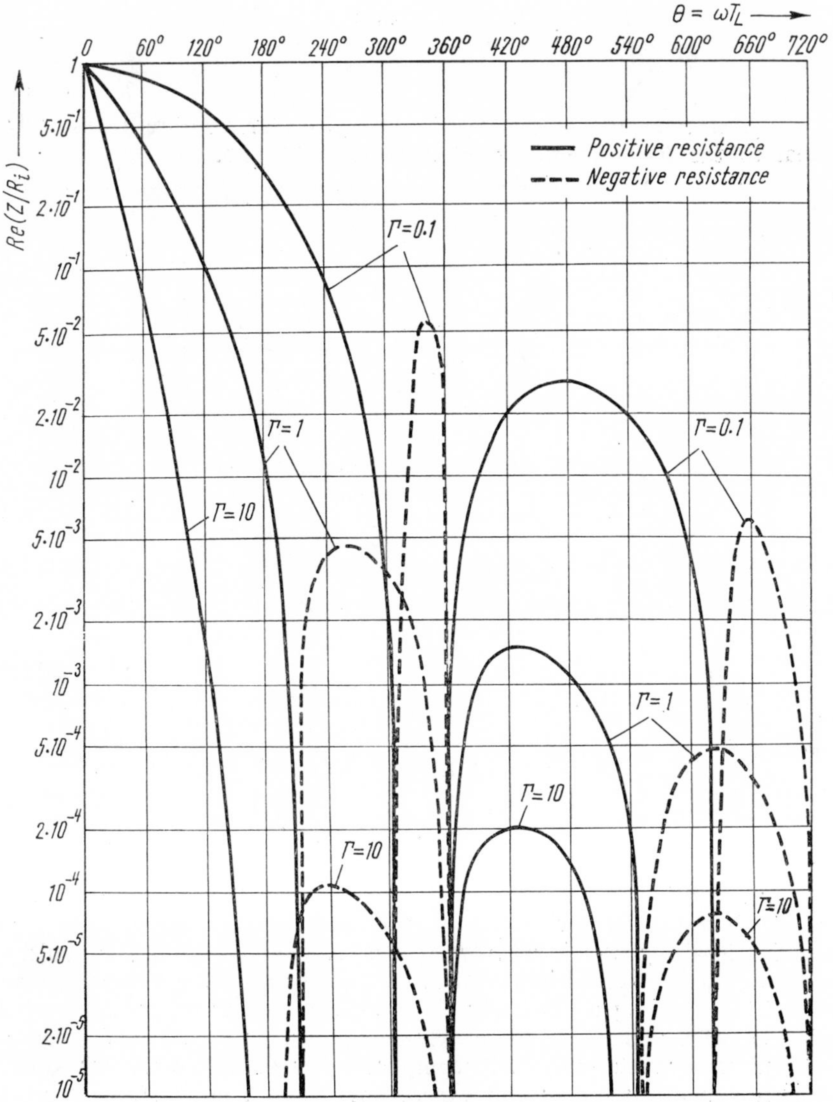

Figure 6.2. The frequency dependence of the series resistance, equation (6.30) with Γ as parameter.

The bias dependence*) of Γ is shown in Figure 6.3 and has a maximum for $J_0 = eJ_R = 2.71828 \ldots J_R$. The maximum value which Γ can have for Schottky-emission

$$\Gamma_{max} = 0.737 \frac{\varepsilon v_l E_R}{L J_R} \tag{6.33}$$

and for a given semiconductor material and a given temperature, is inversely proportional to the Richardson current and sample length.

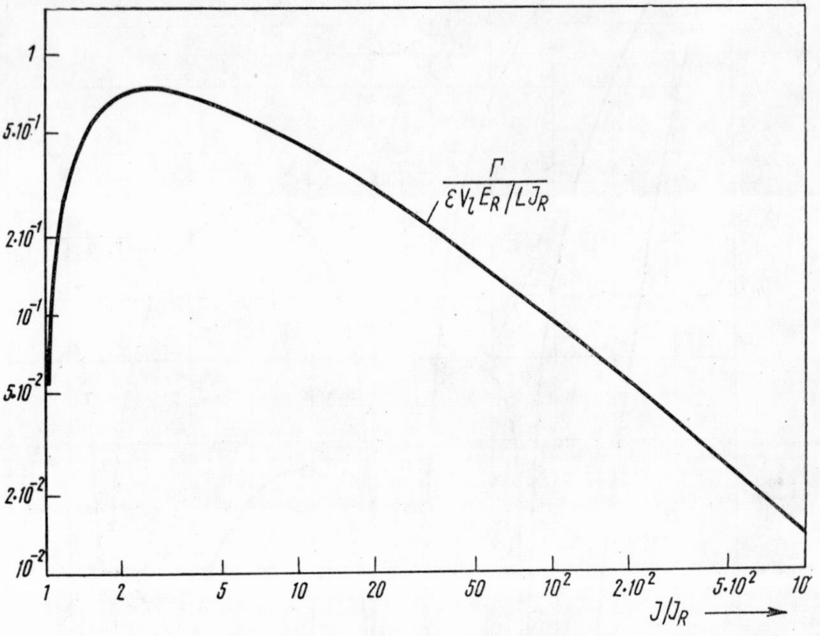

Figure 6.3. Bias dependence of Γ.

Really surprising is the fact that the a.c. impedance of this device does not depend at all upon the device doping, and moreover, it does not matter if we inject holes in a p-type or in an n-type material. Given the intrinsic semiconductor properties ε, v_l (assumed independent of doping); the pure-thermionic-emission current of the injecting contact, J_R; the sample length L; and the temperature, the small signal impedance is completely determined by the signal frequency and the injected steady-state current. A complete study of the device behaviour and a comparison between theory and experiment requires, however, the knowledge of the steady-state $J - V$ characteristic.

*) $J_0 = J_R$ leads to $\Gamma = 0$ and $\sigma_1(0) \to \infty$ which is the situation encountered for SCLC emission. We shall soon see that $J_0 = J_R$ is the boundary between the SCLC region and the Schottky-emission region on the steady-state characteristic.

Sometimes, one assumes tacitly (see, for example, Weller [155]) the fact that the electric field into the semiconductor bulk is uniform and equal to the bias field V_0/L. Consequently, the steady-state characteristic $J_0 = J_0(V_0)$ follows directly from equation (6.31) where $E_c = V_0/L$. Alternatively, the voltage drop V_0 may be expressed as a function of the bias current J_0 as follows

$$V_0 = LE_R(\ln J_0/J_R)^2. \tag{6.34}$$

However, if the mobile injected charge and the fixed space charge are taken into account, the total voltage drop should be*)

$$V_0 = -\frac{eNL^2}{2\varepsilon} + \frac{J_0 L^2}{2\varepsilon v_l} + E_R L\,(\ln J_0/J_R)^2. \tag{6.35}$$

We note the fact that by setting $E_R = 0$, one obtains formally the SCLC characteristic of the saturated velocity diode

$$J_0 = \frac{2\varepsilon v_l}{L^2}\left(V_0 + \frac{eNL^2}{2\varepsilon}\right). \tag{6.36}$$

The fact that by using the zero field at the cathode as the boundary condition, one re-obtains equation (6.36) may be easily checked by direct calculation. For minority carrier injection $N < 0$, the characteristic does exhibit a threshold (or punch-through voltage) $V_{PT} = \dfrac{e|N|L^2}{2\varepsilon}$. It seems that certain experimental devices [128], [156] do indeed show a $J - V$ characteristic of this type. The slope of the characteristic (linear) coincides with the differential (low frequency) resistance which is given by equation (6.29) for $\Gamma = 0$. For $N > 0$, equation (6.36) should be used with caution. Clearly, J_0 cannot be positive for $V_0 \leqslant 0$, as this equation indicates, and equation (6.36) will be acceptable only for V_0 positive and sufficiently large. As a matter of fact, equation (6.36) is not correct because the SCLC boundary condition (zero cathode field) is incompatible with the assumption of saturated velocity. This equation can still be used if the applied voltage is sufficiently large to provide a nearly constant (saturated) drift velocity throughout almost the entire diode length (except a very narrow cathode region). We note the fact that the voltage drop may be also written

$$V_0 = \frac{L^2}{2\varepsilon v_l}(J_0 - eNv_l) \tag{6.37}$$

where eNv_l is the neutral current. This equation should be valid as far as the current is not saturated, i.e. $J_0 < J_R$.

Now let us discuss equation (6.35) which is valid for $J_0 > J_R$. This is the Schottky-emission region of the $J - V$ characteristic. It joins smoothly the SCLC char-

---

*) We assume saturated velocity ($J_0 = ep_0 v_l$) and a Schottky-type boundary condition, equation (6.31). The Poisson equation (3.31) is also used.

acteristic ($J_0 < J_R$) at $J_0 = J_R$*). By comparing equations (6.34) and (6.35), it becomes clear that the first and the second term in the right-hand-side of equation (6.35) correspond, respectively, to the mobile space charge and the fixed space charge. These terms cancel each other only for neutrality, when we have $J_0 = eNv_l$. However, the semiconductor is not, in general, neutral. This is evident for minority carrier injection. For majority carrier injection we can have either accumulation or depletion, as shown in Section 3.10. We shall show below that sometimes the electric field cannot be assumed uniform. Equation (6.35) will be written (see also equation 6.32)

$$V_0 = \frac{J_0 L^2}{2\varepsilon v_l}\left(1 - \frac{eNv_l}{J_0} + \Gamma \ln \frac{J_0}{J_R}\right). \tag{6.38}$$

Let us assume for simplicity

$$J_0 \geqslant J_R \gg eNv_l \tag{6.39}$$

and the second term in parenthesis in equation (6.38) is negligible. The third term has a maximum reached for $J_0 = e^2 J_R$ which is approximately $2\varepsilon v_l E_R / L J_R$. It is now clear that if this quantity is smaller or comparable to unity

$$\frac{2\varepsilon v_l E_R}{L J_R} \lesssim 1 \tag{6.40}$$

the effect of the *mobile* injected charge (first term in equation (6.38)) cannot be neglected, and equation (6.38) cannot be approximated by equation (6.34). For silicon, at room temperature, and for a typical length $L = 10\ \mu$m, equation (6.40) requires approximately $J_R > 100$mA mm^{-2} †).

For illustration, we give here normalized $J - V$ characteristics of the unipolar diode with Schottky-barrier emission.

The *velocity-field-dependence* is taken into account as follows: the mobility is assumed constant and equal to the zero-field mobility μ up to a field E_l (which is taken equal to 20 kVcm^{-1} for *holes in silicon* at room temperature). For $E > E_l$ the velocity is assumed saturated at the value μE_l. The boundary condition, equation (6.31), is used in calculations ($E_R \simeq 53$ kVcm^{-1} for silicon at room temperature).

The normalization is identical to that used in Section 3.9 (see equations (3.77) and (3.88)). Two normalized parameters should be given also. First, we should specify the normalized Richardson current, equation (3.85). Secondly, the doping-length product NL is given (see, for example, equation (3.84)).

Figure 6.4 shows the $J - V$ characteristics for the majority carrier injection [57]. The characteristics coincide in the low current region as long as the mobility

---

*) It may be checked easily that both the $J - V$ characteristic and the incremental resistance (the first-order derivative) are continuous in $J = J_R$.

†) The condition (6.39) should be also satisfied. If, for example, $J_R = 1.6$ A mm^2 (which is a quite usual value [154], [155]), $J_R \gg eNv_l$ will require $N \ll 10^{14}$cm^{-3} ($v_l \simeq 10^7$ cm/sec), i.e. the semiconductor resistivity should be sufficiently high.

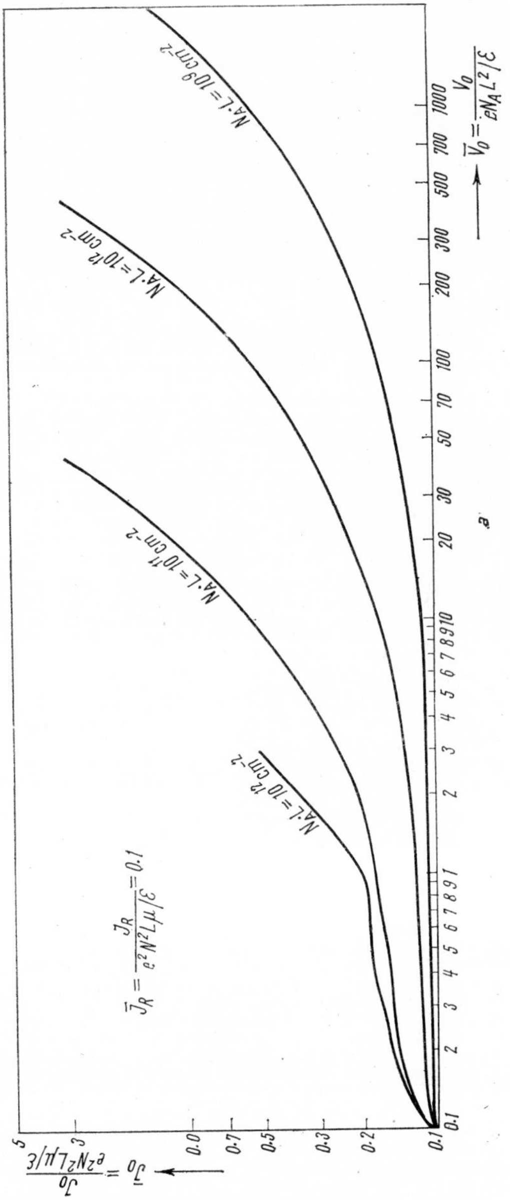

Figure 6.4 a

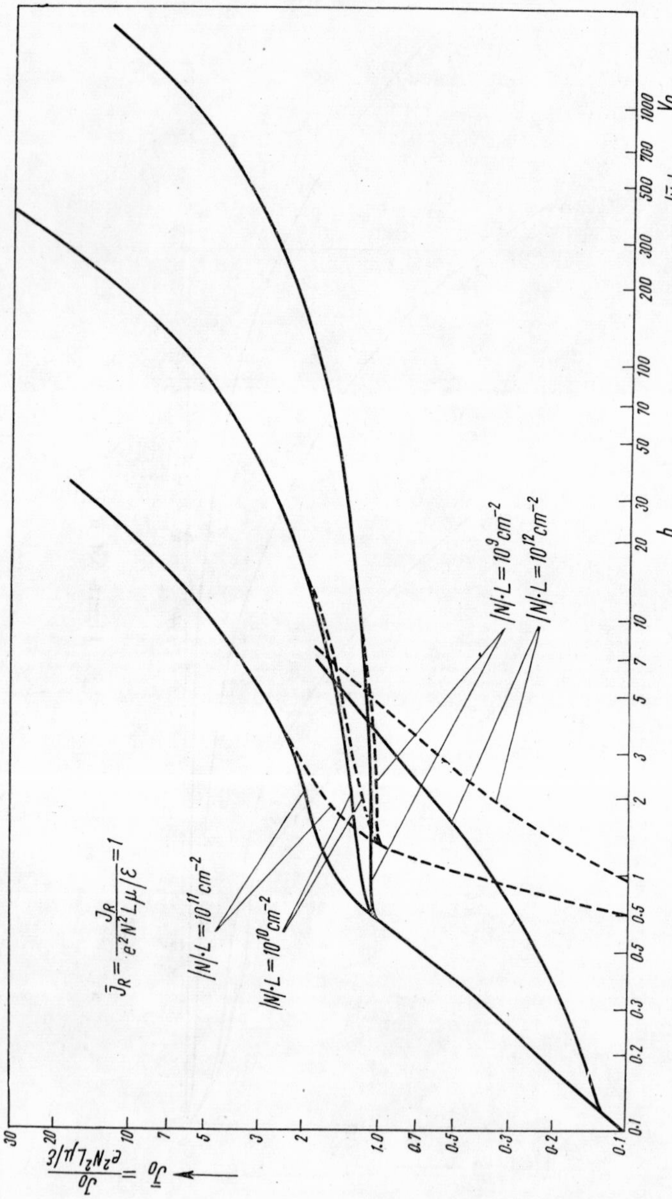
$\bar{J}_R = \frac{J_R}{e^2 N^2 L \mu / \varepsilon} = 1$
$|N| \cdot L = 10^{11} cm^{-2}$
$|N| \cdot L = 10^{10} cm^{-2}$
$|N| \cdot L = 10^{9} cm^{-2}$
$|N| \cdot L = 10^{12} cm^{-2}$
$\bar{J}_0 = \frac{J_0}{e^2 N^2 L \mu / \varepsilon}$
b
$|\bar{V}_0| = \frac{V_0}{e|N|L^2/\varepsilon}$
0.1
0.2
0.3
0.5
0.7
1.0
2
3
5
7
10
20
30
0.1
0.2
0.3
0.5
0.5
1
2
3
5
7
10
20
30
50
70
100
200
300
500
700
1000

Figure 6.4 b

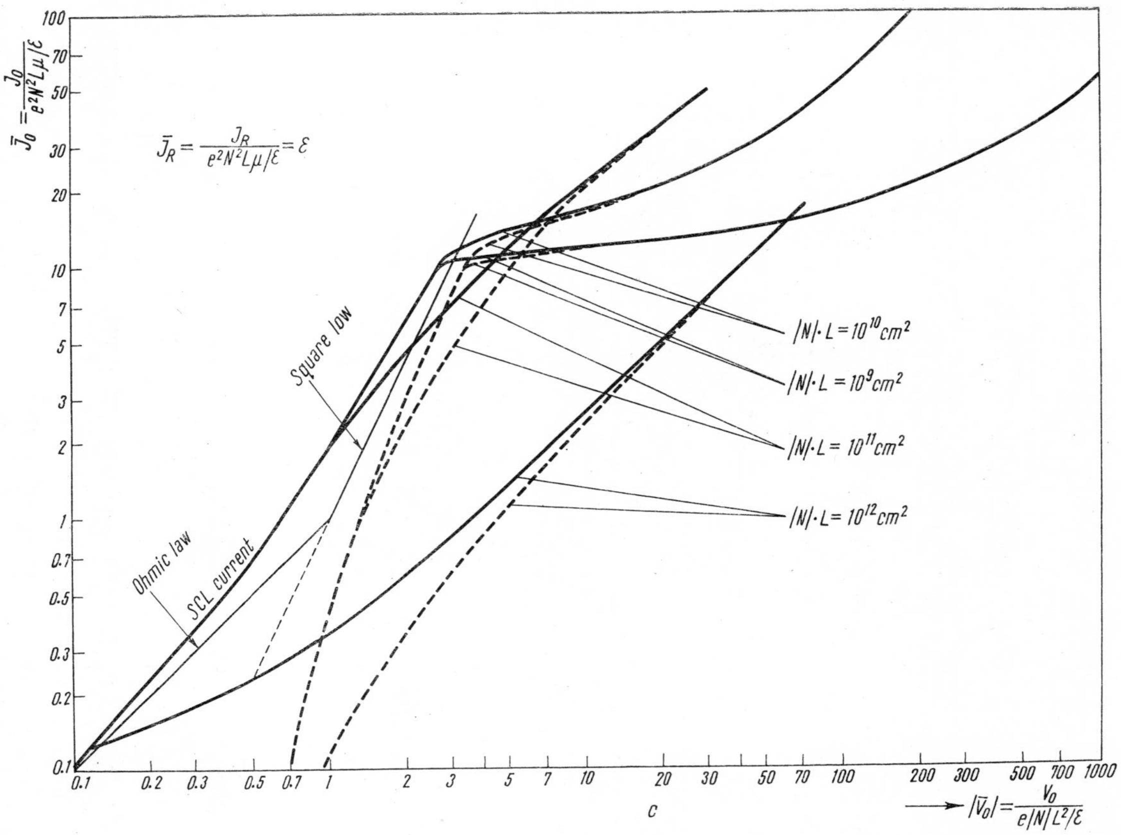

Figure 6.4. Steady-state $J - V$ characteristic of a Schottky-barrier-emission silicon diode, with holes injected as majority carriers (solid lines) or minority carriers (dashed curves) [57].

is constant and the bias current is below the Richardson current. Detailed computations show, for example, that the electric field is largely non-uniform. Figure 6.4c also shows the $J-V$ characteristic for minority carrier injection. By compar-

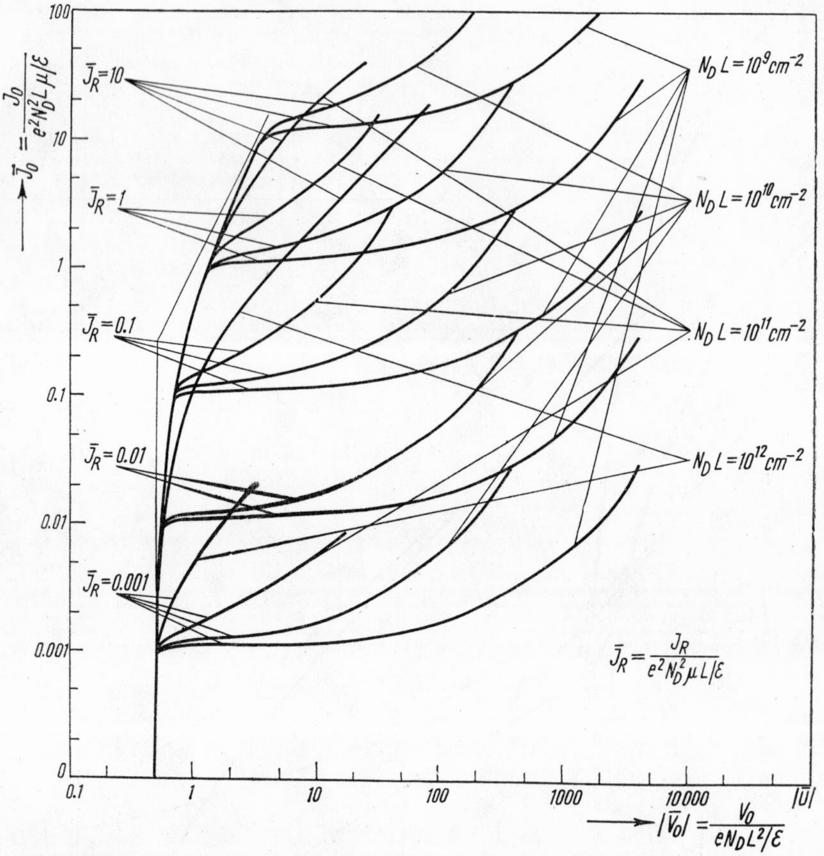

Figure 6.5. Normalized $J-V$ characteristics of a metal-semiconductor-metal diode made from n silicon and having a hole-injecting Schottky-barrier-emission cathode [57].

ing the curves traced for majority and minority carrier injection, respectively, we can see the effect of fixed space charge. These curves coincide indeed at very large applied voltages, where the density of fixed space charge becomes negligible with respect to the injected mobile charge.

The normalized $J-V$ characteristics for the minority carrier injection are shown separately in Figure 6.5 [57]. These characteristics spread themselves out of the SCLC characteristic (punch-through diode) as the current exceeds J_R.

The a.c. parameter Γ can be calculated now as follows:

$$\Gamma = 2E_l E_R \left(\frac{\varepsilon}{eNL}\right)^2 \frac{\ln(\overline{J_0}/\overline{J_R})}{\overline{J_0}} \tag{6.41}$$

and for a given semiconductor, at a given temperature, it depends upon $\overline{J_R}$, NL and $\overline{J_0}$ (the last parameter is bias-dependent). We recall the fact that the expression (6.28) for the a.c. impedance (and the value of Γ, equation (6.41)) are valid only if the carrier velocity is saturated ($v = \mu E_l = v_{sat}$) [57].

6.6. Emitter-current-limited Injection in Negative-mobility Semiconductors

The simplest example of emitter-current-limited (ECL) injection is the pure thermionic-emission limited injection, i.e. the case when the current flow through the device is 'saturated' and equal to J_R = const.,

$$J_0 = J_R = \text{const.} \tag{6.42}$$

(this case can be formally obtained by introducing $E_R \to \infty$ in equation (6.31)). The steady-state current does not depend upon the applied voltage but can be varied by modifying the temperature.

The injected current can be electrically controlled in a three-terminal structure. Such a device was first suggested by Yoshimura [17]. The basic material was silicon and the mobility was considered positive and equal to the low field mobility.

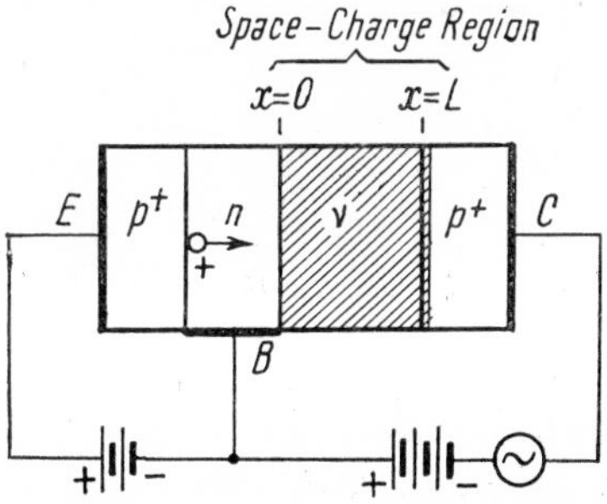

Figure 6.6. $p^+ n \nu p^+$ structure suggested by Yoshimura [17] to operate as an ECL injection device when biased as shown.

Later Atalla and Moll [121] described a similar structure whose operation was based upon the negative mobility of a semiconductor like GaAs. In the structure represented in Figure 6.6, the carrier injection is provided by a p—n junction, the so called emitter-base junction. Almost the entire emitter current is injected into the drift region ν. The external voltage applied between the collector and base terminals is practically equal to the voltage drop V_0 on the ν region. Therefore, the electric field and carrier density distribution inside the drift region should be completely determined by J_0 and V_0. In the language used in

Chapter 3, $J_0 =$ const. corresponds to a *horizontal control characteristic*. The slope of this characteristic is zero, therefore $\sigma_1(0) = 0$ and $\Gamma \to \infty$.

The d.c. and a.c. properties of such a drift region can be studied by using the results obtained in Chapter 3 and 4. The basic assumptions of our model are: (a) one-dimensional analysis, (b) negligible diffusion, (c) constant doping*); and (d) constant differential mobility.

By using the $J_0 =$ const. and $V_0 =$ const. curves plotted in the $\Delta - \theta_d$ plane (Figures 3.24 — 3.27) we can immediately find the normalized transit-time, θ_d, and the injection level Δ. These parameters will be used to determine the steady-state distribution of the electric field, etc. and the frequency characteristics.

The a.c. impedance can be calculated by introducing $\Gamma \to \infty$ into equation (6.6) or any equivalent formula for the general impedance of ICC devices. The result may be written

$$Z = \frac{1}{i\omega C_0} + R_d \bar{J}_0 \frac{\theta_d^3[1 - (\theta_d + i\theta) - \exp - (i\theta + \theta_d)]}{i\theta(\theta_d + i\theta)^2}. \tag{6.43}$$

Therefore the impedance is a series combination of a capacitor having the value of the geometrical capacitance and an impedance which is proportional to the injected current. The reader may easily verify that

$$Z|_{\omega \to 0} = \frac{1}{i\omega C}, \; C \gtrless C_0 \tag{6.44}$$

i.e. the zero-frequence impedance is infinite, as expected (horizontal $J - V$ characteristic). We shall therefore normalize the impedance as follows [106]

$$\bar{Z} = \frac{Z}{T_L/C_0} = \frac{1}{i\theta} + \bar{J}_0 \frac{\theta_d^2[1 - (\theta_d + i\theta) - \exp - (i\theta + \theta_d)]}{i\theta(\theta_d + i\theta)^2}. \tag{6.45}$$

We recall the fact that $\bar{J}_0 = \bar{J}_0(\Delta, \theta_d)$ (see equations of Appendix to Section 3.10) and thus $\bar{Z} = \bar{Z}(i\theta; \Delta, \theta_d)$. However, it seems more convenient to have $\bar{J}_0$ and θ_d as parameters. We have shown in Section 3.10 that the $V_0 =$ const. curves represented in the $\Delta - \theta_d$ plane are almost horizontal (at least that part which is of interest to us). Therefore, for a given voltage the parameter θ_d changes very little with the injected current, J_0. According to equation (3.81) $\theta_d = T_L/\tau_d$ is, roughly, inversely proportional to the applied voltage, whereas $\bar{J}_0$ is proportional to the injected current, J_0, and does not depend upon the voltage drop, V_0.

We are interested in the possibility of the existence of a negative resistance. The series resistance of the impedance, equation (6.45) is

$$R = \mathcal{Re} Z = \frac{T_L}{C_0} \bar{J}_0 \theta_d^2 \mathcal{Re} \frac{1 - (\theta_d + i\theta) - \exp - (i\theta + \theta_d)}{i\theta(\theta_d + i\theta)^2}. \tag{6.46}$$

*) See Problem 3.7 for non-uniform doping.

By simply inspecting this expression of R we can get the following important results:

(a) The existence of a negative resistance depends upon the *value* of θ_d and upon the *sign* of $\bar{J}_0$, which is the sign of μ_d (see Appendix to Section 3.10).

(b) The series resistance, R, is, roughly, proportional to the injected current. Atalla and Moll assumed [121] d.c. neutrality, and thus the injected current was approximately given by

$$J_0 \simeq eNv_0\left(\frac{V_0}{L}\right) \tag{6.47}$$

($\Delta = 1$ and $\bar{J}_0 = 1/\theta_d$ in equation (6.46), according to Appendix 3.10). However, we shall prefer to increase the absolute value of the negative resistance by increasing $\bar{J}_0$ above this value.

(c) The presence of space charge effects ($\Delta \neq 1$, positive) does not modify the frequency dependence of the series resistance, except by a constant factor, because R depends upon Δ only through $\bar{J}_0$*).

For a systematic investigation of the possiblity of a negative resistance we have to calculate the fraction occurring in equation (6.46) as a function of both θ_d and θ

$$F(\theta, \theta_d) = \mathcal{R}e\,\frac{1 - (\theta_d + i\theta) - \exp - (i\theta + \theta_d)}{i\theta(\theta_d + i\theta)^2}. \tag{6.48}$$

It is interesting to mention the fact that we can find the sign of $F(\theta, \theta_d)$ without any new calculation. We note that the series resistance of the SCLC diode practically coincides with equation (6.46)†). There are only minor differences. For SCLC injection we have $\Delta \to \infty$ in $\bar{J}_0 = \bar{J}_0(\theta_d, \Delta)$ and $\mu_d \to \mu > 0$. Consequently, the sign of $F(\theta, \theta_d)$ should be the sign of the SCLC series resistance. For $\theta_d > 0$, $F(\theta, \theta_d)$ is always positive (Figure 5.13). If $\theta_d < 0$, $F(\theta, \theta_d)$ will be negative for certain values of θ, provided that $|\theta_d|$ is sufficiently large (say $|\theta_d| > 3$) (see Figure 5.16). As shown above, for ECL injection the sign of $R(\omega)$ is the sign of $\mu_d F(\theta, \theta_d)$. The possibility of a negative resistance is indicated in Table 6.2.

Table 6.2

N	μ_d	θ_d	$R(\omega)$
> 0 maj. carrier	> 0	> 0	positive for all θ and θ_d
< 0 min. c.	> 0	< 0	negative for $\|\theta_d\| \gtrsim 3$ and $\theta \approx 2\pi$, etc.
> 0 maj. c.	< 0	< 0	negative for any θ if $\|\theta_d\| \lesssim 3$
< 0 min. c.	< 0	> 0	negative for all θ and θ_d

*) We note, however, the fact that for V_0 = const., the frequency-dependence $R = R(\omega)$ slightly changes with J_0, because of T_L which is current dependent.

†) A similar situation was indicated by Llewellyn and Peterson for the vacuum diode [84]. We suggest that the reader verifies that the real part of equation (5.49) reduces to equation (6.48).

Therefore, ECL injection in negative mobility materials*) opens the possibility of a negative resistance in the almost entire frequency spectrum: from very low frequencies (where this resistance occurs in series with a capacitance) up to extremely high frequencies, much above the transit-time frequency. We stress the fact that a similar situation can be obtained by using a non-ohmic injecting contact instead of ECL injection (as shown by Hariu *et al.* [153], see also Figure 6.1). We shall refer to this later.

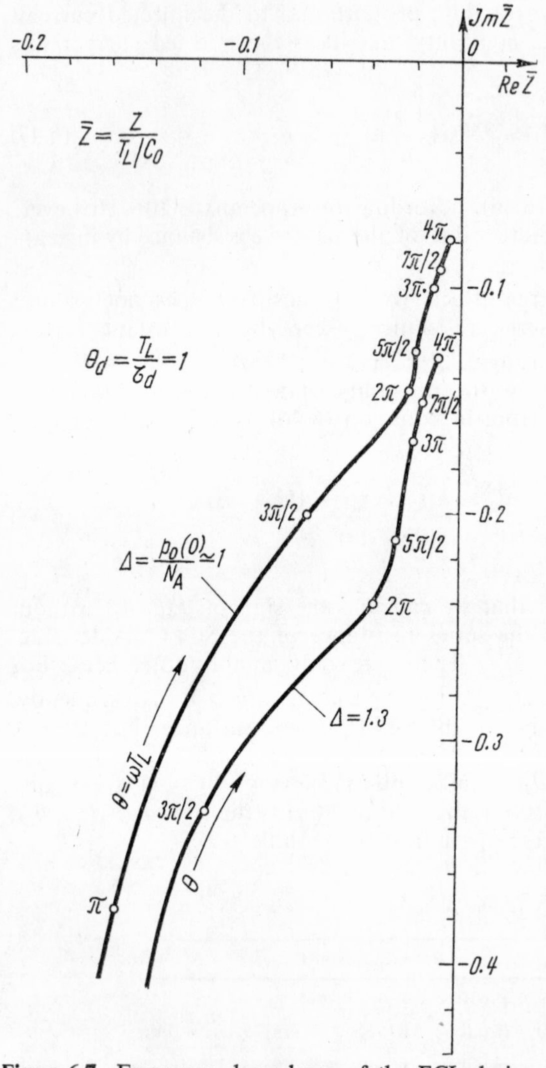

Figure 6.7. Frequency dependence of the ECL device impedance.

The frequency characteristics for ECL injection in GaAs were derived by Atalla and Moll [121] for d.c. neutrality $\Delta = 1$ (also taking diffusion into account), and by Dascalu [161] for an intrinsic drift region i.e. $N \to 0$ and $\theta_d \to 0$. The effect of fixed space-charge (majority and minority carrier injection) was studied in [162]. We shall discuss here only the majority carrier injection in negative mobility materials, i.e. $N>0$, $\mu_d<0$ and $\theta_d < 0$ ($\bar{J}_0 < 0$).

Figure 6.7 shows the frequency-dependence of the device impedance for several values of the normalized bias current and a given θ_d. The real part of this impedance is always negative, as expected. The parallel conductance is plotted in Figure 6.8 as a function of the transit-angle, θ. This diagram indicates an almost constant conductance at frequencies much above the transit-time frequency. We shall demonstrate by direct calculation that such a *constant* conductance does indeed exist.

---

*) Yoshimura [17] derived frequency characteristics for ECL minority carrier injection in positive mobility semiconductors and discussed the negative resistance effect.

By assuming $\theta \gg 1$ and $|\theta_d| \lesssim 1$ the impedance, equation (6.43), becomes

$$Z \simeq \frac{1}{i\omega C_0}\left((1 + i\frac{\bar{J}_0\theta_d^2}{\theta}\right), \theta \gg 1,\ |\theta_d| \lesssim 1 \tag{6.49}$$

where the second term in parentheses is very small as compared to unity. The admittance may be written

$$Y \simeq i\omega C_0\left(1 + \frac{\bar{J}_0\theta_d^2}{i\theta}\right), \ \theta \gg 1,\ |\theta_d| \lesssim 1\cdot \tag{6.50}$$

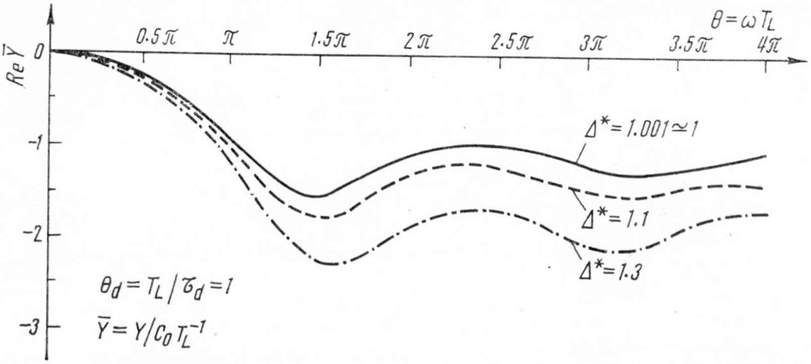

Figure 6.8. Frequency dependence of the parallel conductance.

The v.h.f. equivalent circuit consists in a parallel RC combination. The capacitance is just the geometrical capacitance $C_0 = \dfrac{\varepsilon}{L}$. The parallel conductance is

$$G_{p\infty} = C_0\bar{J}_0\theta_d^2/T_L = (C_0/\tau_d^2)\bar{J}_0 T_L. \tag{6.51}$$

By using the equations of Section 3.9, 3.10 and the corresponding Appendix one obtains

$$G_{p\infty} = \frac{J_0 T_L}{eNLR_d} = \frac{1}{R_d}\frac{Q_{\text{mob}}}{eNL} \tag{6.52}$$

where Q_{mob} is the total mobile charge per unit of electrode area and eNL is the total fixed charge per unit area. For d.c. neutrality $(\Delta = 1)$ $Q_{\text{mob}} = eNL$ and $G_{p\infty} = 1/R_d$. An identical expression for the parallel conductance was derived by Atalla and Moll [121] by assuming that the sample length is equal to an integral number of wavelengths $\lambda = v_0/f$ (where v_0 is the d.c. carrier velocity and $f = \omega/2\pi$ the signal frequency).

Equation (6.52) can be also written

$$G_{p\infty} = Q_{\text{mob}}\,\mu_d/L^2 \tag{6.53}$$

and this expression clearly points out that $G_{p\infty}$ has the sign of μ_d. We stress the fact that equations (6.49) – (6.53) are valid for both majority and minority carrier injection in positive or negative mobility materials*). Let us define a normalized conductance

$$\bar{G}_{p\infty}(\theta_d, \Delta) = \frac{G_{p\infty}}{G_{p\infty}(\theta_d, 1)} = \bar{J}_0\theta_d \tag{6.54}$$

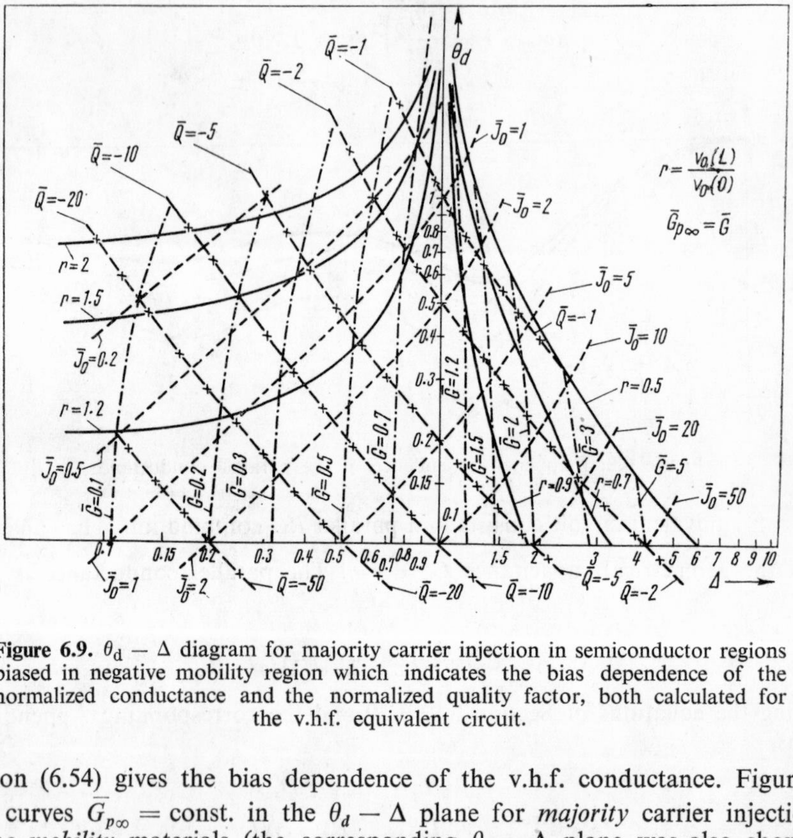

Figure 6.9. θ_d – Δ diagram for majority carrier injection in semiconductor regions biased in negative mobility region which indicates the bias dependence of the normalized conductance and the normalized quality factor, both calculated for the v.h.f. equivalent circuit.

Equation (6.54) gives the bias dependence of the v.h.f. conductance. Figure 6.9 shows curves $\bar{G}_{p\infty} = \text{const.}$ in the $\theta_d - \Delta$ plane for *majority* carrier injection in *negative mobility* materials (the corresponding $\theta_d - \Delta$ plane was also shown in Figure 3.25).

We shall calculate the negative Q of the v.h.f. circuit. This is

$$Q = \frac{\omega C_0}{G_{p\infty}} = \frac{\theta C_0/T_L}{C_0\bar{J}_0\theta_d^2/T_L} = \frac{\theta}{\bar{J}_0\theta_d^2}. \tag{6.55}$$

---

*) Because equation (6.43) is also valid for the SCLC impedance, the v.h.f. conductance (6.53) yields $G_{p\infty} = \varepsilon E(L)\mu L^{-2} = \varepsilon v(L)L^{-2}$ for $\mu = \text{const.}$ and $N = 0$, a result which coincides with that given in Section 5.4.

The device Q normalized to the transit angle θ is

$$\overline{Q} = \frac{Q}{\theta} = \frac{1}{\overline{J}_0 \theta_d^2} = \frac{1}{\overline{G}_{p\infty} \theta_d} \cdot \tag{6.56}$$

Fig. 6.9 also shows curves $\overline{Q}$ = const.

The device Q and the v.h.f. conductance may be modified by adjusting the steady-state bias. In particular, if the device is used for small-signal broad-band amplification*) the parallel conductance should be electrically controlled such as to compensate the losses of the circuit.

Figure 6.9 allows us to calculate the bias dependence of the v.h.f. equivalent circuit. A numerical example is given in Appendix. It is interesting to note the fact that the above results on the v.h.f. circuit are also valid for an arbitrary injecting contact, provided that we have simultaneously

$$|\theta_d| \lesssim 1, \ \theta \gg 1, \text{ and } \theta\Gamma \gg 1 \tag{6.57}$$

or, according to, equation (6.19),

$$\omega T_L \gg 1, \ \omega|\tau_d| \gg 1, \text{ and } \omega\tau_c \gg 1 \tag{6.58}$$

(the transit-time, the cathode relaxation time and the bulk relaxation time, are much longer than the signal period). The proof of this theorem is very simple and is left to the reader†).

6.7 Example of Device with Delayed Injection

Figure 6.10 shows the minority carrier delay diode suggested by Shockley [14]. This is a p^+np^+ structure biased as shown. The left-hand junction (emitter) injects holes into the neutral base (low injection levels). These holes first diffuse through the base and then are swept out rapidly by the high electric field existing in the barrier region of the collector. We assume [14], [16] that: (a) the emitter efficiency is unity; (b) the a.c. electric field in the neutral base region is negligible (due to relatively high electron concentration); and (c) the external a.c. voltage occurs practically on the collector-base barrier region, because the a.c. voltage drop on the forward biased emitter junction is also negligible.

*) The operation frequency is 'cavity-controlled' [153] i.e. is tuned by adjusting the external circuit. The upper operating frequency is limited by the intervalley scattering time [26], [29].

†) By replacing $1 + i\theta\,\Gamma \simeq \theta\,\Gamma \gg 1$ in equation (6.6), one obtains, formally the same result as introducing $\Gamma \to \infty$ (ECL injection).

The a.c. hole current at the emitter junction (which is practically equal to the total current $\underline{\mathbf{J}}_1$) flows through the neutral base and is attenuated and delayed as indicated by the transport factor

$$\beta = |\beta| \exp - i\theta_\beta. \tag{6.59}$$

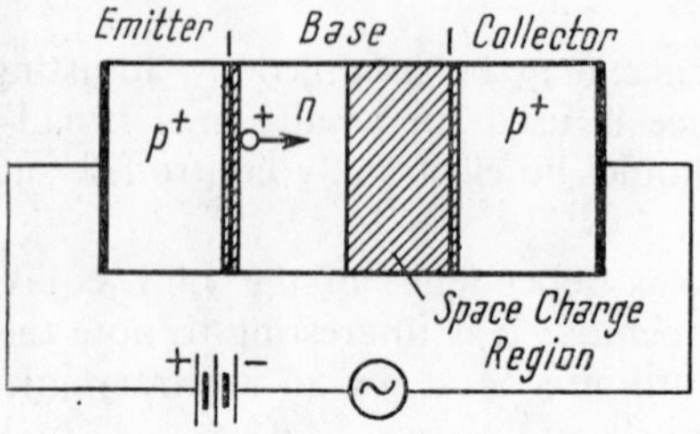

Figure 6.10. A minority carrier delay diode, as suggested by Shockley [14].

This factor is frequency dependent and was calculated by Shockley [14], [16], Wright [87], etc. Shockley [14] also neglected the transit-time of carriers in the collector barrier region. We shall assume that $v = v_{sat}$ and by applying Equs. (4.122) — (4.125) to this region and assuming $\theta \to 0$, one obtains $C_{11} = - C_{12} = = \dfrac{1}{i\omega C_0}$ (where C_0 is the barrier capacitance of the collector region). Because the particle current entering the barrier region is

$$\underline{\mathbf{j}}_1(0) = \beta \underline{\mathbf{J}}_1 \tag{6.60}$$

the total device impedance is, approximately,

$$Z = C_{11} + \beta C_{12} \simeq \frac{1}{i\omega C_0}(1 - \beta) \tag{6.61}$$

The real part of this impedance is

$$R(\omega) = \mathcal{R}_e Z = \frac{|\beta|}{\omega C_0} \sin \theta_\beta \tag{6.62}$$

and takes a negative value if the delay experienced by the holes in the neutral base satisfies the following equation:

$$\pi < \theta_\beta < 2\pi. \tag{6.63}$$

If the transit-time in the barrier collector region cannot be neglected, the impedance is

$$Z = \frac{1}{i\omega C_0}\left[1 - \frac{1 - \exp(-i\theta)}{i\theta} |\beta| \exp - \theta_\beta\right]. \tag{6.64}$$

The effect of transit-time in both base and collector regions was studied by Wright [86], [87]. His 'transistor transit-time oscillator' has three d.c. terminals but its a.c. operation is almost identical to that described above.

6.8 Conclusion

There are a large number of transit-time devices which can be studied by using the a.c. theory developed here. We give other few examples.

(a) The a.c. stability of transferred-electron devices with 'imperfect cathode boundary conditions' [52]. The case studied by Kroemer [52] can be derived directly from equation (6.6) where $\Delta \approx 1$, $\Gamma =$ real, positive and $\theta_d < 0$ (due to $\mu_d < 0$).

(b) Yu *et al.* [154] indicated the importance of transit-time effects a.c. in operation of negative mobility diodes with Schottky barrier contact. They made also a simple analytical estimation by assuming uniform electric field and taking diffusion into account. If one neglects the diffusion, it will be possible to take into account the electric field nonuniformity, by using the formulae given here in Section 6.3. The general theory of Chapter 4 can be used for direct computer calculation of the device impedance*).

(c) The drift region of the avalanche-generation transit-time devices can be studied by use of equations (4.122) — (4.125) in conjunction with proper boundary conditions relating the injected current to the applied electric field [29].

Problems

6.1. Calculate the ECL impedance by assuming d.c. neutrality and negligible diffusion.

Hint: Use either equation (6.27) or the basic equations of the unipolar current flow in solids.

6.2. The impedance of the saturated-velocity transit-time diode is given by equation (6.28). Determine the a.c. equivalent circuit at frequencies much higher than the transit-time frequency.

Hint: The results should be discussed by taking Γ as a parameter.

6.3. Verify that the low-frequency resistance (6.29) is identical with the differential resistance resulting from the steady-state characteristic.

6.4. Calculate the low-frequency capacitance for ECL injection, which is defined by equation (6.44). Discuss the bias dependence of this capacitance.

---

*) Yu *et al.* [154] simulated the device behaviour starting from the basic equations, which is much more difficult.

6.5. Show that the small-signal impedance for ECL injection in the limit of zero ion density is given by

$$Z = \frac{T_L}{C_0} \frac{1}{i\theta} \left\{ 1 + \frac{p}{(i\theta)^2} [\exp(-i\theta) - 1 + i\theta] \right\} \tag{P.6.1}$$

where

$$p = \frac{2[v_0(0) - v_0(L)]}{v_0(0) + v_0(L)}. \tag{P.6.2}$$

7

Carrier Inertia Effects and High-frequency Negative Resistance

7.1 Carrier Inertia

In this chapter we put forward a method of detailed analysis of transit-time effects in charge-controlled electron devices. In these devices there exist capacitive effects due to charge accumulation, as well as delays due to the finite time required for charge transport.

The current flowing through a unipolar device has two components: the particle (drift) current and the displacement current. The particle current is due to the transport of charge carriers and is proportional to both carrier density and carrier velocity. We have already shown the fact that at frequencies comparable to the transit-time frequency, the a.c. particle current is non-uniform in space and is delayed with respect to the applied voltage. This delay increases with increasing frequency. If the delay of the *total* current lies between $-\dfrac{\pi}{2}$ and $-\dfrac{3\pi}{2}$, the device exhibits a negative resistance, i.e. generates a.c. power and may be used for signal amplification or generation.

Such an h.f. negative resistance is known for both vacuum and solid-state devices. 'However*), the mode of operation for semiconductive diodes differs markedly from that for vacuum diodes. This difference arises from the fact that in the case of vacuum devices the charge carriers, electrons, move with conservation of momentum so that the speed at any instant is dependent upon their past history. In contrast, electrons or holes in a semiconductor suffer collisions at a frequency of about 10^{12} times per second; as a result they continue to move in the direction of an applied field only as long as the field exerts a force upon them. It is possible to achieve negative resistance operation in semiconductive devices even though the momentum of the charge carriers is not conserved, by effecting the functions necessary for negative resistance in a manner and with means not available in vacuum diodes'' [16].

For the sake of analysis, we distinguish here two kinds of carrier inertia †). First, we have the inertia of carriers in transport, which is simply the property

---

*) Here is a long quotation from Shockley's pioneering work [16].

†) We are concerned here with the inertia of *free* carriers and do not discuss the properties of bounded electrons or carriers trapped in defect centres.

that the carrier drift from the injecting to the collecting electrode requires a finite time. Secondly, there is the velocity inertia, i.e. the property that the carrier velocity does not respond instantly to the variations of the local electric field acting upon that carrier. The first property is general and is intrinsic to the charge-controlled device concept itself. The second property is specific to the conductive medium.

If the conductive space is the vacuum, the one-dimensional form of the motion law for an electron in an electric field is

$$\mathrm{d}v(t)/\mathrm{d}t = -(e/m)\,E(t) \tag{7.1}$$

where E is the electric field acting at the time t upon the electron with the velocity v. If $E(t)$ *changes rapidly*, the electron velocity will be delayed with respect to $E(t)$. Assume now that $E(t)$ is equal to a d.c. component plus a small a. c. perturbation. This electric field acts upon the electron travelling from the injecting plane to the collecting plane. The transit-time is approximately equal to the d. c. transit-time T_L and it is this time interval T_L which has to be taken as a reference interval when we discuss the rapid or slow variation of the electric field. This happens because $v(t)$ will experience a considerable delay with respect to $E(t)$ when the signal period becomes comparable to the time interval the electron is subjected to the electric field and the latter is just T_L. The fact that the delay of $\underline{\mathbf{v}}_1$ with respect to $\underline{\mathbf{E}}_1$ becomes appreciable at frequencies comparable to the transit-time frequency, was confirmed by detailed calculations [142]. Therefore, at transit-time frequencies, the electron in vacuum exhibits a velocity inertia.

For a charge carrier (say a hole) moving in a solid, the 'motion equation' can be approximated by

$$m\frac{\mathrm{d}v}{\mathrm{d}t} + \nu m v = eE \tag{7.2}$$

where ν is the collision frequency. This equation was discussed in Chapter 2. If the collisions are very frequent during the carrier passage in the interelectrode space, the motion equation (7.2) may be approximated by

$$v = \mu E, \qquad \mu = e/\nu m \tag{7.3}$$

and v is 'in phase' with E — no velocity inertia exists. This relation holds up to very high frequencies (of the order of 100 GHz). At higher electric fields the mobility is field dependent. By differentiating equation (7.3) one obtains

$$\underline{\mathbf{v}}_1 = \left(\frac{\mathrm{d}v}{\mathrm{d}E}\right)_0 \underline{\mathbf{E}}_1 \tag{7.4}$$

where $\mu_d = (\mathrm{d}v/\mathrm{d}E)$ is the differential mobility. As far as the frequency of the applied signal is sufficiently low, the carrier velocity still follows the field variations.($\mathrm{d}v/\mathrm{d}E$ is a real quantity) and no velocity inertia exists.

The a.c. mobility μ_d can be, however, negative (as in GaAs InP, etc.) and in this case the velocity exhibits a 180° delay with respect to the electric field. This is a kind of 'velocity inertia' specific to solids and finds application in transferred-electron (or Gunn-type) devices. Strictly speaking, we do not have here an inertia: the velocity responds instantly to the field variations. The carrier velocity is an average value and decreases with the increasing electric field because of carrier redistribution between two or more sets of energy levels which have different carrier mass [28], [29].

Both 'transport' inertia and 'velocity' inertia are useful in providing the desired phase lag between voltage and current and may be identified as 'sources' of a.c. power generation. Except for the negative mobility-case, these 'sources' come into action at frequencies comparable to the transit-time frequency. Therefore, an analysis of transit-time effects is indispensable if we wish to understand the details of the a.c. power generation process.

7.2 Carrier-density Modulation and Carrier-velocity Modulation

An important step towards the understanding of the transit-time processes in electron vacuum devices was the famous Llewellyn's monograph 'Electron Inertia Effects' [163]. The early workers (Benham, Müller, Llewellyn) were mostly concerned with the physical understanding of the basic electronic equations. About 20 years ago, Shockley [16] investigated the possibility of achieving a negative resistance due to the transit-time effects in solids. He developed the impulsive-impedance method which allows a good insight into the transient response at small perturbations.

We suggested elsewhere [82], [125], [142], [143], [164] a *quantitative* analysis of transit-time effects upon the frequency characteristics. This method is based upon the transport coefficients defined in Chapter 3 and is suitable for computer analysis.

The starting point is the observation that the small-signal a.c. particle current has two components (see Chapter 4)

$$\underline{\mathbf{j}}_1 = e p_0 \underline{\mathbf{v}}_1 + e \underline{\mathbf{p}}_1 v_0 \tag{7.5}$$

namely the carrier density modulation component

$$\underline{\mathbf{J}}_{1p} = e v_0 \underline{\mathbf{p}}_1 \tag{7.6}$$

and the carrier velocity modulation component

$$\underline{\mathbf{J}}_{1v} = e p_0 \underline{\mathbf{v}}_1. \tag{7.7}$$

$\underline{\mathbf{J}}_{1p}$ is due to the modulation of the concentration of the carrier moving with the d.c. velocity, whereas $\underline{\mathbf{J}}_{1v}$ is the result of the modulation of the velocity of the carriers injected by the steady-state bias. The third component is the displacement current

$$\underline{\mathbf{J}}_{1d} = i\,\omega\,\varepsilon\,\underline{\mathbf{E}}_1. \tag{7.8}$$

The total alternating current density

$$\underline{\mathbf{J}}_1 = \underline{\mathbf{J}}_{1p} + \underline{\mathbf{J}}_{1v} + \underline{\mathbf{J}}_{1d} \tag{7.9}$$

must be independent of position x. The above separation was also mentioned by other authors [8], [17], [110]. What is really new in our analysis is the systematic quantitative examination of the device response. We define three 'f' functions

$$f_p = \frac{\underline{\mathbf{J}}_{1p}}{\underline{\mathbf{J}}_1},\quad f_v = \frac{\underline{\mathbf{J}}_{1v}}{\underline{\mathbf{J}}_1},\quad f_d = \frac{\underline{\mathbf{J}}_{1d}}{\underline{\mathbf{J}}_1}. \tag{7.10}$$

These are complex quantities which should satisfy (according to equation (7.9)) the equation:

$$f_p + f_v + f_d = 1. \tag{7.11}$$

Let us now assume that it is possible to introduce F such as

$$\underline{\mathbf{E}}_1(x) = F\underline{\mathbf{J}}_1. \tag{7.12}$$

As an example we shall calculate this function F for the particular case when it is possible to define the cathode conductivity $\sigma_1(0) = \underline{\mathbf{j}}_1(0)/\underline{\mathbf{E}}_1(0)$. By taking into account equations (4.16), (4.17) and (4.64) one obtains

$$F = K_{21x} + \frac{K_{22x}}{\sigma_1(0) + i\omega\varepsilon} = F(x). \tag{7.13}$$

It can be shown that F depends upon x only through the transit-time T_x (from $x = 0$ to the distance x). Therefore, we have $F = F(T_x)$.

Then, it is permissible to write

$$f_d = \frac{i\omega\varepsilon\underline{\mathbf{E}}_1}{\underline{\mathbf{J}}_1} = i\omega\varepsilon F(T_x) \tag{7.14}$$

$$f_p = \frac{e\mathbf{p}_1 v_0}{\underline{\mathbf{J}}_1} = \frac{v_0\varepsilon}{\underline{\mathbf{J}}_1}\frac{\mathrm{d}\underline{\mathbf{E}}_1}{\mathrm{d}x} = \frac{\varepsilon}{\underline{\mathbf{J}}_1}\frac{\mathrm{d}\underline{\mathbf{E}}_1}{\mathrm{d}T_x} = \varepsilon\frac{\mathrm{d}F}{\mathrm{d}T'_x} \tag{7.15}$$

$$f_v = 1 - f_p - f_d = 1 - i\omega\varepsilon F - \varepsilon\frac{\mathrm{d}F}{\mathrm{d}T_x}\cdot \tag{7.16}$$

The complex quantities f_p, f_v, f_d calculated for the SCLC solid-state diode with the square law steady-state characteristic, equation (4.85), are represented in Figure 7.1. Let us consider for example the curve

$$f_p(\theta_x) = \frac{\underline{\mathbf{J}}_{1p}(x, \omega)}{\underline{\mathbf{J}}_1(\omega)},\qquad \theta_x = \omega T_x. \tag{7.17}$$

Because we can write

$$\frac{\underline{\mathbf{J}}_{1p}(x, \omega)}{\underline{\mathbf{J}}_{1p}(0, \omega)} = \frac{f_p(\theta_x)}{f_p(0)}, \tag{7.18}$$

the curve f_p shown in Figure 7.1 should indicate the variation of the $\underline{\mathbf{J}}_{1p}$ component with the distance from the cathode, for a given frequency ω. We must know the

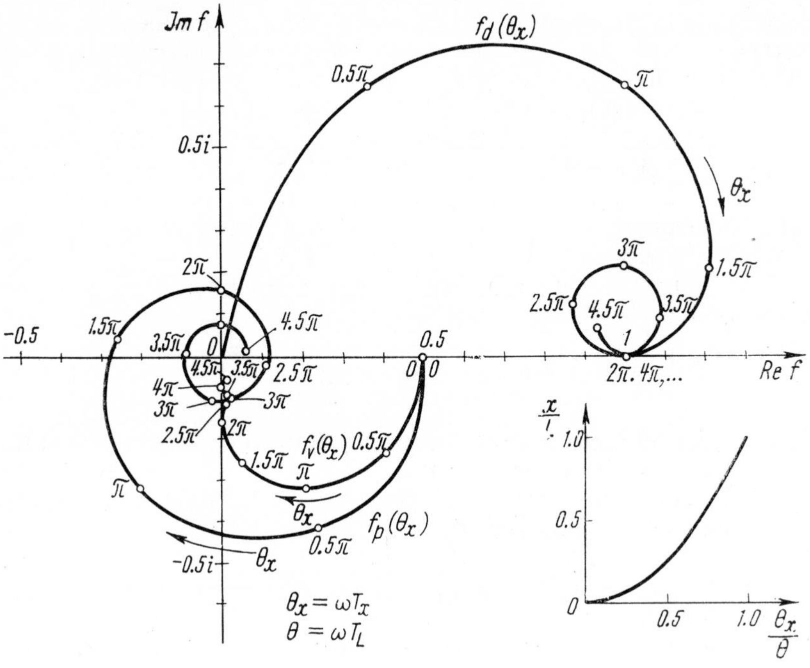

Figure 7.1. The "f" functions for the ideal (square-law) SCLC solid-state diode [125].

dependence $T_x = T_x(x)$. For the particular case discussed here (Figure 7.1) one obtains

$$\frac{T_x}{T_L} = \frac{\theta_x}{\theta} = \left(\frac{x}{L}\right)^{1/2} \tag{7.19}$$

and the gradation in transit angles can be converted in a gradation in distances. The limiting point of the f_p curve is given by $\theta_x = \theta = \omega T_L$, where ω is given. The same operation may be done for f_v and f_d.

We stress the fact that each of the alternating current components 'bears' the variation in time of a physical quantity, namely the carrier density ($\underline{\mathbf{J}}_{1p}$), carrier velocity ($\underline{\mathbf{J}}_{1v}$) and electric field ($\underline{\mathbf{J}}_{1d}$). For example, $\underline{\mathbf{J}}_{1p}$ is in phase with $\underline{\mathbf{p}}_1$ and its amplitude is proportional to the amplitude of $\underline{\mathbf{p}}_1$. Hence, the curves plotted in Figure 7.1 can be used to get an intuitive representation of the mechanism of the device response.

7.3 Space-charge Waves

We are particularly interested in the variation of the $\underline{\mathbf{J}}_{1p}$ component, which gives us information about the motion of the 'space-charge wave' injected at the cathode. We shall characterize the propagation of the space charge wave by the ratio (7.18). The phase of this quantity is exactly the phase delay of $\underline{\mathbf{p}}_1(x)$ with respect to $\underline{\mathbf{p}}_1(0)$. By inspecting Figure 7.1 we note that the variation of the a.c. space-charge density is delayed by an angle smaller than the transit angle θ_x. This shows the space-charge wave injected at the cathode moves at a velocity greater than the d.c. velocity, v_0, although, by definition, $\underline{\mathbf{J}}_{1p}$ is due to the density modulation of the carrier moving with a velocity v_0*).

We shall explain this situation as follows. The electric field at the distance, x, is given by the relationship

$$\underline{\mathbf{E}}_1(x) = \frac{e}{\varepsilon} \int_0^x \underline{\mathbf{p}}_1(x)\,\mathrm{d}x \tag{7.20}$$

(obtained by integrating the Poisson equation with the boundary condition $\underline{\mathbf{E}}_1(0) = 0$, SCLC injection) and is determined by the injected space-charge contained at a given instant between the cathode and the plane x. A variation of the space charge induces a variation of the electric field, and, consequently, a variation of the carrier velocity

$$\underline{\mathbf{v}}_1(x) = \mu \underline{\mathbf{E}}_1(x). \tag{7.21}$$

On turn, the modulation of the carrier velocity determines an additional (local) displacement of the charge injected by the d.c. bias and this represents a perturbation of the space-charge wave which propagates from the cathode to the anode. Because $\underline{\mathbf{E}}_1(x)$, and consequently $\underline{\mathbf{v}}_1(x)$ anticipate the variation of charge density ($\underline{\mathbf{p}}_1(x)$) (see the above equations) it is clear why the delay of $\underline{\mathbf{p}}_1(x)$ with respect to $\underline{\mathbf{p}}_1(0)$ is smaller than θ_x.

We suggest (see the demonstration below) that an SCLC diode with negligible velocity modulation should exhibit an unperturbed space-charge wave, (i.e. the delay is exactly θ_x), or, by definition

$$\frac{f_p(\theta_x)}{f_p(0)} = \frac{\underline{\mathbf{J}}_{1p}(x, \omega)}{\underline{\mathbf{J}}_{1p}(0, \omega)} = exp\,(-\,i\theta_x) \tag{7.22}$$

*) The phase velocity of the 'space charge wave' is greater than the steady-state velocity. We suggest that the reader evaluates and discusses the phase velocity of the wave represented by the particle current.

An example of such an SCLC diode is the insulator or dielectric diode with a high density of shallow traps. This device was studied in Section 5.10. We assumed slow traps (see equation 5.78) of a high density ($\delta \ll 1$). Above a certain frequency ($\omega \gg 1/\tau$) the velocity modulation is negligible, as indicated by equation (5.93).

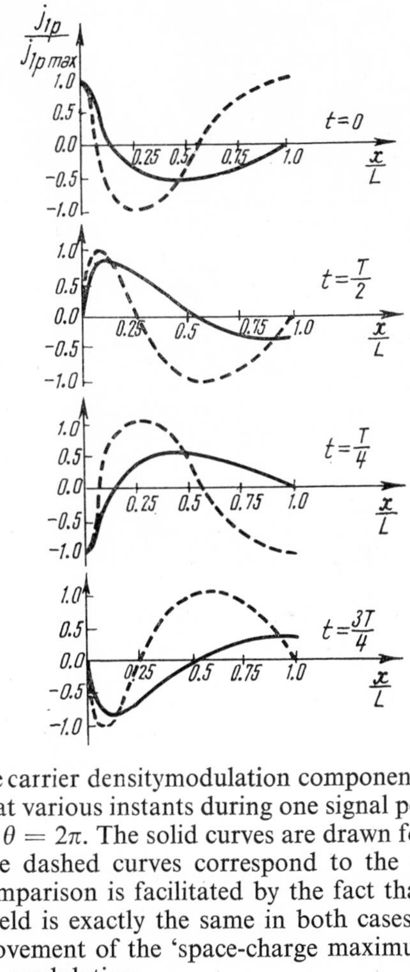

Figure 7.2. Variation of the carrier density modulation component with respect to x, at a few time moments in square-law SCLC diodes with $\delta = 1$ (full lines) and $\delta \simeq 0$ (dashed curves). The operating frequency is equal to the transit-time frequency ($\theta = 2\pi$). The total alternative current flowing through the diode is $j(t) = j_{\max} \cos \omega t$, i.e. at $t = 0$ it is maximum, and a maximum injection takes place at the cathode (where the displacement current is always equal to zero) [125].

Figure 7.2 shows the variation of the carrier density modulation component with respect to the distance from the cathode, at various instants during one signal period. The transit angle has the particular value $\theta = 2\pi$. The solid curves are drawn for the square-law SCLC diode ($\delta = 1$) and the dashed curves correspond to the diode with slow shallow traps ($\delta = 0$). The comparison is facilitated by the fact that the distribution of the steady-state electric field is exactly the same in both cases. The diagrams of Figure 7.2 show that the movement of the 'space-charge maximum' is faster in the presence of carrier velocity modulation.

Figure 7.3. shows (at an arbitrary scale) the variable space charge inside the device at various time-instants. These data were obtained by using the $\underline{\mathbf{J}}_{1p} = e\underline{\mathbf{p}}_1 v_0$ curves from Figure 7.2 and by taking into account the $v_0 = v_0(x)$ dependence. Positive values indicate an excess of charge density with respect to the steady-state value at the same point, and negative values — a deficit of charge. This figure visualises the displacement and the attenuation of the space-charge density maximum, in transit through the device (see Appendix).

Let us now consider minutely the case of negligible velocity modulation ($f_v \simeq 0$). There are at least three situations of this kind, namely: — the saturation of the carrier velocity, — the punch-through injection at very low current intensities ($J_0 \to 0$), and — the effect of high-density slow shallow traps at relatively high frequencies.

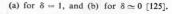

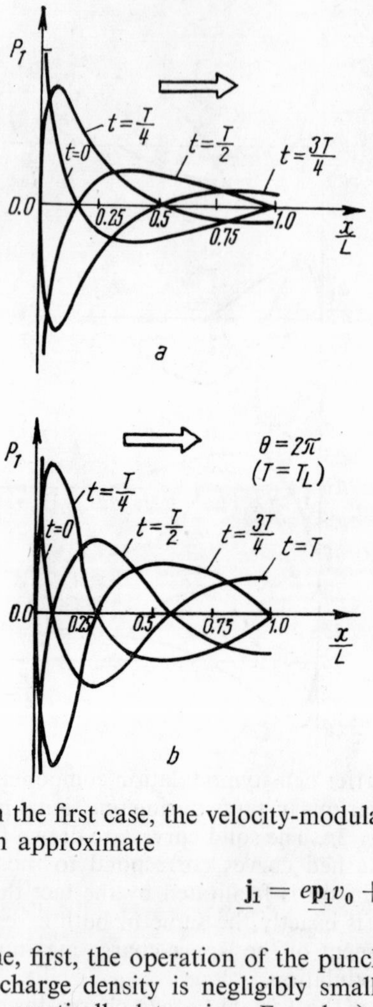

Figure 7.3. Variation of the carrier density distribution, calculated by using the previous figure:
(a) for $\delta = 1$, and (b) for $\delta \simeq 0$ [125].

In the first case, the velocity-modulation is zero ($\mathbf{J}_{1v} = 0$). In the last two cases we can approximate

$$\underline{\mathbf{j}}_1 = e\underline{\mathbf{p}}_1 v_0 + e p_0 \underline{\mathbf{v}}_1 \simeq e v_0 \underline{\mathbf{p}}_1. \tag{7.23}$$

Assume, first, the operation of the punch-through diode. For $J_0 \to 0$ the d.c.mobile space charge density is negligibly small as compared to the fixed charge density (which practically determines E_0 and v_0). Hence, it is permissible to write

$$\frac{\underline{\underline{\mathbf{j}}}_1}{J_0} = \left(\frac{\underline{\underline{\mathbf{p}}}_1}{p_0}\right) + \left(\frac{\underline{\underline{\mathbf{v}}}_1}{v_0}\right) \simeq \frac{\underline{\underline{\mathbf{p}}}_1}{p_0} \tag{7.23'}$$

because, formally, $p_0 \to 0$ and the first term becomes dominant. Equation (7.23′) proves equation (7.23).

Figure 7.4 shows the frequency characteristics of the SCLC punch-through diode calculated with and without velocity modulation, the second result is approximate. This approximation becomes acceptable at very large values of $|\theta_r|$ (J_0 small, Figure 5.15).

In the third case a similar reasoning applies. We do not discuss here this situation but refer the reader to equations (5.88), and (5.93) which speak by themselves.

If $\underline{\mathbf{J}}_{1v}$ is negligible, by introducing $f_v = 0$ in equation (7.16) one obtains a differential equation in $F = F(T_x)$, which yields

$$F(T_x) = F(0) + \frac{1}{i\omega\varepsilon}[1 - \exp(-i\theta_x)]. \tag{7.24}$$

By assuming $\underline{\mathbf{j}}_1(0) = \sigma_1(0)\,\underline{\mathbf{E}}_1(0)$ we obtain

$$F(0) = \frac{\underline{\mathbf{E}}_1(0)}{\underline{\mathbf{J}}_1} = \frac{1}{\sigma_1(0) + i\omega\varepsilon}. \tag{7.25}$$

Then by using the notations (6.2), (6.7) and (6.8) one obtains ($\Gamma = \varepsilon/\sigma_1(0)\,T_L$)

$$\frac{f_p(T_x)}{f_p(0)} = \frac{\underline{\mathbf{J}}_{1p}(x, \omega)}{\underline{\mathbf{J}}_{1p}(0, \omega)} = [\exp(-i\theta_x)] - i\theta\Gamma[1 - \exp(-i\theta_x)]. \tag{7.26}$$

For SCLC injection $\sigma_1(0) = \infty$ and $\Gamma = 0$. Therefore equation (7.26) becomes identical to equation (7.18) which is the definition of the unperturbed space-charge wave*). This relation may be also written

$$\underline{\mathbf{j}}_1(x) = [\exp(-i\theta_x)]\,\underline{\mathbf{j}}_1(0). \tag{7.27}$$

We stress the fact that the above relation is no longer valid if the injecting contact is *non-ohmic* (ICC injection), i.e. $\Gamma \neq 0$, then equation (7.26) will be used.

7.4 A.C. Power Dissipation inside the Device

We have shown in Section 7.1 that there exist two aspects of carrier inertia: inertia of charge transport and inertia of carrier velocity with respect to the electric field variations. The former may be associated with the space-charge wave and the carrier-density modulation and the latter with the velocity modulation. We wish to see what is the rôle played by these inertia effects in determining the occurrence of an h.f. negative resistance. One possibility [125] is to calculate the active a.c. power dissipated by each component of the alternating current. These amounts of power are

*) If the drift velocity is saturated we obtain immediately $\underline{\mathbf{n}}_1(x) = [\exp(-i\theta_x)]\underline{\mathbf{n}}_1(0)$, where $\theta_x = \omega T_x = \omega x/v_{sat}$.

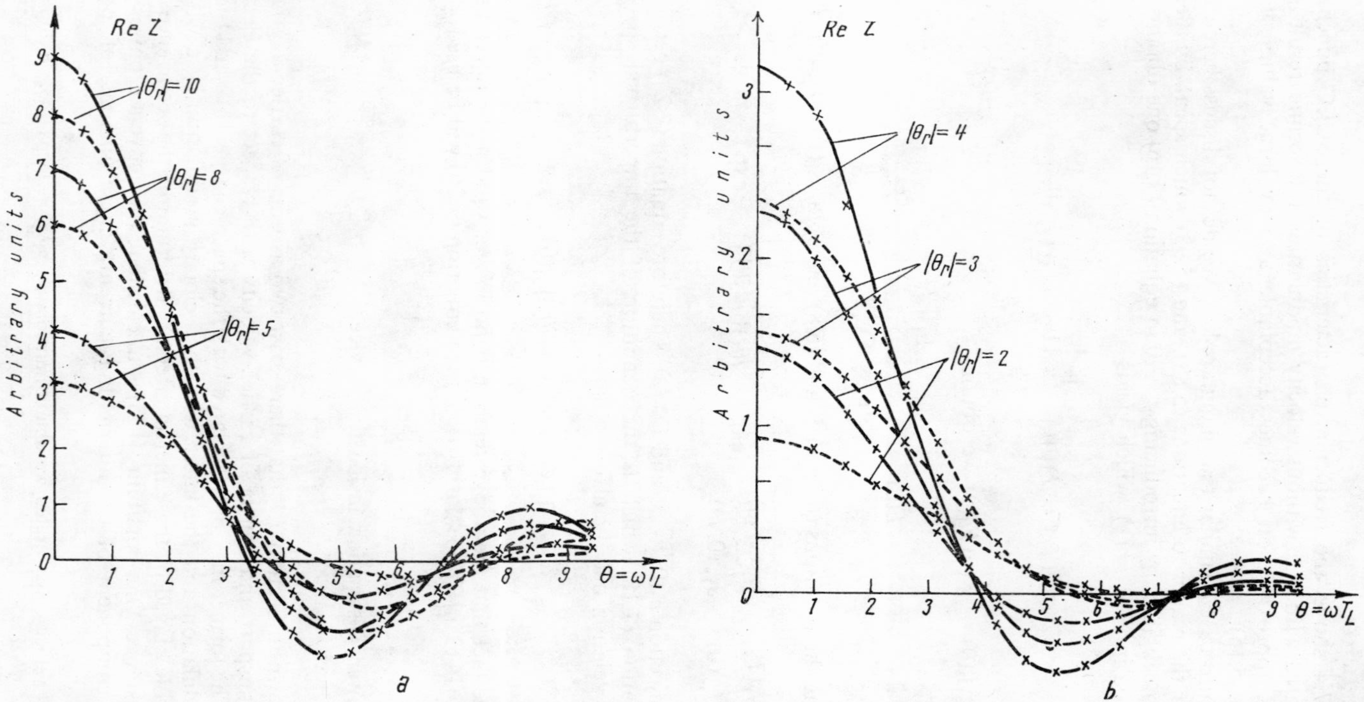

Figure 7.4. The frequency characteristics (series resistance versus transit-angle) of the punch-through diode calculated with (full curves) and without (dashed curves) velocity modulation (communicated by N. Marin).

denoted by $P_{(a)p}$, $P_{(a)v}$ and $P_{(a)d}$ respectively. By definition we have (the asterisk denotes the conjugate of a complex quantity)

$$P_{(a)p} = \mathcal{Re}\, P_p(\omega) = \mathcal{Re}\int_0^L \underline{\mathbf{J}}_{1p}^* \underline{\mathbf{E}}_1 \, \mathrm{d}x \tag{7.28}$$

$$P_{(a)v} = \mathcal{Re}\, P_v(\omega) = \mathcal{Re}\int_0^L \underline{\mathbf{J}}_{1v}^* \underline{\mathbf{E}}_1 \, \mathrm{d}x \tag{7.29}$$

$$P_{(a)d} = \mathcal{Re}\, P_d(\omega) = \mathcal{Re}\int_0^L \underline{\mathbf{J}}_{1d}^* \underline{\mathbf{E}}_1 \, \mathrm{d}x = \mathcal{Re}\int_0^L (-i\omega\, \underline{\mathbf{E}}_1^*)\, \underline{\mathbf{E}}_1 \, \mathrm{d}x. \tag{7.30}$$

The active power dissipated by the displacement current is zero, as expected. We shall use again the transport coefficients introduced in chapter 4. By taking into account equations (7.12), (7.15), (7.16), (7.28) and (7.29) one obtains

$$P_{(a)p} = \varepsilon\, |\underline{\mathbf{J}}_1|^2\, \mathcal{Re}\int_0^L F \frac{\mathrm{d}F^*}{\mathrm{d}T_x} \mathrm{d}x \tag{7.31}$$

$$P_{(a)v} = |\underline{\mathbf{J}}_1|^2\, \mathcal{Re}\int_0^L \left(1 + i\omega\varepsilon F^* - \varepsilon \frac{\mathrm{d}F^*}{\mathrm{d}T_x}\right) F\mathrm{d}x = |\underline{\mathbf{J}}_1|^2\, \mathcal{Re}\int_0^L \left(1 - \varepsilon \frac{\mathrm{d}F^*}{\mathrm{d}T_x}\right) F\mathrm{d}x\,. \tag{7.32}$$

Here, F is a simple function of the transport coefficients, as shown for example by equation (7.13) (which is valid for injection from a non-ohmic cathode).

We shall consider again the 'ideal' SCLC solid-state diode (with a square-law $J - V$ characteristic). After some algebraic manipulations one obtains (Problem

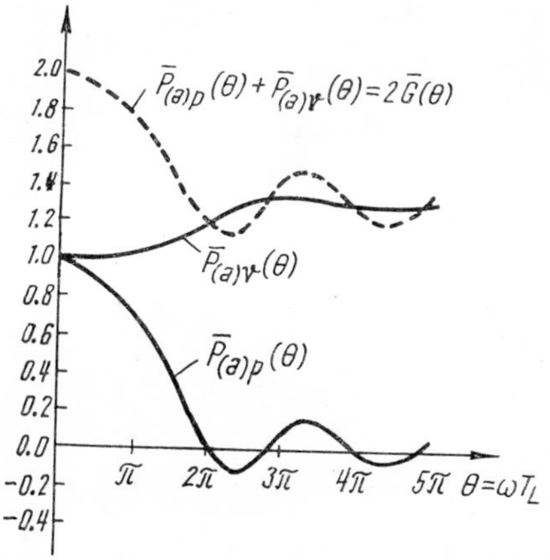

Figure 7.5 Active power dissipation inside an ideal SCLC solid-state diode ($N = 0$, $\mu =$ const.) as a function of frequency. Full curves represent frequency dependence of the power dissipated by carrier density modulation and carrier velocity modulation. The dashed curve is the sum of the above two, corresponds to the total power and is proportional to the high frequency conductance of the device [125]

7.2) $P_{(a)p}$ and $P_{(a)v}$. These functions are plotted versus θ by assuming that as ω changes the a.c. voltage amplitude is held constant. A normalization to the low-frequency ($\theta \to 0$) value was also made. The results are shown in Figure 7.5. We note the fact that the sum $P_{(a)p} + P_{(a)v}$ is the total a.c. power dissipated and should by proportional to the device (parallel) *conductance* ($\underline{\mathbf{V}}_1 =$ fixed). We find now that the oscillations of $G = G(\omega)$ are due to the carrier density modulation (or to the propagation of the space-charge wave). These oscillations of $G = G(\omega)$ for an SCLC diode were discussed in Section 5.4. Kroemer [83] demonstrated that if diffusion is included in calculations of this kind, these oscillations will be strongly attenuated. Now we are able to understand that diffusion has a major effect upon the propagation of the space-charge wave. In Appendix to Section 5.6 it was shown that if we consider a finite carrier density at the cathode*), (which is a more realistic boundary condition for SCLC injection), the carrier density modulation at the cathode should vanish. Although carrier density modulation will exist in the bulk, we have no longer an *injected* space charge wave ($\underline{\mathbf{J}}_{1p}(0) = 0$) in the true sense of the word.

Another observation is that the diode conductance at frequencies much higher than the transit-time frequency is approximately due to the carrier velocity modulation and is almost constant. In the square-law diode ($v = \mu E$) the velocity modulation will always dissipate power ($\underline{\mathbf{v}}_1$ in phase with $\underline{\mathbf{E}}_1$) and the v.h.f. conductance is always positive. We may expect a v.h.f. negative conductance for certain negative mobility devices and this is indeed confirmed by the theory developed in Section 6.6. We shall return to this later. The reader has noticed, of course, the fact that $P_{(a)v}$ is not constant, but varies with frequency. This effect is due to the redistribution of the a.c. electric field.

There is another interesting result disclosed by Figure 7.5. There are frequency ranges where the carrier density modulation *generates* a.c. power ($P_{(a)p} < 0$). This is, of course, a transit-time effect and takes place when $\underline{\mathbf{p}}_1$ acquires a sufficient phase delay with respect to the electric field, equation (7.20). This fact is very important and deserves a special discussion. We shall go in more detail and represent the *distribution* of power dissipated by $\underline{\mathbf{J}}_{1p}$ and $\underline{\mathbf{J}}_{1v}$ inside the conduction space. Such a distribution is plotted in Figure 7.6 for the SCLC square-law diode, at a transit angle of 2π. The total power generated by each component is proportional to the area between the curve and the axis (the integral). The density modulation *generates* a.c. power in the anode region†) but dissipates in the cathode region (where the delay of the space-charge wave is still small). At $\theta = 2\pi$ these amounts of power balance each other and $\underline{\mathbf{J}}_{1p}$ dissipates no power. By increasing the frequency, the anode region will increase while the cathode region will decrease and will absorb little power: consequently, $\underline{\mathbf{J}}_{1p}$ will generate a.c. power. But this is only an *internal* effect: the overall conductance is positive due to the losses accompanying velocity modulation. If we want an external negative-resistance, we will try to suppress this velocity modulation.

At frequencies much higher than the transit-time frequency, the $\underline{\mathbf{J}}_{1p}(x)$ component will alternately dissipate and generate power as x increases from cathode to anode.

---

*) The amplitude of free carrier density variation at the cathode will be also finite, in contrast to the situation depicted in Figure 7.3.

†) In a certain sense we have here a mechanism similar to that observed in a travelling-wave tube but, in contrast to this case, the charge carriers in these solid-state devices interact with the electric field given by their own space charge.

The total power is likely to oscillate and decrease in amplitude with increasing frequency (see above).

At this point it is interesting to consider the behaviour of the vacuum diode*). This device *does* exhibit an external negative resistance (Figure 4.2b) [74], [84], [163]. It is a common place in literature the assertion that this is an effect of velocity inertia

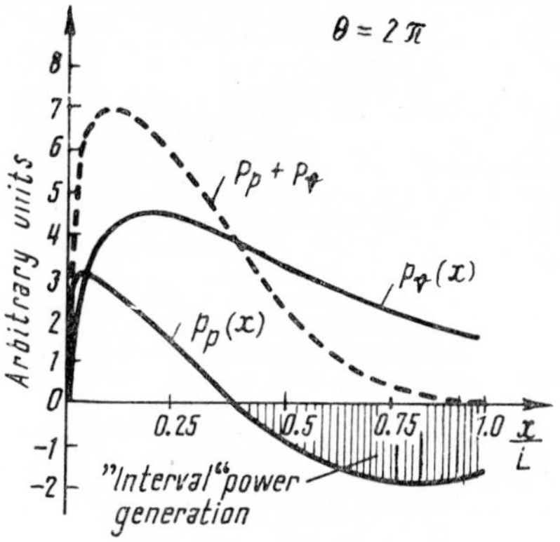

Figure 7.6. Distribution of power dissipation with the distance from the cathode in the ideal SCLC solid-state diode, for $\theta = 2\pi$ (the signal period is equal to the transit-time). Full curves represent the power distribution for the $\underline{\mathbf{J}}_{1p}$ component (p_p) and the $\underline{\mathbf{J}}_{1v}$ component (p_v). The dashed curve is the sum of the above two and gives (in arbitrary units) the distribution of the total active power [125].

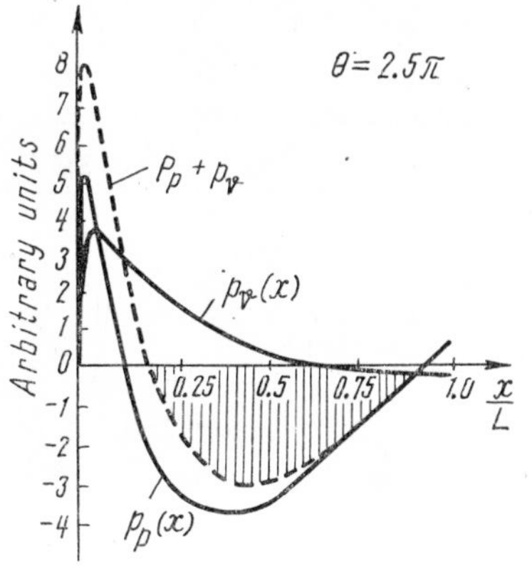

Figure 7.7. Distribution of power dissipation in the planar SCLC vacuum diode, at $\theta = 2.5\pi$ [125] (see also the previous figure).

with respect to the electric field variations [10], [16]. Because such an inertia does not exist in solids (at least in materials like Si and Ge, at usual frequencies and temperatures), we do not have special reasons to expect an h.f. negative resistance in unipolar SCLC diodes. Shockley [14], [16] tried to overcome this difficulty by imagining more complex devices in which it is possible to acquire the desired delay between the current and the voltage.

Now if we represent the distribution of power dissipated inside the vacuum diode by $\underline{\mathbf{J}}_{1p}$ (however, for electrons instead of holes) and $\underline{\mathbf{J}}_{1v}$ at the transit-angle $\theta = 2.5\,\pi$ (maximum absolute value of the series negative resistance), the situation shown in Figure 7.7 will be obtained. We see that the almost entire power genera-

*) For a detailed comparison between the h.f. operation of the square-law SCLC solid-state diode and the planar vacuum diode, see [82] and [142].

tion inside the device is due to the carrier density modulation and not to the carrier velocity modulation! In other words, the negative resistance of the SCLC vacuum diode ($\theta = 2.5\,\pi$) is practically the result of the phase shift introduced by carrier transport and not the result of the velocity inertia. However, it is not quite so. If we compare Figure 7.6 and Figure 7.7, we shall notice the fact that, due to electron

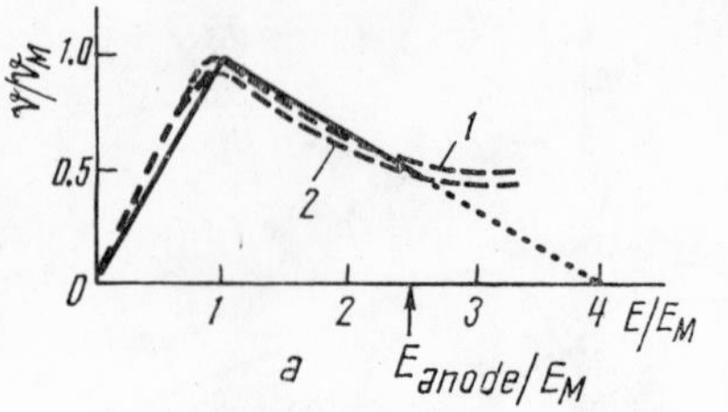

Figure 7.8. (a) Piecewise linear velocity field dependence as an aproximation of experimental (1) and theoretical (2) data (reproduced by Kroemer, IEEE Spectrum, vol. 5, pp. 47 – 56, 1968) for electrons in GaAs. Computations for (b) and (c) are performed for a maximum field (E_{anode}) equal to 2.5 times the threshold value (E_M), (b) Distribution of the active a.c. power dissipated inside the device by the $\underline{J}_{1p}$ component (p_p), the $\underline{J}_{1v}$ component (p_v) and the total particle current ($p_p + p_v$) at very low frequencies; (c) The same for $\theta = 2\pi$ (at the transit-time frequency).

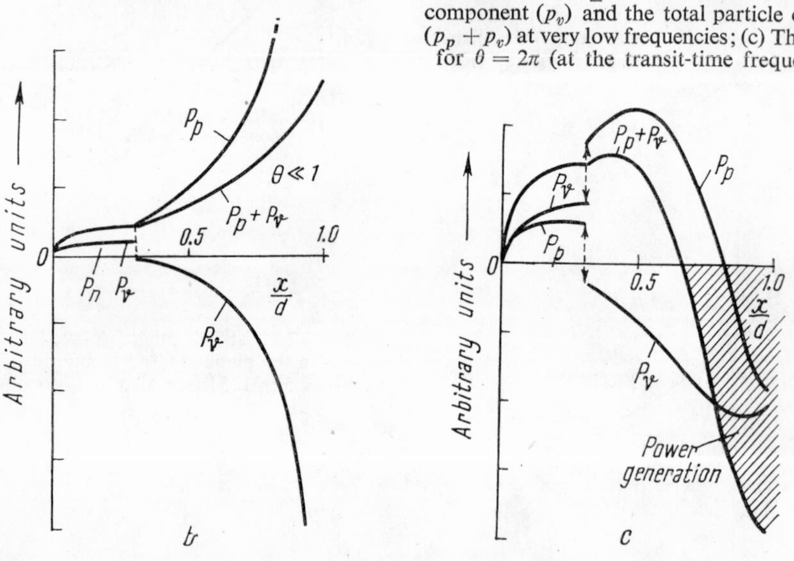

velocity inertia, the amount of power dissipated by $\mathbf{J}_{1v}$ decreases considerably and therefore this inertia effect is also important for the h.f. power generation.

We shall give another example illustrating the importance of transit-time effects in general and of the carrier density modulation in particular. Let us consider the negative mobility amplifier with zero ion density studied by Kroemer [83] (Section 5.4). The low-frequency resistance, R_i, is positive even when the device is biased in the negative-mobility region of the $v = v(E)$ characteristic. If V_0 increases the carrier velocity decreases but the mobile charge increases and J_0 still increases ($R_i > 0$). If the applied signal changes sufficiently fast, the injected mobile charge cannot follow the signal and a negative resistance effect is possible. This qualitative explanation is due to Kroemer [117]. A quantitative analysis published subsequently [164] substantiates these simple arguments. The power distribution inside the device at $\theta \to 0$ (Figure 7.8b) indicates power generation by $\underline{\mathbf{J}}_{1v}$ in the anode region, where

$\mu_d = \frac{dv}{dE} < 0$, as expected. The $v = v(E)$ characteristic is approximated as shown in Figure 7.8a (for GaAs) and the anode field is 2.5 times the threshold field. Figure 7.8c represents the power distribution calculated for the transit-time frequency ($\theta = 2\pi$), where the device resistance is negative. The total power dissipated by carrier density modulation has decreased (as compared to $\theta \to 0$). There is even a region (near the anode) where $\underline{\mathbf{J}}_{1p}$ generates a.c. power. Note also that $p_v = p_v(x)$ has changed its form, which denotes a redistribution of the a.c. field. To conclude, the external negative resistance of this device is a combination of negative mobility and transit-time effects.

7.5 Negative Resistance Devices

The preceding analysis demonstrated the complexity of the small-signal behaviour of unipolar injection currents in solid-state regions. The total alternating current, $\underline{\mathbf{J}}_1$, has three components, each of them depending upon the other two (through the continuity equation) and upon the steady-state quantities. The mechanism of the diode response can be analysed and the sources of a.c. power dissipation (or generation) can be identified, thus explaining the h.f. negative resistance effects.

If the velocity variation is in phase with the electric field

$$\underline{\mathbf{v}}_1 = \mu_d \underline{\mathbf{E}}_1 \quad , \quad \mu_d > 0 \tag{7.33}$$

the velocity modulation dissipates a.c. power and is, therefore, undesirable for negative resistance devices. If there is a sufficient delay between $\underline{\mathbf{v}}_1$ and $\underline{\mathbf{E}}_1$ $\left(\text{from } -\frac{\pi}{2} \text{ to } +\frac{\pi}{2}\right)$ $\underline{\mathbf{J}}_{1v}$ generates a.c. power. In both cases, active power may be generated *at certain frequencies*, by the carrier-density modulation.

Let us first discuss the SCLC solid-state diodes. We shall take again as reference the square-law diode. Here, the fixed space-charge is zero and the carrier mobility is constant. Let us first discuss the effect of the fixed space charge.

(a) Majority carrier injection (the SCLC resistor, $p^+\pi p^+$ or $n^+\nu n^+$). At high bias currents the $J - V$ characteristic is parabolic ($J_0 \propto V_0^2$). Both $\underline{\mathbf{J}}_{1v}$ and $\underline{\mathbf{J}}_{1p}$ are present (at $\theta = 0$ these components are equal). As the bias current decreases, the mobile charge injected *in excess* of the thermal equilibrium charge *decreases*. It can be shown that $\underline{\mathbf{J}}_{1p}$ also decreases*). At vanishingly small bias currents, the conduction is ohmic an no carrier-density modulation exists. We have shown above that the oscillations of $G = G(\omega)$ are due to carrier-density modulation. Figure 5.13 shows indeed

*) The a.c. carrier density decreases because the transit-time increases and becomes comparable to the dielectric relaxation time.

that the oscillations of the parallel conductance decrease in amplitude and then disappear as the bias current decreases and becomes very small.

(b) Minority carrier injection (the punch-through diode). The high-current region of the steady-state characteristic is parabolic again. We have already shown that at low currents, the mobile charge becomes negligible and therefore the carrier-velocity modulation becomes negligible (see equation (7.23) and Figure 7.4). This is a favourable effect because $\underline{\mathbf{J}}_{1v}$ is a source of the a.c. power loss. The h.f. resistance becomes negative due to the a.c. power generated by the $\underline{\mathbf{J}}_{1p}$ component. Thus, the negative resistance of the punch-through diode is a typical transit-time effect.

(c) Slow shallow trapping (see Sections 3.13 and 5.10). The $J - V$ characteristic is of the form $J \propto \delta V^2$ where δ is a trapping factor. If δ is small as compared to unity, $\underline{\mathbf{J}}_{1v}$ is small as compared to $\underline{\mathbf{J}}_{1p}$ (Section 7.3). The dynamic resistance becomes negative again, due to the a.c. power generated by the carrier-density modulation. We stress the fact that the negative resistance of the SCLC diode with shallow traps is a pure transit-time effect. Trapping by itself does not represent a 'power source' and plays only an indirect rôle [92], [143].

We shall now discuss the effect of mobility field-dependence. We assume for simplicity a zero fixed space-charge ($N = 0$). If the a.c. mobility μ_d is positive, the SCLC solid-state has a dynamic resistance which is always positive*). However, the resistance may become negative if the differential mobility is negative. As shown at the end of the previous Section, the a.c. power generation is due to both transit-time and velocity inertia effects. The velocity-modulation component generates power in that region of the semiconductor bulk where μ_d is negative. Nevertheless this generation is not efficient because $\underline{\mathbf{J}}_{1v}$ dissipates power in the cathode region, where the steady-state ranges from zero (at the cathode) to the threshold field. This region can either be made thinner by using a non-uniformly doped semiconductor diode (Section 5.8) or it can be suppressed by using a non-ohmic injecting cathode (the cathode field may now exceed the threshold field).

We have already shown that if the SCLC semiconductor diode ($N \neq 0$, positive or negative) is biased at a very high steady-state field and the carrier velocity is saturated, the real part of the impedance will be non-negative (theoretically it reaches zero for $\theta = 2\pi, 4\pi, \ldots\ldots$). The $\underline{\mathbf{J}}_{1p}$ component generates (at high θ) a.c. power in a certain part of the semiconductor but dissipates in the rest and the *total* dissipated power is positive or zero. However, a net generation of a.c. power is possible in a high-field (saturated velocity) semiconductor diode, as demonstrated by recent experiments [127], [128], [156]. This discrepancy may be due to the incomplete velocity saturation (in practical punch-through diodes) as well as to the effect of diffusion which ought to be important at low bias currents. The low field cathode region of this device should be more accurately described from the theoretical point of view [128]. The additional delay between the a.c. current and voltage (delay required to explain the negative resistance), may originate from this cathode region. A detailed analysis of the a.c. power distribution should indicate more power loss in the vicinity

---

*) However, the punch-through diode ($N < 0$) with positive field-dependent mobility μ_d exhibits a negative resistance effect.

of the cathode. However, the overall effect of a more realistic description of the cathode region may be favourable to a negative resistance effect. The explanation lies in the modification of the phase relation between the a.c. carrier density and the a.c. electric field and therefore in the modification of the a.c. power dissipated by $\underline{\mathbf{J}}_{1p}$ (density modulation).

An important redistribution of the a.c. power dissipated by $\underline{\mathbf{J}}_{1p}$ can be achieved by using a non-ohmic (or imperfect) cathode, instead of an Ohmic cathode, which provides SCL injection. This can be shown analytically for a diode with negligible velocity modulation by using equations (7.24), (7.25) and (7.31). The total a.c. power is equal to the power dissipated by $\underline{\mathbf{J}}_{1p}$ and may be written as

$$P_{(a)p} = |\underline{\mathbf{J}}_1|^2 \left\{ \int_0^L \frac{\sin \omega T_x}{\omega \varepsilon} \,\mathrm{d}x + \mathcal{R}e \int_0^L F(0) \left[\exp\left(i\,\omega\, T_x\right)\right] \mathrm{d}x \right\}. \tag{7.34}$$

The second term of the above sum is zero for SCL injection ($\sigma_1(0) \to \infty$, $F(0) \to 0$) and, therefore, it represents the effect of the finite a.c. resistance of the injecting contact*). Let us refer again to the saturated-velocity transit-time diode. Here $\mathrm{d}x = v_l \mathrm{d}T_x$ and the first term of the a.c. power equation (7.34), cannot be negative. On the other hand, the second term of equation (7.34) (where $F(0)$ is complex, see equation (7.25)) can be negative at certain frequencies.

We have also shown that a negative resistance effect can be obtained by using a non-uniform doping (Section 5.8) or a convergent geometry (Section 5.9). The details of the device behaviour may be studied by using the method given in this chapter. In both cases the a.c. power is generated by density modulation and is a pure transit-time effect.

Therefore, a *transit-time* negative resistance can be achieved in positive mobility and negative mobility devices with ohmic or non-ohmic injecting contacts. This negative resistance occurs (or has a maximum absolute value) at frequencies comparable to the transit-time frequency. The transit-time effects connected to the propagation of 'space-charge wave' are essential. The a.c. power generation associated with this wave is strongly frequency dependent. This generation process is not efficient, because the electric field is determined, at least in part, by the a.c. component of the injected charge. We shall explain this below. The $\underline{\mathbf{J}}_{1p}$ component generates a.c. power if a sufficient phase delay is provided between the a.c. component of the mobile charge and the local electric field (see equation (7.28)). The electric field is†)

$$\underline{\mathbf{E}}_1(x) = \underline{\mathbf{E}}_1(0) + \frac{e}{\varepsilon} \int_0^x \underline{\mathbf{p}}_1(x)\,\mathrm{d}x \tag{7.35}$$

---

*) For $\sigma_1(0) \to \infty$, we have $\underline{\mathbf{J}}_1 = \underline{\mathbf{j}}_1(0)$. Because for a non-ohmic contact $\sigma_1(0) =$ finite, the second term in equation (7.34) may be also considered as the effect of the delay occurring between the injected particle current $\underline{\mathbf{j}}_1(0)$ and the total alternating current.

†) This relation is obtained by integrating the Poisson equation. For SCLC injection we re-obtain equation (7.20), because $\underline{\mathbf{E}}_1(0) = 0$.

and has two components (hence two components of equation (7.28) can be distinguished). The first one is the cathode field whereas the second component is the result of the a.c. mobile charge. The required phase delay between $\underline{\mathbf{J}}^*_{1p} = e\, v_0\, \underline{\mathbf{p}}^*_1(x)$ and the second component of equation (7.35) cannot be provided in the entire interelectrode space, even if the frequency is properly chosen (in the immediate vicinity of the cathode this component will always give a power loss). In this sense the above transit-time mechanism is not too efficient. If the current is SCL, it is only this second term of (7.35) which exists (see Section 7.4). The opposite case

$$\underline{\mathbf{E}}_1(x) \simeq \underline{\mathbf{E}}_1(0) \quad , \quad \underline{\mathbf{v}}_1 \simeq \underline{\mathbf{E}}_1(0) \, . \, L \tag{7.36}$$

(negligible effect of the injected charge) is also possible. If the space-charge wave is injected with a suitable delay and experiences little modification (small transit-time effects in *this* region), then the density modulation can generate power everywhere in the considered region. This is the principle of the structures called here 'with delayed injection' (Section 6.7).

In positive mobility materials the velocity modulation dissipates a.c. power and is, therefore, undesirable. Several devices which reduce the effect of this component have been discussed previously. On the other hand, the same component generates a.c. power in negative mobility semiconductor regions. The negative resistance of a negative mobility diode with either ohmic or non-ohmic injecting contact is the result of both transit-time and negative mobility effects [164].

At frequencies higher than the transit-time frequency, the transit-time effects become less important. We have shown in Section 6.6 that if the signal period is much shorter than the carrier transit-time, the bulk relaxation time and the cathode relaxation time, then the device exhibits a constant (frequency independent) conductance, which is shunted by the geometrical capacitance $C_0 = \varepsilon/L$. If the differential mobility, μ_d may be approximated as being constant, this v.h.f. conductance will be given by equation (6.53), and will be proportional to μ_d. We have already shown (Section 7.4) for the square law SCLC diode that this v.h.f. conductance is due precisely to the velocity modulation. We can obtain a v.h.f. *negative* conductance due to the velocity modulation in a negative mobility diode, if the semiconductor is properly biased. This idea is exemplified in Section 6.6, by studying ECL injection.

We do not discuss here the effect of the mobility relaxation time or of the intervalley scattering time. The same method of detailed analysis can be applied if these effects are taken into account.

To conclude, the method of analysis developed in this chapter allows a better understanding of the a.c. response mechanism of charge-controlled devices. We have seen how the injected space-charge wave produces a modulation of the bulk electric field and therefore a modulation of the carrier velocity. On the other hand, the velocity modulation determines a 'supplementary' displacement of the electric charge, and this displacement 'perturbs' the space-charge wave, etc. By separating the two components of the particle current it is possible to get an intuitive representation of transit-time and carrier-inertia effects.

An important result of this analysis is the detailed explanation of the h.f. negative resistance calculated for a number of devices. A good understanding of all phenomena taking place in these negative resistance elements is necessary in order to exploit the various capabilities of solid-state devices, as well as to improve the efficiency of the already existing structures.

Problems

7.1. Show that the 'f' functions for the square-law unipolar diode are given respectively by

$$f_p = \frac{1}{\theta_x}\left[\sin\theta_x - \frac{1}{\theta_x}(1 - \cos\theta_x)\right] + i\frac{1}{\theta_x}\left(\cos\theta_x - \frac{\sin\theta_x}{\theta_x}\right) \tag{P.7.1}$$

$$f_v = \frac{1}{\theta_x^2}(1 - \cos\theta_x) + i\frac{1}{\theta_x}\left(\frac{\sin\theta_x}{\theta_x} - 1\right) \tag{P.7.2}$$

$$f_d(\theta_x) = 1 - \frac{\sin\theta_x}{\theta_x} + i\frac{1}{\theta_x}(1 - \cos\theta_x) \tag{P.7.3}$$

where $\theta_x = \omega T_x$ is the 'partial' transit angle.

7.2. Calculate the a.c. power dissipated by the $\underline{\mathbf{J}}_{1p}$ and $\underline{\mathbf{J}}_{1v}$ components in the square-law diode (Figure 7.5).

7.3. The small-signal admittance may be imagined as a parallel combination of three 'partial' admittances, corresponding respectively to the components of the alternating current. For the vacuum diode these admittances are

$$Y_p(\theta) = \frac{f_p(\theta)}{Z(\theta)} = \frac{2}{\theta^2}\left(2\cos\theta + \theta\sin\theta - \frac{2\sin\theta}{\theta}\right) + i\frac{2}{\theta^2}\times$$

$$\times\left[\frac{2}{\theta}(1 - \cos\theta) + \theta\cos\theta - 2\sin\theta\right] \tag{P.7.4}$$

$$Y_v(\theta) = \frac{f_v(\theta)}{Z(\theta)} = \frac{2}{\theta^2}\left[\frac{2\sin\theta}{\theta} - 1 - \cos\theta\right] + i\frac{2}{\theta_x^2}\left[\sin\theta + \frac{2(\cos\theta - 1)}{\theta}\right] \tag{P.7.5}$$

$$Y_d(\theta) = \frac{f_d(\theta)}{Z(\theta)} = 1 + \frac{2}{\theta}\left(\frac{1 - \cos\theta}{\theta} - \sin\theta\right) + \frac{2i}{\theta^2}(\sin\theta - \theta\cos\theta)\,. \tag{P.7.6}$$

Discuss the high-frequency negative resistance of this device by using the 'partial' admittances defined above.

7.4. Explain the discontinuity of curves plotted in Figure 7.8, b and c.

7.5. Calculate the components of the a.c. particle current for ECL injection, by assuming a negligible fixed space-charge. Discuss the following special cases: a) Very small bias current; b) Saturated velocity.

7.6. Calculate and plot the distribution of the active a.c. power dissipated in a saturated-velocity diode and show the effect of the a.c. boundary condition (i.e. of the particular injection mechanism involved).

8

Transit-time Effects in Field-effect Devices

8.1 Simplified Model of the Insulated-gate Field-effect Transistor (IGFET)

A field-effect device cannot be described by a one-dimensional model. Consequently, the transport coefficients derived in Chapter 4 are not suitable for studying the transit-time effects in such devices.

Due to the lack of space we cannot discuss here all the details of the problem indicated by the title of this chapter. One possibility is to discuss briefly the most important models for various devices (junction-gate and insulated-gate field-effect triodes and tetrodes). Another possibility is to concentrate on a single model and discuss it as completely as possible. We have chosen the second alternative. This chapter deals with the small-signal theory of a simplified model of an IGFET*). An exact mathematical solution for the small-signal problem will be found and used to derive equivalent circuits and frequency characteristics. For the junction-gate FET the analytical approach and the final results are similar.

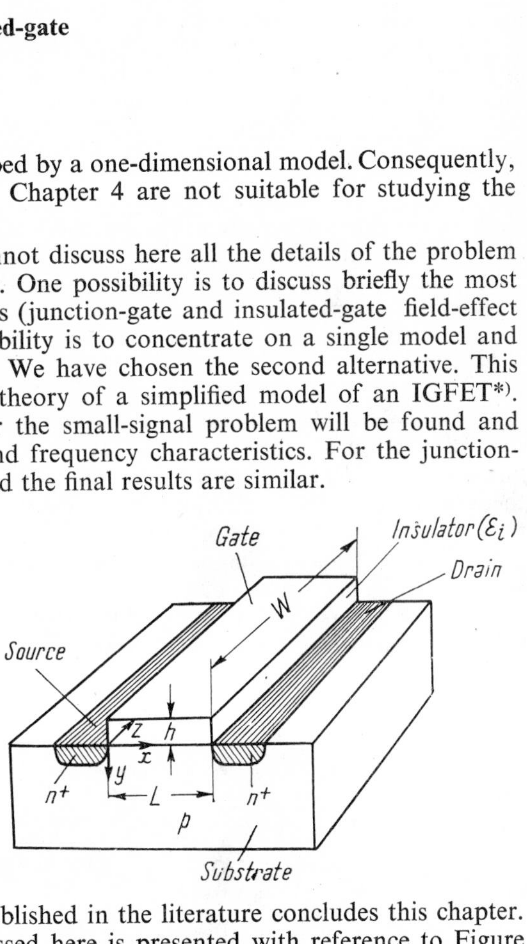

Figure 8.1. Sketch of an IGFET structure indicating the notations used in this chapter.

A brief review of the results published in the literature concludes this chapter.

The mathematical model discussed here is presented with reference to Figure 8.1. The mobile carriers are injected by the source and collected by the drain, which

*) The same device is discussed briefly in Chapter 9 (noise properties) and Chapter 12 (transient behaviour).

is properly biased with respect to the source (at ground potential). The electrical conduction takes place at the semiconductor surface. The magnitude of the mobile charge and, therefore, the source-drain current is controlled by the potential applied on the third electrode, called the gate, which is separated from the semiconductor surface by a thin insulating layer.

We assume the fact that the conductive layer at the semiconductor surface (the channel) is extremely thin as compared to the insulator thickness h. Therefore, the entire structure resembles a plane-parallel capacitor, with a capacitance

$$C_i = \varepsilon_i / h \tag{8.1}$$

per unit of gate-electrode area (ε_i is the insulator permittivity). This is really the case if the source-drain space L is much larger than h

$$L \gg h \tag{8.2}$$

because the drain potential $V_D(t)$ does not distort appreciably the electric field into the above-mentioned capacitor. In other words, the longitudinal electric field E_x may be assumed to be very small as compared to the transversal component E_y i.e.

$$E_x(t) \ll E_y(t) \; . \tag{8.3}$$

This is Shockley's 'gradual approximation' [12], which avoids an exact mathematical solution of the two-dimensional Poisson-equation (for both semiconductor and insulator). E_y and E_x will be determined separately, the former from the Gauss theorem applied to the metal-insulator-semiconductor system, the latter by writing the continuity equation along the channel.

Let $V(x, t)$ be the channel potential at the distance x from source $(x = 0)$, and $V_G(t)$ the gate potential (the source-electrode is taken as the reference). We denote by σ_s the surface density of the mobile charge in the channel, and by σ_{s_I} the fixed space-charge per unit of interface area. By the Gauss theorem we have

$$\operatorname{div}_s \mathbf{D} = \sigma_s + \sigma_{sI} \tag{8.4}$$

or

$$E_y\big|_{y=0+} - E_y\big|_{y=0-} = \sigma_s + \sigma_{sI} \tag{8.5}$$

where $E_y(0^+) = 0$ and $E_y(0^-) = \varepsilon_i (V_G - V)/L$.

Therefore

$$\sigma_s = -C_i [V_G(t) - V_T - V(x, t)] \tag{8.6}$$

where

$$V_T = -\frac{\sigma_{sI} L}{\varepsilon_i} \tag{8.7}$$

is a threshold voltage. Assume for example an n-type conduction ($\sigma_s < 0$) in the presence of surface traps ($\sigma_{sI} < 0$). No current flows unless $V_G > V_T > 0$ because the channel contains no mobile charge. V_T is a threshold voltage of the same nature as that which occurs in insulator diodes with deep traps. If we inject electrons in an n-type thin (surface) channel, V_T is negative (because of $\sigma_{sI} > 0$, ionized donors). The second situation occurs for an IGFET with a built-in channel. The former ($V_T >$ > 0) describes either the thin-film transistor (TFT) with a large amount of traps, or the metal-oxide-semiconductor (MOS) transistor. In the n-type MOS transistor the n-channel ($\sigma_s < 0$) is an inversion layer, σ_{sI} is also negative (the ionized acceptors of the p-substrate) and $V_T > 0$. If the interface and the oxide charge have to be taken into account, these effects can be included in V_T, provided that these charges do not depend upon the applied voltages and the injected charge and do not vary with time*).

The particle current in the channel I_c is given by

$$I_c(x, t) = v(x, t)\, W \sigma_s(x, t) = \mu E_x W \sigma_s \tag{8.8}$$

and because

$$E_x = -\frac{\mathrm{d}V}{\mathrm{d}x} \tag{8.9}$$

and σ_s is given by equation (8.7), one obtains

$$I_c(x, t) = \frac{\mu \varepsilon_i W}{h} [V_G(t) - V_T - V(x, t)] \frac{\partial V(x, t)}{\partial x} \tag{8.10}$$

where I_c, which flows from the drain ($x = L$) to the source ($x = 0$), is positive by convention. Hence, the continuity equation should be†)

$$\frac{\partial(-I_c)}{\partial x} + W\frac{\partial \sigma_s}{\partial t} = 0 \tag{8.11}$$

or

$$\frac{\partial I_c(x, t)}{\partial x} + \frac{W\varepsilon_i}{h}\frac{\partial}{\partial t}[V_G(t) - V(x, t)] = 0 \tag{8.12}$$

Relationships (8.10) and (8.12) are the basic equations of our problem. They should be completed with the postulate of the drain-current saturation, as follows:

*) An accurate description of the MOST should take into account the 'fixed' space-charge of the depletion layer which occurs between the surface inversion layer and the semiconductor bulk (substrate).

†) We implicitly neglect the displacement current along the x axis (see equation (8.3)). The second term in equations (6.7) and (6.8) may be visualised as a current which charges the gatechannel capacitor of infinite small area $W\mathrm{d}x$, at the distance x from the source.

For steady-state operation ($V_D = V_{D,0}$, $V_G = V_{G,0}$) the particle current is uniform, and by integrating equation (8.10) where $I_c = I_{D,0} =$ const. from source ($x = 0$, $V = 0$) to drain ($x = L$, $V = V_{D,0}$) one obtains

$$I_{D,0} = \frac{\mu \varepsilon_i W}{hL}\left[(V_{G,0} - V_T)\, V_{D,0} - \frac{V_{D,0}^2}{2}\right]. \tag{8.13}$$

The above relationship holds only if

$$V_{D0} \leqslant V_{G0} - V_T = V_{D\text{sat}}. \tag{8.14}$$

In the opposite case the channel is depleted of mobile charge at the drain end, and equation (8.6) is no longer valid. We postulate the fact that for $V_{D,0} > V_{G,0} - V_T$ the drain current $I_{D,0}$ remains constant at

$$I_{D\text{sat}} = I_D\,(V_{G,0},\, V_{D\text{sat}}) = \frac{\mu \varepsilon_i W}{2Lh}(V_{G,0} - V_T)^2. \tag{8.15}$$

This saturation of the $I_D = I_D\ (V_D)$ characteristic ($V_G =$ const.) is not obvious, nor is it satisfied rigorously by actual devices. The saturation of the drain current is extensively discussed in the literature [165] — [169], but does not concern us here.

8.2 Small-signal Alternating-current Theory of IGFET

By introducing

$$\Psi(x,\ t) = V_G(t) - V_T - V(x,\ t) \tag{8.16}$$

equations (8.10) and (8.12) become, respectively,

$$I_c = -\ \frac{\varepsilon_i \mu W}{h}\ \Psi\ \frac{\partial \Psi}{\partial x} \tag{8.17}$$

$$\frac{\partial I_c}{\partial x} = -\ \frac{\varepsilon_i\, W}{h}\ \frac{\partial \Psi}{\partial t} \tag{8.18}$$

and may be combined to give

$$\frac{\partial}{\partial x}\left(\Psi\ \frac{\partial \Psi}{\partial x}\right) = \frac{1}{\mu}\,\frac{\partial \Psi}{\partial t} \tag{8.19}$$

or

$$\frac{\partial^2 \Psi}{\partial x^2} = \frac{2}{\Psi} \frac{\partial \Psi}{\partial t} \tag{8.20}$$

which is a non-linear differential equation with partial derivatives.

In order to solve the small-signal problem we shall use a perturbation method similar to that discussed in Section 4.5. Let us introduce

$$\Psi(x, t) = \Psi_0(x) + \underline{\Psi}_1(x) \exp i\omega t, \ |\underline{\Psi}_1| \ll \Psi_0 \tag{8.21}$$

(where $\underline{\Psi}_1$ is a small, first-order quantity) into equation (8.19) and then let us equalize separately the zero-order and first-order quantities. One obtains, respectively,

$$\frac{\mathrm{d}}{\mathrm{d}x}\left(\Psi_0 \frac{\mathrm{d}\Psi_0}{\mathrm{d}x}\right) = 0 \tag{8.22}$$

$$\frac{\mathrm{d}}{\mathrm{d}x}\left(\underline{\Psi}_1 \frac{\mathrm{d}\Psi_0}{\mathrm{d}x} + \Psi_0 \frac{\mathrm{d}\underline{\Psi}_1}{\mathrm{d}x}\right) = \frac{i\omega}{\mu} \underline{\Psi}_1 . \tag{8.22'}$$

The d.c. or steady-state solution is known already from the preceding Section. We do need, however, $\Psi_0 = \Psi_0(x)$ from equation (8.22), to introduce it into equation (8.22′).

Equation (8.22) may be integrated twice to give

$$\Psi_0^2 = C_1 x + C_2. \tag{8.23}$$

The integration constants C_1 and C_2 should be determined by using the following boundary conditions

$$x = 0, \quad \Psi_0(0) = V_{G,0} - V_T \tag{8.24}$$

$$x = L, \quad \Psi_0(L) = V_{G,0} - V_T - V_{D,0}. \tag{8.25}$$

After some algebraic manipulation one obtains

$$\Psi_0(x) = (V_{G,0} - V_T)(1 - ax/L)^{1/2} \tag{8.26}$$

where

$$a = \frac{I_{D,0}}{I_{D\,\mathrm{sat}}} = a(V_{G,0}, V_{D,0}) \tag{8.27}$$

(see equations 8.13 and 8.15).

The 'a.c.' equation, equation (8.22), becomes

$$(1 - ax/L)^{1/2} \underline{\Psi}_1'' - \frac{a}{L}(1 - ax/L)^{-1/2} \underline{\Psi}_1' - \left[\frac{a^2}{4L^2} (1 - ax/L)^{-3/2} + \frac{i\omega T_0}{L^2} \right] \underline{\Psi} 0_1 = \quad (8.28)$$

where

$$T_0 = \frac{L^2}{\mu (V_{G,0} - V_T)} \quad (8.29)$$

and $\underline{\Psi}_1'$, $\underline{\Psi}_1''$ are the first and second-order derivatives of

$$\underline{\Psi}_1 = \underline{\Psi}_1(x).$$

The general solution of the above equation may be written (see Appendix)

$$\underline{\Psi}_1 = K_1 \mathbf{J}_{2/3} \left(\frac{4}{3a} \sqrt{S'}\, z^{3/4} \right) + K_2 \mathbf{J}_{-2/3} \left(\frac{4}{3a} \sqrt{S'} z^{3/4} \right) \quad (8.30)$$

where $\mathbf{J}_n$ are modified Bessel functions of n-order [170], then

$$z = 1 - a\,(x/L) \quad (8.31)$$

$$S' = i \frac{\omega}{\omega_0}, \quad \omega_0 = \frac{1}{T_0} \quad (8.32)$$

and K_1 and K_2 are arbitrary constants. The solution to our problem will be determined by suitable boundary conditions, namely*)

$$x = 0,\ z = 1, \quad \underline{\Psi}_1 = \underline{\mathbf{V}}_{G1} \quad (8.33)$$

$$x = L,\ z = 1 - a,\ \underline{\Psi}_1 = \underline{\mathbf{V}}_{G1} - \underline{\mathbf{V}}_{D1}. \quad (8.34)$$

The constants in equation (8.30) should be given by

$$K_1 = A_1 \underline{\mathbf{V}}_{G,1} + B_1 \underline{\mathbf{V}}_{D,1} \quad (8.35)$$

$$K_2 = A_2 \underline{\mathbf{V}}_{G,1} + B_2 \underline{\mathbf{V}}_{D,1} \quad (8.36)$$

*) All electrical variables may be written as a d.c. term plus an a.c. (complex) term, i.e.

$$V_G(t) = V_{G,0} + \underline{\mathbf{V}}_{G,1} \exp i\omega t, \text{ etc.}$$

(see Chapter 4).

where A_1, A_2, B_1 and B_2 are complex quantities which depend upon the normalized frequency ω/ω_0 and the normalized bias current $I_{D,0}/I_{D\,sat}$ (through Bessel functions).

It can be shown (see Appendix) that the a.c. particle current $\underline{\mathbf{I}}_{c1}$ may be expressed as

$$\underline{\mathbf{I}}_{c1} = g_{ms}\sqrt{S'}\,z^{1/4}\left\{K_1\,\mathbf{J}_{-1/3}\left(\frac{4}{3a}\sqrt{S'}\,z^{3/4}\right) + K_2\,\mathbf{J}_{1/3}\left(\frac{4}{3a}\sqrt{S'}\,z^{3/4}\right)\right\} \tag{8.37}$$

where

$$g_{ms} = \frac{\mu\,\varepsilon_i\,W}{hL}(V_{G,0} - V_T) \tag{8.38}$$

is the device transconductance computed in saturation. Here $\underline{\mathbf{I}}_{c,1} = \underline{\mathbf{I}}_{c,1}(x)$ is determined completely as a function of the external voltages $\underline{\mathbf{V}}_{G,1}$, $\underline{\mathbf{V}}_{D,1}$, the signal frequency and the d.c. bias. We can now calculate the circuit currents $\underline{\mathbf{I}}_{D,1}$, $\underline{\mathbf{I}}_{G,1}$, $\underline{\mathbf{I}}_{s,1}$, where (Figure 8.2)

$$\underline{\mathbf{I}}_{G,1} = \underline{\mathbf{I}}_{S,1} - \underline{\mathbf{I}}_{D,1} \tag{8.39}$$

and we have

$$x = 0, \quad z = 1, \qquad \underline{\mathbf{I}}_{c,1}(0) = \underline{\mathbf{I}}_{S,1} \tag{8.40}$$

$$x = L, \quad z = 1 - a, \quad \underline{\mathbf{I}}_{c,1}(L) = \underline{\mathbf{I}}_{D,1} \tag{8.41}$$

The above relations determine the external currents as a function of the applied a.c. voltages. It is then possible to determine any small-signal parameter of the IGFET.

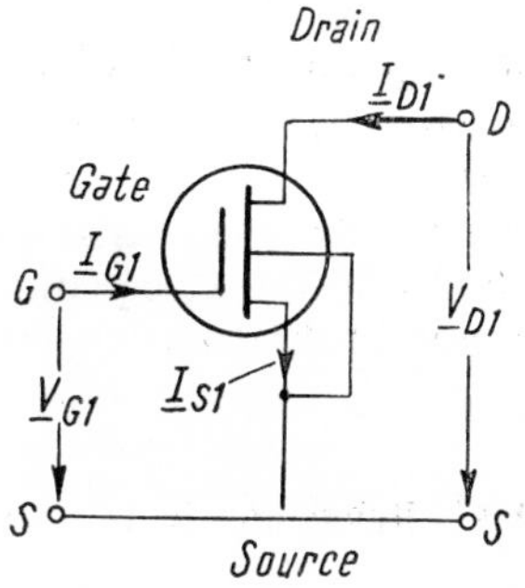

Figure 8.2. A.c. voltages and currents of the IGFET.

These a.c. parameters can be normalized with reference to the transconductance in saturation g_{ms} (see above). A suitable normalization of the angular frequency is indicated by equation (8.22). The normalized frequency characteristics should depend upon the parameter a, which is the drain current *versus* the saturation current. In saturation, $a = 1$ and the normalized frequency characteristics no longer depend upon the applied bias.

The time-constant T_0 multiplying the frequency ω in equations (8.30) and (8.37), may be written as the product of the low-current conductance

$$g_{d,0} = \left.\frac{\partial I_{D,0}}{\partial V_{D,0}}\right|_{\substack{V_{G,0} = \text{ct.} \\ V_{D,0} \to 0}} = \frac{\mu \varepsilon_i W}{hL}(V_{G,0} - V_T) = g_{ms} \tag{8.42}$$

and the total gate-channel capacitance

$$C_{gc} = WLC_i = \varepsilon_i WL/h. \tag{8.43}$$

In the theory of one-dimensional devices developed in Chapters 4—7, we found that the transit-time of charge-carriers is a very important parameter. First, the transit-time T_x was used as the independent variable when we solved the basic differential equations of the problem. However, T_x does not appear to be a suitable variable for the differential equation (8.22). This fact can be understood if one takes into account that the mobile charge injected in our transistor does not move in the electric field created by its own space-charge, as was the case in one-dimensional devices.

Secondly, the transit-angle (the angular frequency multiplied by the total transit-time T_L) which occurred systematically in the high-frequency theory of the preceding chapters, occurs no longer here. This can be proved by calculating the d.c. transit-time according to our model (Problem 8.2)

$$T_L = T_0 \frac{4}{3a^2}[1 - (1 - a)^{3/2}] = T_L(V_{G,0}, V_{D,0}). \tag{8.44}$$

The time constant T_0, used for frequency normalization in equation (8.32), is $\left(\frac{3}{4}\right)$ of the transit-time in the saturation region ($a = 1$, see equation 8.27). From equation (8.37) we see that ω *does not* appear multiplied by the transit-time (except for $a = 1$). It is the time constant $T_0 = C_{gc}\, g_{d,0}^{-1}$ (and not the transit-time T_L) which characterizes the dynamic behaviour of the IGFET. This observation may be connected with the fact that equations (8.17) and (8.18) are the same as the equations of a non-linear transmission line [171], [172].

8.3 The A.C. Behaviour in the Saturation Region

If the IGFET is biased in the saturation region, then because of equation (8.27) we shall have

$$a = 1. \tag{8.45}$$

The boundary condition (8.34) introduced in equation (8.30), requires $K_2 = 0$ because $\mathbf{J}_{-2/3}$ becomes infinite when the argument tends to zero ($x = L$, $z = 0$), and $\underline{\Psi}_1$ should be finite. K_1 is determined by the boundary condition (8.33), and $\underline{\Psi}_1$ becomes

$$\underline{\Psi}_1 = \underline{V}_{G,1} \frac{\mathbf{J}_{2/3}\left(\frac{4}{3}\sqrt{S'}\, z^{3/4}\right)}{\mathbf{J}_{2/3}\left(\frac{4}{3}\sqrt{S'}\right)}. \tag{8.46}$$

Then, equation (8.37) yields

$$\underline{I}_{c,1} = \frac{\underline{V}_{G,1}\, g_{ms}}{\mathbf{J}_{2/3}\left(\frac{4}{3}\sqrt{S'}\right)} z^{1/4}\sqrt{S'}\, \mathbf{J}_{-1/3}\left(\frac{4}{3}\sqrt{S'}\, z^{3/4}\right). \tag{8.47}$$

The high-frequency transconductance can be calculated by use of the boundary condition (8.41) and relationship (8.2.7). The result is

$$Y_{21} = \left.\frac{\underline{I}_{D,1}}{\underline{I}_{G,1}}\right|_{\text{sat}} = g_{ms} \frac{(S')^{1/3}}{(2/3)^{1/3}\, \Gamma\left(\frac{2}{3}\right) \mathbf{J}_{2/3}\left(\frac{4}{3}\sqrt{S'}\right)} \tag{8.48}$$

(see equations 8.2.7 and 8.2.8). The low-frequency ($\omega \to 0$, $S' = \omega T_0 \to 0$) series development is

$$Y_{21}(S') \simeq \frac{g_{ms}}{1 + \frac{4}{15} S' + \frac{(S')^2}{45} + \frac{8(S')^3}{8910} + \cdots}. \tag{8.49}$$

The input admittance Y_{11} will be calculated as follows

$$Y_{11} = \left.\frac{\underline{I}_{G,1}}{\underline{V}_{G,1}}\right|_{\text{sat}} = \frac{\underline{I}_{S,1} - \underline{I}_{D,1}}{\underline{V}_{G,1}} \tag{8.50}$$

where $\underline{I}_{S,1}$, $\underline{I}_{D,1}$ are given by equation (8.47) and by the boundary conditions (8.40) and (8.41). The other two Y parameters*) are zero, i.e.:

$$Y_{12} = 0 \quad , \quad Y_{22} = 0. \tag{8.51}$$

*) These parameters are defined by $\underline{I}_{G,1} = Y_{11}\underline{V}_{G,1} + Y_{12}\underline{V}_{D,1}$ and $\underline{I}_{D,1} = Y_{21}\underline{V}_{G,1} + Y_{22}\underline{V}_{D,1}$. Here we discuss the common-source connection.

This result is due to the fact that a variation of the drain-source voltage cannot affect the device currents if the device is biased in the saturation region.

The normalized frequency characteristics are represented in Figures 8.3 and 8.4, after Burns [171]. The input susceptance is capacitive. The output current $\underline{I}_{D1}$ is delayed with respect to the input voltage.

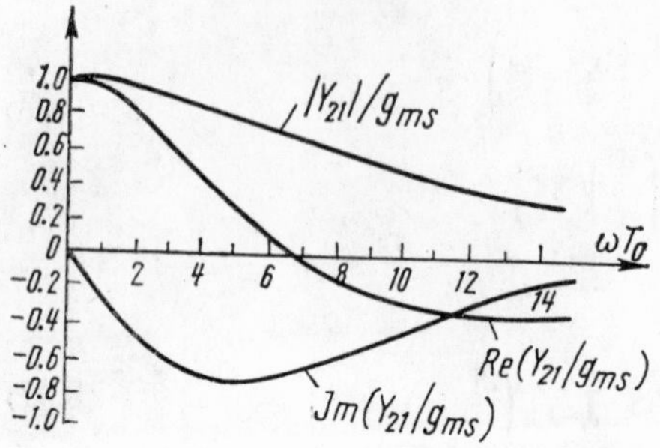

Figure 8.3. Frequency dependence of the normalized transconductance in saturation, calculated by Burns [171].

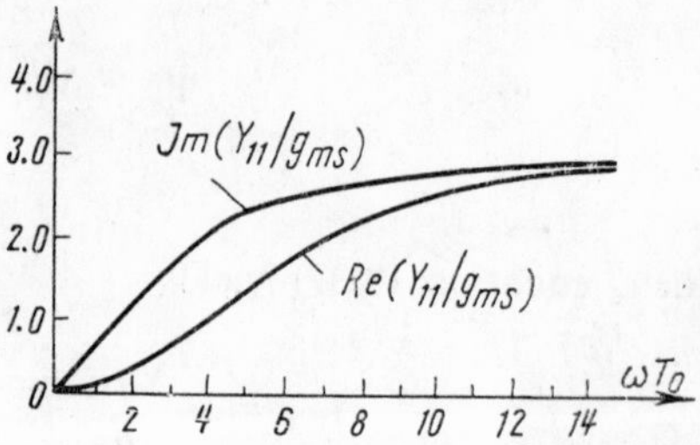

Figure 8.4. Frequency dependence of the normalized input admittance in saturation, calculated by Burns [171].

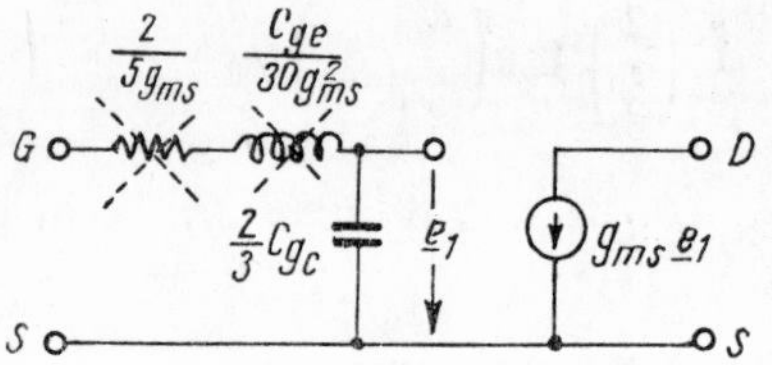

Figure 8.5. Small-signal low-frequency equivalent circuit for a device biased in the saturation region [171].

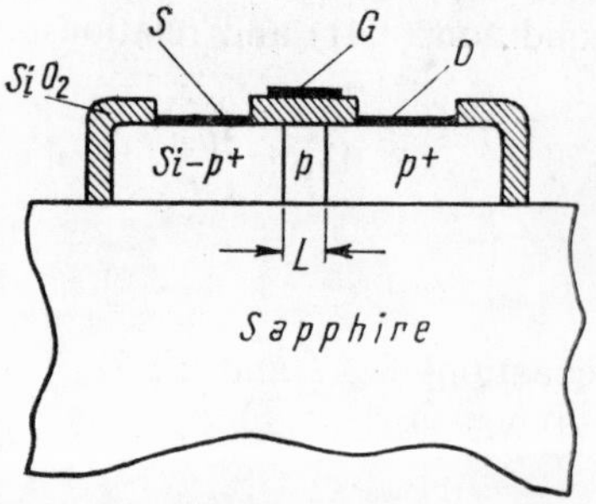

Figure 8.6. Experimental transistor structure, made from silicon grown on sapphire [173].

The low-frequency equivalent circuit can be easily obtained by calculating Y_{11} and Y_{21} for $S' \to 0$ $(\omega \to 0)$. A possible form of this circuit is shown in Figure 8.5.

It may be shown (Problem 8.4) that the input capacitance $\frac{2}{3} C_{gc}$ *) is exactly the quasi-stationary capacitance

$$C_{st} = \left. \frac{\partial Q_{mob}}{\partial V_G} \right|_{V_D = \text{const.}} \tag{8.52}$$

calculated in saturation.

*) The series resistance and series inductance in Figure 8.5 have a negligible effect at very low frequencies.

Burns [173] also demonstrated the fact that the above theory is well verified by experiment. A special experimental device was constructed (Figure 8.6). This was a silicon MOS transistor, grown epitaxially on sapphire [173]. Thus the parasitic channel-substrate and drain-substrate capacitances were eliminated. The

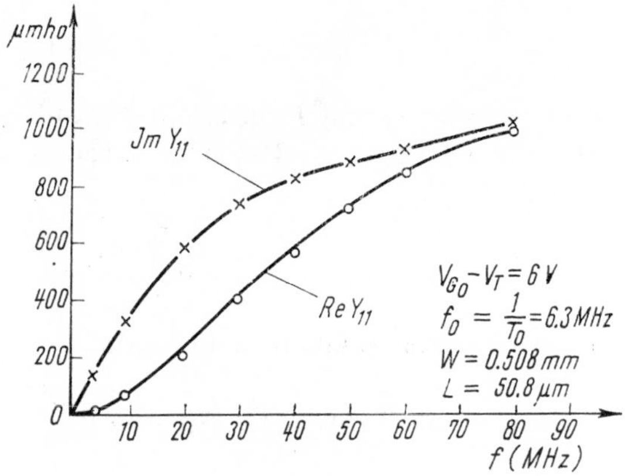

Figure 8.7. Comparison between the theoretical input impedance (Fig. 8.4) and the experimental results measured on the transistor of Figure 8.6 [173]. — = Theoretical; o,x= Experimental.

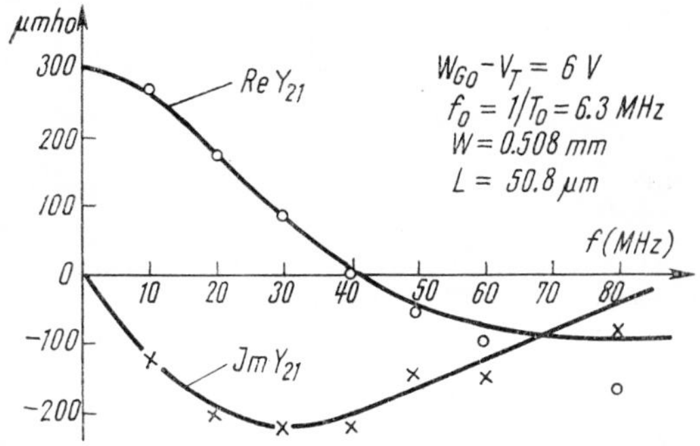

Figure 8.8. Comparison between the theoretical transconductance (Figure 8.3) and the experimental results measured on the transistor of Figure 8.6. [173]. — = Theoretical; o, x = = Experimental.

channel length L is also unusually large (25 and 50 μm) and thus the time constant equation (8.29) has a very large value as compared to the values normally encountered in commercial devices. Therefore, ωT_0 is comparable to unity at relatively low frequencies and the effect of parasitic elements such as the lead inductance and transistor-case capacitance can be neglected and accurate measurements of the *internal* reactive effects can be performed. The result of the experiments is compared with the theory in Figures 8.7 and 8.8 [173].

8.4 Small-Signal Operation in the Linear Region

We shall consider now the linear region of the $I_D - V_D$ characteristic equation (8.13), which is characterized by $I_{D,0} \to 0\,(V_{D,0} \to 0)$ and

$$a \to 0. \tag{8.53}$$

The a.c. parameters may be determined easily starting from the a.c. equation (8.28), where a is set equal to zero. This equation reduces to

$$\frac{\underline{\Psi}_1}{\mathrm{d}x^2} = \frac{i\omega\, T_0}{L^2}\underline{\Psi}_1 \tag{8.54}$$

and has a general solution of the form

$$\underline{\Psi}_1(x) = C_1\,[\exp(bx/L)] + C_2\,[\exp(-\,bx/L)] \tag{8.55}$$

where

$$b = \sqrt{i\hat{\omega}} = \sqrt{\frac{\omega}{2}}(1+i),\ \ \hat{\omega} = \omega T_0. \tag{8.56}$$

Another form of $\underline{\Psi}_1(x)$ is

$$\underline{\Psi}_1(x) = D_1 \sinh\frac{bx}{L} + D_2\cosh\frac{bx}{L}\,. \tag{8.57}$$

The integration constants D_1, D_2 can be determined as follows

$$x = 0\ \ ,\ \ \underline{\Psi}_1(0) = \underline{\mathbf{V}}_{G1} = D_2 \tag{8.58}$$

$$x = L\ \ ,\ \ \underline{\Psi}_1(L) = D_1\sinh b + D_2\cosh b = \underline{\mathbf{V}}_{G,1} - \underline{\mathbf{V}}_{D,1}\,. \tag{8.59}$$

Therefore

$$\underline{\Psi}_1(x) = \frac{1}{\sinh b}\left\{[\underline{\mathbf{V}}_{G,1} - \underline{\mathbf{V}}_{D,1}]\sinh\frac{bx}{L} - \underline{\mathbf{V}}_{G,1}\sinh b\left(\frac{x}{L} - 1\right)\right\} \tag{8.60}$$

and $\underline{\mathbf{I}}_{c1}$ (calculated as shown in the Appendix to Section 8.2) becomes

$$\underline{\mathbf{I}}_{c1} = \frac{g_{ms}\,L}{\sinh b}\left\{\frac{b}{L}\underline{\mathbf{V}}_{G,1}\cosh b\left(\frac{x}{L} - 1\right) - (\underline{\mathbf{V}}_{G,1} - \underline{\mathbf{V}}_{D,1})\frac{b}{L}\cosh\left(b\frac{x}{L}\right)\right\}. \tag{8.61}$$

The external currents will be calculated as shown by equations (8.39) to (8.41). After some algebraic calculations we get the following parameters (common-source connection):

$$Y_{11} = \frac{2\,g_{ms}\,b}{\sinh b}\,[\cosh b - 1] \tag{8.62}$$

$$Y_{12} = Y_{21} = \frac{g_{ms}\,b}{\sinh b}\,[1 - \cosh b] \tag{8.63}$$

$$Y_{22} = \frac{g_{ms}\,b}{\sinh b}\,\cosh b. \tag{8.64}$$

Since the matrix of the Y coefficients is reciprocal, the device behaviour should be that of a passive element*). We note the fact that the transconductance

$$g_m = \left.\frac{\partial I_{D,0}}{\partial V_{G,0}}\right|_{V_{D,0} = \text{const.}} = \frac{\mu\varepsilon_i\,W}{L\,h}\,V_{D,0} \tag{8.65}$$

vanishes indeed for vanishingly small currents ($I_{D,0} \to 0,\ V_{D,0} \to 0$).

The low-frequency ($\omega \to 0$) equivalent circuit of the IGFET in the linear region ($I_{D,0} \to 0$) is simply the conductance $g_{d,0}$ given by equation (8.43) ($g_{d,0} = g_{ms}$). The high-frequency behaviour is described by equations (8.62) — (8.64). We shall examine here two special cases.

(A) *The FET resistor.* The gate is a.c. short-circuited to source. The device behaves between the drain and source terminals as a complex admittance given by

$$Y_{d,0} = \frac{g_{d,0}\,b}{\operatorname{tgh} b},\quad b = \sqrt{i\hat{\omega}}\,,\ \hat{\omega} = \omega T_0\,. \tag{8.66}$$

At relatively low frequencies, the series expansion of tanh b may be used and one obtains

$$Y_{d,0} \simeq g_{d,0}\left(1 + \frac{b^2}{3}\right) = g_{d,0} + i\omega\,\frac{C_{g,c}}{3} \tag{8.67'}$$

(because $g_{d,0}\,T_0$ is equal to C_{gc}, according to equations (8.29), (8.42) and (8.43), i.e. the low frequency conductance is shunted by a capacitance which is one third of the gate-to-channel capacitance. This result could not be anticipated without a detailed calculation.

*) This can be clearly seen by calculating the admittances Y_a, Y_b, Y_c, Y_m of the π-type equivalent circuit. One obtains $Y_m = 0$.

(B) *The gate-to-channel admittance.* This is 'the MOS-transistor $V_D = 0$ admittance' calculated by Pierret [174]*), i.e. the gate-to source admittance with drain connected to source ($V_D = 0$ and $\underline{\mathbf{V}}_{D,1} = 0$). This admittance should be Y_{11} from equation (8.62)

$$Y_{11} = i\omega C_{gc}\left[\frac{2(\cosh b - 1)}{b \sinh b}\right] \tag{8.67}$$

By expanding in series the hyperbolic sin and cos, one obtains the low-frequency admittance

$$Y_{11}\,|_{\omega\to 0} \simeq i\omega C_{gc} + \frac{\omega^2 C^2{}_{gc}}{12\, g_{d,0}},\ \omega T_0 \ll 1. \tag{8.68}$$

The v.h.f. admittance may be calculated as follows ($\omega T_0 \gg 1$)

$$Y_{11}\,|_{\omega\to\infty} \simeq \frac{2i\omega C_{gc}}{b} = (1+i)\,(2\omega C_{gc}\, g_{d,0})^{1/2}. \tag{8.69}$$

The parallel conductance is proportional to ω^2 at low frequencies and to $\sqrt{\omega}$ at very high frequencies ($\omega T_0 \gg 1$). The same result was obtained by Pierret [174] and was also verified experimentally. In fact, the structure is a MOS capacitor, provided with lateral injecting contacts (source and drain) which supply minority carriers to the surface inversion layer. The lateral flow of carriers may explain, at least in part, the frequency-dependence of the equivalent capacitance of the structure, which have been sometimes considered as due entirely to the capture and releasing of mobile carriers by surface states [174].

8.5 General Frequency Characteristics

The theory of Section 8.2 yields the external currents $\underline{\mathbf{I}}_{G,1}$, $\underline{\mathbf{I}}_{D,1}$, $\underline{\mathbf{I}}_{S,1}$ as a function of the applied voltages. The Y parameters may be normalized to g_{ms} and should be of the form

$$\overline{Y}_{ij} = Y_{ij}/g_{ms} = \overline{Y}_{ij}(\hat{\omega}, a);\ i, j = 1, 2 \tag{8.70}$$

where

$$\hat{\omega} = \omega T_0,\ T_0 = \frac{C_{gc}}{g_{ms}} = \frac{C_{gc}}{g_{d,0}} = \frac{L^2}{\mu(V_{G,0} - V_T)} \tag{8.71}$$

*) Pierret [174] calculated this admittance by using a local equivalent circuit derived from the transmission-line model. He took into account the effect of the substrate, which is neglected here.

is the normalized frequency, and a is a bias-dependent parameter (see equations (8.13), (8.15), (8.27)), repeated here for convenience

$$a = \frac{I_{D,0}}{I_{Dsat}} = \frac{2V_{D,0}}{V_{G,0} - V_T} - \left(\frac{V_{D,0}}{V_{G,0} - V_T}\right)^2. \tag{8.72}$$

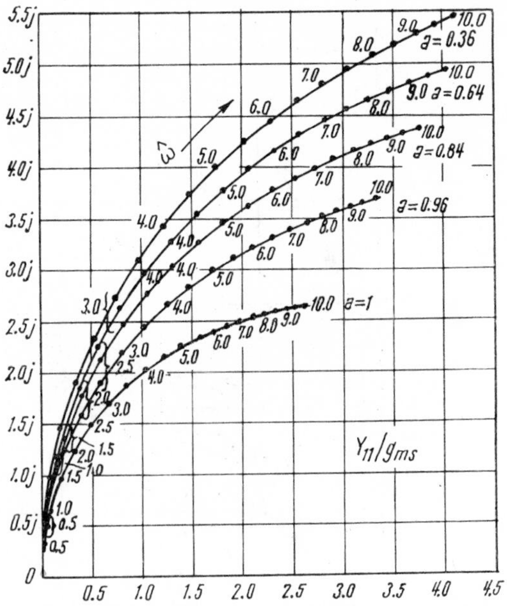

Figure 8.9. Input admittance with short-circuited output (common source connection) as a function of frequency for a few operating points [175].

We studied above two special cases: the saturation region ($a = 1$) and the linear region ($a \to 0$). The intermediate case $0 < a < 1$ will be presented here.

Figures 8.9 – 8.12 reproduce after Geurst and Nunnink [175] the frequency-dependence of the admittance parameters Y_{ij} in common source connection. Each curve corresponds to a value of the parameter a, given by equation (8.72).

Figures 8.9 shows that the input admittance (short-circuited output) has a capacitive imaginary part, as expected. For a given $V_{G,0}$ the input *low-frequency* capacitance decreases with increasing $V_{D,0}$ (this capacitance should be equal to C_{gc} at low currents and (2/3) C_{gc} in the saturation region, as shown above). The output short-circuit admittance Y_{22} (Figure 8.11) also exhibits a capacitive susceptance. Figure 8.10 shows that the output current is delayed with respect to the input voltage. This delay is attributed to the transit-time effects in the channel.

A π-type circuit like that shown in Figure 5.13 is, sometimes, more convenient. The transformation relations are indicated by equations (5.117) – (5.120). By use of the above figures we find that the output capacitance (see equation (5.119) and Figures 8.11, 8.12) of the π-type circuit is *negative*. In a similar manner we find that Y_m exhibits an inductive behaviour. Both these effects should be understood intuitively as due to carrier-transport inertia. However, this transport is accompanied by the distributed capacitive control of the gate electrode.

For practical purposes, an equivalent-circuit with frequency-independent parameters is also useful. Such a circuit may be derived for low frequencies ($\hat{\omega} \ll 1$). The calculations can be performed analytically by expanding in series of powers

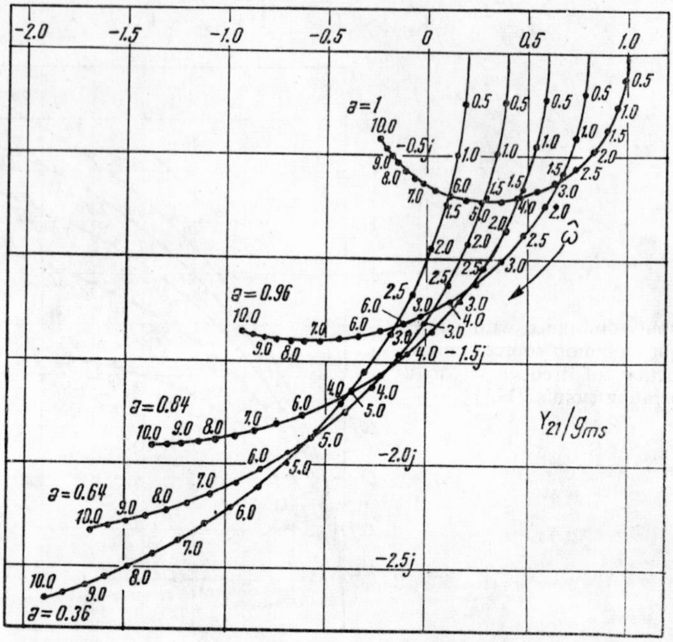

Figure 8.10. Transconductance (common sources connection) as a function of frequency for a few operating points [175].

of $i\hat{\omega}$ the Bessel functions appearing in Section 8.2, and retaining only the first terms*).

The final result should be of the form

$$Y_{11} = i\omega C_{11} \tag{8.73}$$

$$Y_{12} = -\,i\omega C_{12} \tag{8.74}$$

$$Y_{21} = g_m - i\omega C_{21} \tag{8.75}$$

$$Y_{22} = g_d + i\omega C_{22} \tag{8.76}$$

---

*) We shall point out a difficulty encountered in these calculations. For $\omega \to 0$ ($S' \to 0$) the Bessel function of the order $-2/3$ in equation (8.30)) becomes infinite and K_2, apparently, should be equal to zero in order to keep $\underline{\Psi}_1$ = finite (see also Section 8.3). However, K_2 cannot be identically equal to zero, because the boundary conditions (8.33) and (8.34) would not be satisfied simultaneously. This difficulty may be avoided as follows: K_1 and K_2 introduced in equation (8.30) are constants with respect to z (or x) and may depend upon ω. In particular, K_2 should vanish for $\omega \to 0$. More precisely, for $\omega \to 0$, K_2 should vary as $(S')^{1/3}$ (or $\omega^{1/3}$). Therefore, K_1 and K_2 will be calculated as a sum of the first terms of a series expansion with respect to ω.

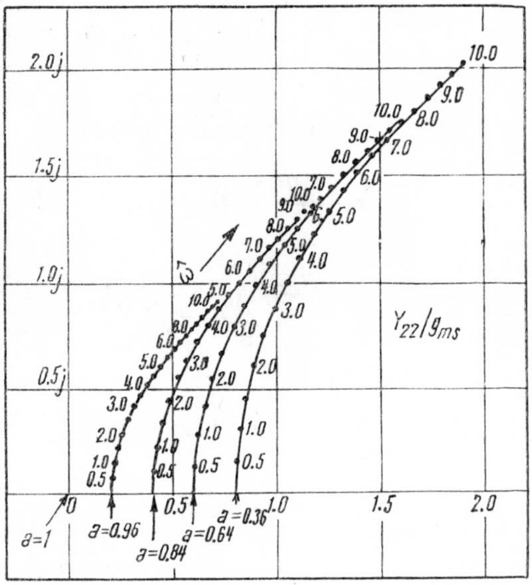

Figure 8.11. Output admittance with short-circuited input (common source connection), for a few operating points [175].

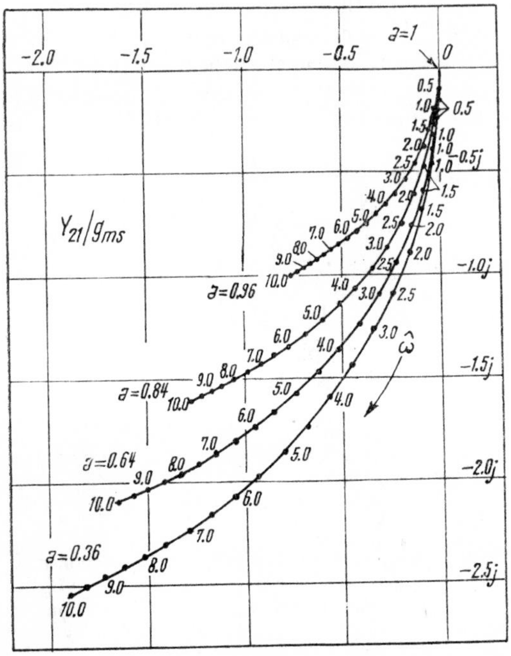

Figure 8.12. Reverse transconductance Y_{12} (common source connection), for a few operating points [175].

where all capacities C_{ij} are positive, as well as the transconductance g_m and the drain conductance g_d. The low-frequency 'π-type' circuit is shown in Figure 8.13 ($C_{22} < C_{12}$ and $C_{21} > C_{12}$).

A similar low-frequency circuit was derived by Richer [176] for the junction gate FET(JGFET). Richer [176] pointed out that part of the parameters occurring in such a circuit can be found from quasi-stationary calculations. So do g_m, g_d and the capacitances C_{11} and C_{12} which are, respectively, the input capacitance with short-circuited output and the transfer (or feedback) drain-gate capacitance, re-

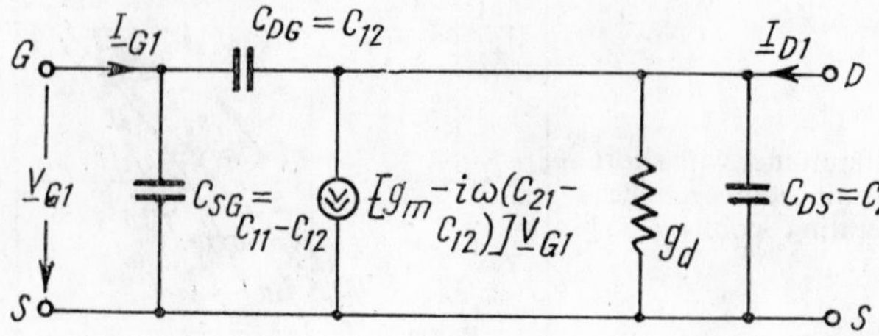

Figure 8.13. π-type low frequency equivalent circuit for the field effect transistor.

spectively. Both these capacitances may be calculated by differentiating the electric charge contained in the channel. The calculations are long and tedious, except the case of the device biased in saturation (when $C_{11} = (2/3)\ C_{gc}$ and $C_{12} = 0$ for the IGFET described by the model studied in this chapter).

Geurst's theory [175], [177], was verified by a.c. measurements on CdSe thin-film transistors [178]. However, large differences between theory and experiment were reported for the saturation region [176]. This occurred because the measured transistors did not exhibit a pronounced saturation, whereas the theory assumed perfect saturation.

A similar a.c. theory was developed recently by Cherry [179].

8.6 Small-signal Transient Response of an IGFET

The small-signal transient response can be found by replacing the complex variable for $i\omega$ in the formulae derived for the a.c. régime (see Sections 4.5 and 4.6).

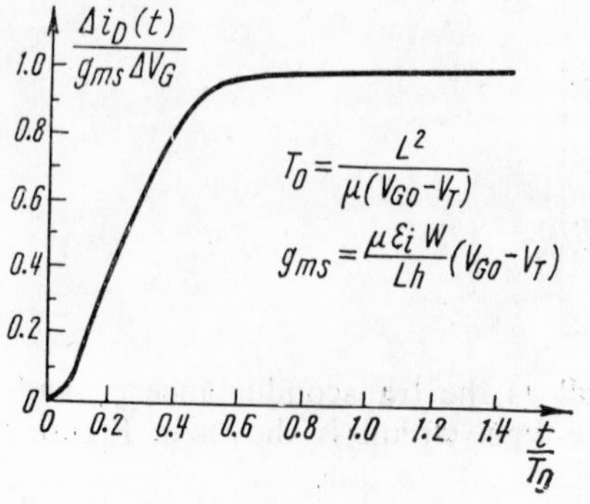

Figure 8.14. Small-signal transient response of the drain current to a voltage step ΔV_G [171]. The transistor is biased in saturation.

Burns [171] calculated the transient drain current when a small voltage step is applied between gate and source terminals. A formula equivalent to (8.47) was used to calculate the response. By use of the theory of Laplace transform, Burns [171] found

$$\Delta i_D(t) = g_{mS}\,\Delta V_G\,[1 - 0.99\exp(-6.406\,t/T_0) + 1.07\,(\exp -24.1\,t/T_0) - \ldots]. \quad (8.77)$$

The response is plotted in Figure 8.14. The 10—90% rise time is approximately 0.37 T_0.

8.7 Conclusion

The problems encountered in calculating the a.c. parameters of JGFET are similar. We refer below to both IGFET and JGFET, except where otherwise stated.

We may distinguish two categories of theoretical papers dealing with the a.c. behaviour of field-effect devices. The papers belonging to the first category suggest low-frequency equivalent circuits derived by use of the quasi-stationary or the 'charge-control' approach. The second category of papers allow a correct evaluation of high-frequency effects, starting from the general non-stationary equation of the problem*). It is this second class of papers which are reviewed below.

A solution to the small-signal problem for FE transistors was derived by van der Ziel and Ero [180], Hauser [181], Candler and Jordan [182], Richer [176], Reddy and Tromfimenkoff [183], van Nielen [184], and even Huang [185] †), who used a series expansion in $i\omega$ and retained the first terms.

A closed-form solution of differential equations (accompanied either by series developments or by numerical calculations) were presented by Tromfimenkoff and Treleaven [186], Haslet and Tromfimenkoff [187], Shoji [188], Geurst [177], [189], Geurst and Nunnink [175], Burns [171], Cherry [179].

An interesting approach was put forward by Das [172], [190]. This author started from the analogy of an FET with a non-linear RC transmission line. This analogy is based upon the equations describing the electrical behaviour. The a.c. behaviour is *approximated* by replacing the circuit with distributed constants of a lumped-parameter RC circuit. An extensive study based upon numerical calculations and experimental measurements substantiates this approach [172]. These results [191] are particularly interesting because they allow us to take into account the substrate resistivity effects in the MOS transistor††). The device is considered as a 4-terminal element, in view of the control properties of the substrate. An *exact* analytical solution for this more complex situation would probably be very difficult or even impossible to develop.

---

*) All these theories incorporate the gradual approximation (indicated at the beginning of this chapter). Richer [176] suggested that this approximation might be incorrect at frequencies comparable to the transit-time frequency.

†) Van Nielen [184] used succesive approximations whereas Huang [185] considered a charge-control approach. Both methods are based upon the 'exact' equations of the type (8.10) and (8.11).

††) The theory of Section 8.1 applies to an MOST with infinite substrate resistivity.

Problems

8.1. Show the equivalence between equations (8.17), (8.18) and the equations of a non-linear RC transmission line.

8.2. Show that the d.c. transit-time from source to drain depends upon the d.c. bias as indicated by equation (8.44).

8.3. Calculate the input admittance (8.50) for an IGFET in saturation and expand it in power series, as shown for the complex transconductance (8.49).

8.4. Find the total mobile charge inside the channel as a function of $V_{G,0}$ and $V_{D,0}$ and determine the quasi-stationary input capacitance (common-source connection). Verify that this capacitance decreases from $C_{g,c}$ to $\frac{2}{3} C_{gc}$ as the bias current increases from zero to its saturation value ($V_{G,0} =$ given). Explain intuitively this decrease.

8.5. Derive the upper frequency limit for the validity of the equivalent circuit of Figure 8.5.

8.6. Determine the complete low-frequency equivalent circuit in the linear region of the $I_D - V_D$ characteristics, by using equations (8.62) – (8.64). How can this circuit be modified to include the effect of amplification. Compare this circuit with the results of the charge-control theory [185].

9

Current Fluctuations

9.1. Introduction

This chapter is devoted to the noise properties of unipolar devices. Only the shot noise and the thermal noise are discussed. The main objective of our analysis is to point out the effect of carrier transport upon the basic noise properties, in the transit-time frequency region.

The noise problem of unipolar (single) injection and double injection in solids represents a wide field for both theoretical and experimental research*). The thermal noise in unipolar (SCLC) devices was the subject of numerous investigations. However, its basic properties are still a matter of dispute. We do not attempt to examine these questions in too many details. Instead, we give here examples of calculation of transit-time effects.

We start our discussion with a brief description of thermal noise and shot noise processes†).

(a) *Thermal noise.* The power spectrum (or spectral intensity) S_i of the short-circuit current fluctuations produced by a passive two-terminal element having the admittance $Y = G + iB$ in internal thermal equilibrium at a uniform temperature T, is given by Nyquist's formula

$$S_i = 4kTG. \tag{9.1}$$

(b) *Shot noise.* The full shot noise has the power spectrum

$$S_i = 2eI_0 \tag{9.2}$$

and refers to the fluctuation in a current of mean value I_0; the current consists of a flow of particles charged with the electric charge e.

---

*) See Proceedings of the Workshop '*Fluctuation phenomena in single and double injection devices*', 28—29 May 1970, Basel (Switzerland), published in *Solid State Electronics*, **14**, May 1971.

†) An excellent introduction to the electrical noise problem is represented by Bennett's book [191]. A recent review of noise phenomena was given by van der Ziel [192], [193].

A recent paper due to Robinson [194], studies both thermal and shot-noise processes, on the basis of mechanical statistics and establishes a clear distinction between these two categories of noise. Confusion between these different processes was rather frequent in the earlier literature.

According to Robinson [194], the distinction between a shot noise process and a thermal noise process may be formulated as follows.

'In a shot noise process the region in which the carriers interact with external fields is physically distinct from the region in which the statistical properties of the carriers are established. The carriers interacting with fields have no influence on, nor are influenced by processes in the emitter region' [194]*).

'In a thermal noise process, on the other hand, the interaction region is coincident with the generation region. During their interaction with external fields the carrier are in approximate thermal equilibrium with a lattice...' [194].

Generally, the effect of both shot-noise and thermal noise processes may be identified in the same device. We assume below that these fluctuation phenomena are independent and/or one of them is dominant. We discuss the shot-noise effect in injection-controlled current (ICC) diodes, and the 'thermal noise' properties of SCLC diodes and field-effect transistors.

9.2 Coupling of the Injection Noise to the External Circuit

Assume a two-terminal unipolar semiconductor device consisting of one or several cascaded semiconductor regions. We intend to investigate the effect of fluctuations of the particle current†) injected by the cathode or by a certain emitter region. We can treat each component of the injection noise as a sinusoidal signal. The effect of this signal in the external circuit may be investigated by using the transport coefficients defined and calculated in Chapter 4 (for one-dimensional regions).

For simplicity, we shall assume a device having a single active semiconductor region. The transport coefficients K_{ij} of this region are defined by equations (4.14) and (4.15), rewritten here for convenience

$$\underline{\mathbf{V}}_1 = K_{11}\,\underline{\mathbf{J}}_1 + K_{12}\,\underline{\mathbf{E}}_1(0) \tag{9.3}$$

$$\underline{\mathbf{E}}_1(L) = K_{21}\,\underline{\mathbf{J}}_1 + K_{22}\,\underline{\mathbf{E}}_1(0). \tag{9.4}$$

We shall also state a property of the injecting plane $x = 0$: we assume that the bias current is modulated by the electric field intensity $\underline{\mathbf{E}}_1(0)$ and that an a.c. (par-

*) 'Shot noise processes are usually associated with devices not in thermal equilibrium ··· but the conditions for a pure shot noise process can be present in a system in thermal equilibrium such as an unbiased semiconductor diode. Then and only then, the shot noise formulae must lead to identically the same result as the thermodynamic Johnson noise (i.e. thermal noise) formula' [194].

†) These fluctuations represent the so-called injection noise. A typical example is the shot noise given by equation (9.2). However, this formula is not sufficiently general [194].

ticle) current $\sigma_1(0)\underline{\mathbf{E}}_1(0)$ results. Here $\sigma_1(0)$ is assumed real and it is the slope of the control characteristic (defined in Section 3.6) for the applied bias field. The total alternating current $\underline{\mathbf{J}}_1$ at the 'cathode' ($x = 0$) will be

$$\underline{\mathbf{J}}_1 = \sigma_1(0)\,\underline{\mathbf{E}}_1(0) + i\omega\varepsilon\,\underline{\mathbf{E}}_1(0) + \underline{\mathbf{j}}_i. \tag{9.5}$$

Here, the second therm in the right-hand-side is the displacement current and the third term is the ω-frequency component of the injection noise which consist in random fluctuations of the bias current (which is the average value of the injected current). These fluctuations are sufficiently low to be treated as small-amplitude signals.

We assume now that a load impedance Z_l is connected at the a.c. terminals of our device. We consider a unit-area section device and the total current is equal to the current density $\underline{\mathbf{J}}_1$. The voltage drop on Z_l is

$$\underline{\mathbf{V}}_1 = -\,Z_l\,\underline{\mathbf{J}}_1. \tag{9.6}$$

The alternating current flowing in this circuit is due to the $\underline{\mathbf{j}}_i$ component, which acts as a current generator. By combining equations (9.3) and (9.5) we obtain

$$\underline{\mathbf{V}}_1 = Z\,\underline{\mathbf{J}}_1 - K'_{12}\underline{\mathbf{j}}_i \tag{9.7}$$

where

$$K'_{12} = \frac{K_{12}}{\sigma_1(0) + i\omega\varepsilon} \tag{9.8}$$

and

$$Z = K_{11} + K'_{12}, \tag{9.9}$$

where Z is the diode impedance, equation (4.18). By eliminating $\underline{\mathbf{V}}_1$ between equations (9.6) and (9.7) one obtains

$$\underline{\mathbf{J}}_1 = \frac{K'_{12}}{Z + Z_l}\,\underline{\mathbf{j}}_i\,. \tag{9.10}$$

Consequently, the current $\underline{\mathbf{J}}_1$ in our circuit can be non-zero if either $\underline{\mathbf{j}}_i \neq 0$ (noise) or $Z + Z_l = 0$ (self-oscillations). The mean-square current due to the injection noise should be

$$\overline{J_1^2} = \left|\frac{K'_{12}}{Z + Z_l}\right|^2 \cdot \overline{j_1^2}. \tag{9.11}$$

The short-circuit noise ($Z_l = 0$) is

$$\overline{J_1^2}_{(s.\,c.)} = \left|\frac{K'_{12}}{K_{11} + K'_{12}}\right|^2 \overline{j_i^2} = F_0\,\overline{j_i^2} \tag{9.12}$$

where F_0 is a short-circuit noise factor. The mean-square voltage is, according to equations (9.7) and (9.10),

$$\overline{V_1^2} = \left| \frac{K'_{12} Z_l}{Z_l + Z} \right|^2 \overline{j_i^2} \tag{9.13}$$

and the open-circuit ($Z_l \to \infty$) noise should be

$$\overline{V_{1(o.c.)}^2} = | K'_{12} |^2 \overline{j_i^2} = F_\infty \overline{j_i^2} \tag{9.14}$$

so that

$$\overline{V_{1(o.c.)}^2} = | Z |^2 \overline{J_{1(s.c.)}^2}. \tag{9.15}$$

We shall calculate below the open-circuit noise factor F_∞ (which has the dimensions of ohm²). From equations (9.12), (9.14) and (9.15) F_0 is simply equal to $F_\infty | Z |^{-2}$.

From equations (9.8) and (9.14), F_∞ is given by

$$F_\infty = \left| \frac{K_{12}}{\sigma_1(0) + i\omega\varepsilon} \right|^2 = \left| \frac{K_{12}}{\sigma_1(0)} \right|^2 \frac{1}{1 + \omega^2 \tau_c^2} \tag{9.16}$$

where

$$\tau_c = \frac{\varepsilon}{\sigma_1(0)} \tag{9.17}$$

is the dielectric (differential) relaxation time, equation (6.8), defined for the 'cathode' plane ($x = 0$). This parameter is real and frequency-independent. K_{12} is a complex number and is given by equation (4.58) (where s should be replaced by $i\omega$)

$$K_{12} = \frac{v_0(0)}{1 - \alpha v_0(0)} \int_0^{T_L} [1 - \alpha v_0(T_x)] \exp(-i\omega T_x) \, \mathrm{d}T_x, \quad \alpha = \frac{eN}{J_0}. \tag{9.18}$$

A first integration yields

$$K_{12} = \frac{v_0(0)}{1 - \alpha v_0(0)} \left\{ \frac{1 - \exp(-i\theta)}{i\omega} - \alpha \left[\frac{v_0(0) - v_0(L) \exp(-i\theta)}{i\theta} + \right. \right.$$

$$\left. \left. + \int_0^{T_L} \frac{\mathrm{d}v_0}{\mathrm{d}T_x} \frac{\exp(-i\omega T_x)}{i\omega} \mathrm{d}T_x \right] \right\}. \tag{9.19}$$

$K_{12} = K_{12}(\omega)$ should exhibit an oscillatory behaviour and vanish asymptotically at very high frequencies (a similar problem was met in Section 5.4). A further decrease of F_∞ (attenuation of the injection noise) at very high frequencies is due to the $(1 + \omega^2\tau_c^2)^{-1}$ factor in equation (9.16). The frequency-dependence can be studied better by normalizing $F_\infty(\omega)$ to its low frequency ($\omega \to 0$) value. We have

$$\bar{F}_\infty = \frac{F_\infty(\omega)}{F_\infty(0)} = \left|\frac{K_{12}(\omega)}{K_{12}(0)}\right|^2 \frac{1}{1 + \omega^2\tau_c^2} \tag{9.20}$$

where

$$F_\infty(0) = \frac{|K_{12}(0)|^2}{[\sigma_1(0)]^2}\,. \tag{9.21}$$

We note that τ_c and $\sigma_1(0)$ depend upon the injecting contact, whereas K_{12} is a transport coefficient which characterizes the semiconductor region itself.

To be specific, we shall refer below to the model of Section 4.8, where the differential mobility is assumed constant. By use of equation (4.103) in equation (9.20) we get

$$\bar{F}_\infty = \frac{\theta_d^2\,[1 + \exp(-2\theta_d) - 2(\exp -\theta_d)\cos\theta]}{(1 - \exp - \theta_d)^2\,(\theta^2 + \theta_d^2)\,(1 + \theta^2\Gamma^2)}\,,\quad \Gamma = \frac{\tau_c}{T_L} \tag{9.22}$$

i.e. $\bar{F}_\infty = \bar{F}_\infty(\theta;\, \theta_d, \Gamma)$. Another possible form is ($\theta_d = T_L/\tau_d$)

$$\bar{F}_\infty = \frac{1 + [\exp - 2(T_L/\tau_d)] - 2[\exp(-T_L/\tau_d)\cos\theta]}{(1 - \exp - T_L/\tau_d)^2(1 + \omega^2\tau_d^2)(1 + \omega^2\tau_c^2)}\,. \tag{9.23}$$

The factors $(1 + \omega^2\tau_d^2)^{-1}$ and $(1 + \omega^2\tau_c^2)^{-1}$ determine a reduction of $\bar{F}_\infty$ at high frequencies due to the dielectric relaxation mechanism in the bulk (τ_d) and at the 'cathode' (τ_c) respectively. The remaining quantities in equation (9.23) introduce the transit-time effect. Indeed, if the transit-time were very short as compared to the bulk relaxation time τ_d, the factor $\bar{F}_\infty$ would become simply

$$\bar{F}_\infty \simeq \frac{1}{(1 + \omega^2\tau_d^2)\,(1 + \omega^2\tau_c^2)}\,. \tag{9.24}$$

Assume the opposite case when both τ_d and τ_c are extremely short as compared to T_L. At frequencies comparable to or lower than the transit-time frequency, $\bar{F}_\infty$ becomes

$$\bar{F}_\infty \simeq 1 + 2\,[\exp(-T_L/\tau_d)]\,[1 - \exp(-T_L/\tau_d)]^{-2}\,(1 - \cos\theta) \tag{9.25}$$

and exhibits small oscillations due to the transit-time effect.

Another particular case is characterized by no bulk relaxation effects, i.e.

$$\tau_d = \frac{\varepsilon}{eN\mu_d} \to \infty \tag{9.26}$$

which may occur for either no fixed space-charge ($N = 0$) or for saturated velocity ($\mu_d = 0$). Therefore F_∞ becomes

$$F_\infty = \left[\frac{v_0(0)T_L}{\sigma_1(0)}\right]^2 \frac{2(1 - \cos\theta)}{\theta^2} \frac{1}{1 + \omega^2\tau_c^2} \tag{9.27}$$

The first factor in equation (9.27) is just $F_\infty(0)$. The second and the third factors take into account, respectively, the transit-time effects and the relaxation effects at the cathode. $F_\infty = F_\infty(\omega)$ exhibits attenuated oscillations.

Finally, let us discuss the low-frequency noise. According to equations (4.103) and (9.21) we have

$$F_\infty(0) = \left[\frac{v_0(0)\, T_L}{\sigma_1(0)}\right]^2 \left[\frac{1 - \exp(-T_L/\tau_d)}{T_L/\tau_d}\right]^2 \tag{9.28}$$

or

$$F_\infty(0) = \left(\frac{v_0\tau_c\tau_d}{\varepsilon}\right)^2 \left[1 - \exp(-T_L/\tau_d)\right]^2 \tag{9.29}$$

which indicates the way in which the low frequency open-circuit injection noise depends upon the time constants τ_c, τ_d, T_L. The short-circuit noise factor $F_0(0)$ is, perhaps, more suggestive. This is

$$F_0(0) = \left|\frac{K'_{12}}{K_{11} + K'_{12}}\right|^2_{\omega=0} = \frac{1}{\left|1 + \sigma_1(0)\dfrac{K_{11}(0)}{K_{12}(0)}\right|^2} = \left|\frac{K'_{12}(0)}{Z(0)}\right|^2. \tag{9.30}$$

We note that $F_0(0)$ is zero for SCLC injection ($\sigma_1(0) \to \infty$) and unity for ECL injection ($\sigma_1(0) \to 0$). In the first case, the injected noise is totally suppressed (an idealized situation, of course — see the Appendix) whereas in the second case it is transmitted unmodified in the external circuit ($\overline{J_1^2} = \overline{j_i^2}$). Then, we wish to know if the injected noise at $\omega \to 0$ is reduced or enhanced by the transport process. The discussion proceeds as follows. We show first that both $K_{11}(0)$ and $K_{12}(0)$ are positive. This can be shown by noting that the low-frequency (incremental) resistance is

$$R_i = K_{11}(0) + K_{12}(0)/\sigma_1(0). \tag{9.31}$$

This formula is analogous to equation (3.56): note the fact that $\sigma_1(0) = (\mathrm{d}j_c/\mathrm{d}E)_{E=Ec}$, according to equation (3.54). The first term, $K_{11}(0)$, is the 'bulk' resistance (3.57) and is always positive. The 'cathode' resistance $K_{12}(0)/\sigma_1(0)$ has the sign of $\sigma_1(0)$ because $K_{12}(0)$ is also positive (see equation (3.58)). Therefore, $E_0(0)$ is below unity for positive $\sigma_1(0)$ and above unity if $\sigma_1(0) < 0$ (negative slope of the control characteristic, see Chapter 3). In other words, a negative conductivity of the emitting plane*) determines an 'excess noise', or $\overline{J_1^2} > \overline{j_i^2}$. Of course, this external noise will decrease at higher frequencies, as shown above.

9.3 Frequency-dependence of Noise Factors

In this Section we shall apply the above results to certain particular cases.

Let us first discuss the transit-time saturated-velocity diode (Section 6.5). The open-circuit noise factor F_∞ is given by equation (9.27) because $\mu_d = 0$ and $\tau_d \to \infty$. It may be written in normalized form as follows

$$\overline{F}_\infty = \frac{2(1 - \cos\theta)}{\theta^2(1 + \theta^2\Gamma^2)}, \quad \Gamma = \frac{\tau_c}{T_L}. \tag{9.32}$$

The frequency-dependence is plotted in Figure 9.1 with Γ as parameter. F_∞ decreases markedly at high frequencies due to both transit-time and cathode-relaxation effects.

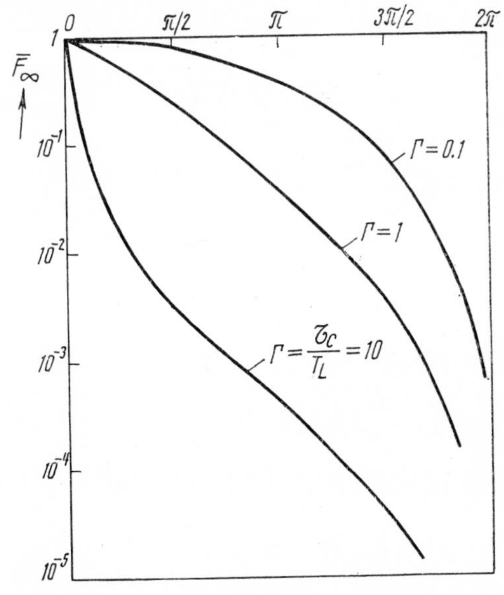

Figure 9.1. Frequency-dependence of the normalized open-circuit noise factor for the injection-limited saturated-velocity diode.

---

*) Such a negative conductivity can be achieved by use of imperfect-cathode injection in negative-mobility semiconductors [52].

The low frequency value of F_∞ is

$$F_\infty(0) = \left(\frac{v_0(0)\, T_L}{\sigma_1(0)}\right)^2 = \left(\frac{L}{\sigma_1(0)}\right)^2 \tag{9.33}$$

whereas $F_0(0)$ can be found by use of equations (6.29), (9.30), and (9.33):

$$F_0(0) = \left(\frac{1}{1 + \dfrac{\sigma_1(0)\, L}{2\varepsilon v_l}}\right)^2 = \left(\frac{1}{1 + \dfrac{T_L}{2\tau_c}}\right)^2 < 1. \tag{9.34}$$

Therefore, if the cathode relaxation time τ_c is much shorter than the transit-time T_L ($\Gamma \ll 1$), the low-frequency short-circuit noise $\overline{J_1^2} = F_0(0)\,\overline{j_i^2}$ is much below the level of the injection noise.

If we identify the injection noise with the full shot-noise, equation (9.2), we obtain (for a device of unit area),

$$\overline{j_i^2} = 2eJ_0\,\Delta f \tag{9.35}$$

where (Δf) is the bandwidth. The open-circuit noise is given by equations (9.14), (9.3), (9.33) and (9.35) and has the value

$$\overline{V_{1\,(o.c.)}^2} = 2eJ_0\,\Delta f \left[\frac{L}{\sigma_1(0)}\right]^2 \frac{2(1 - \cos\theta)}{\theta^2(1 + \theta^2\Gamma^2)}. \tag{9.36}$$

An identical result was obtained by Haus *et al.* [160]. For a more detailed discussion of frequency and bias dependence of the noise measure*) see this reference.

The ECL injection will be examined now. By definition we have $\sigma_1(0) = 0$. Therefore one obtains

$$F_\infty = |K'_{12}|^2 = \left|\frac{K_{12}}{i\omega\varepsilon}\right|^2 \tag{9.37}$$

and, because $K_{12}(0) \neq 0$ (except for the SCLC injection, which is beyond discussion) for $\omega \to 0$, $F_\infty(0)$ tends to infinity. In agreement with this result, the d.c. ($\omega \to 0$) impedance for ECL injection is due to a pure capacitance (see equation (6.44)). The short-circuit noise factor should be finite. We have

$$F_0 = \left|\frac{1}{1 + i\omega\varepsilon\,\dfrac{K_{11}(\omega)}{K_{12}(\omega)}}\right|^2 \tag{9.38}$$

the low-frequency factor being unity (as pointed out above).

*) The noise measure is defined as $\overline{V_{1\,(o.c.)}^2}/4kT_0R\Delta f$, where R is the device resistance. The above-cited authors [160] calculated the noise measure for the maximum absolute value of the negative resistance (Section 6.5).

Let us consider, for example, the special case $N = 0$ and $\mu = \text{const}$. The transport coefficients were calculated in Section 4.7. By replacing equations (4.76) and (4.77) into equation (9.38), one obtains

$$F_0 = \left| \frac{R(i\theta)}{1 + \frac{\zeta}{1-\zeta} i\theta\, P(i\theta)} \right|^2 = F_0(\theta; \zeta) \tag{9.39}$$

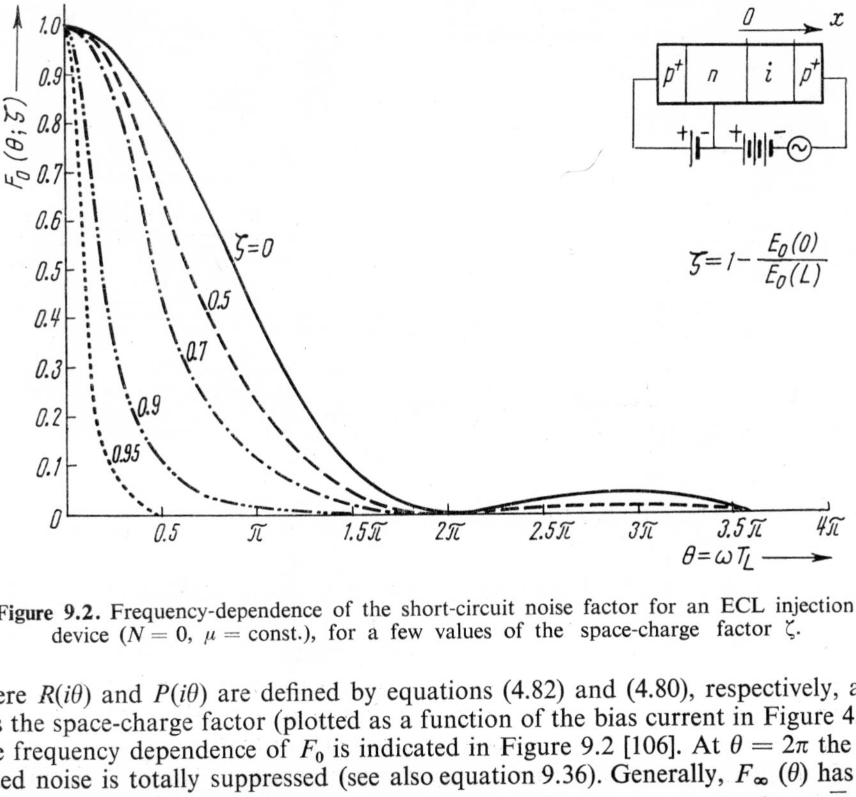

Figure 9.2. Frequency-dependence of the short-circuit noise factor for an ECL injection device ($N = 0$, $\mu = \text{const.}$), for a few values of the space-charge factor ζ.

where $R(i\theta)$ and $P(i\theta)$ are defined by equations (4.82) and (4.80), respectively, and ζ is the space-charge factor (plotted as a function of the bias current in Figure 4.1). The frequency dependence of F_0 is indicated in Figure 9.2 [106]. At $\theta = 2\pi$ the injected noise is totally suppressed (see also equation 9.36). Generally, $F_\infty(\theta)$ has an oscillatory-attenuated behaviour*). At a given transit angle, the noise factor $\overline{F}_0 = \overline{F}_0(\theta)$ decreases with increasing ζ (increasing non-uniformity of the electric field, due to the injected space-charge). This indicates that the coupling of the external circuit to the fluctuations of the injected current is reduced by the presence of the bulk space charge. However, this reduction does not take place in other cases, as shown below.

---

*) A high-frequency attenuation of the injected (shot) noise was predicted theoretically and verified experimentally for the planar vacuum diode (see, for example, Section 6.4 of reference [74]).

We consider now another example of ECL injection, namely ECL injection in negative mobility materials in the limit of zero doping [161], i.e. $\mu_d < 0$ and $N = 0$ in the equations developed in Section 4.8. The final result is

$$F_0 = \overline{F}_0 = \left| \frac{R(i\theta)}{\left(\frac{1+r}{2}\right) + (1-r)\,Q(i\theta)} \right|^2 \tag{9.40}$$

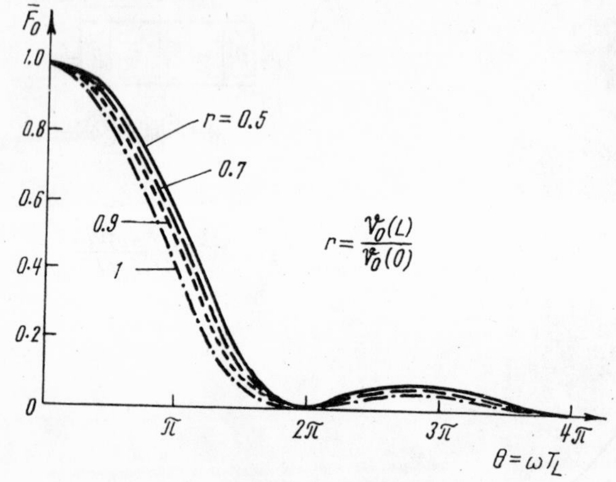

Figure 9.3. Frequency dependence of the short circuit noise factor for ECL injection in a negative-mobility medium ($\mu < 0$, $N = 0$).

where $R(i\theta)$, $Q(i\theta)$ are defined by equation (4.106) and

$$r = \frac{v_0(L)}{v_0(0)} \tag{9.41}$$

where r should be $\leqslant 1$ because $\mu_d < 0$ and the electric field increases from $x = 0$ to $x = L$. Equation (9.40) is plotted *versus* θ in Figure 9.3 [161]. For a given θ, F_0 increases with decreasing r (increasing field non-uniformity).

In an ECL injection structure similar to that depicted in Figure 6.6, the injected current is equal to the emitter current which crosses the 'base' by diffusion. The forward-biased emitter junction yields a full shot-noise. By assuming unit emitter efficiency, we have [192]

$$S_i \simeq 2eI_E \tag{9.42}$$

where I_E is the emitter current. If we consider the injection from a non-ohmic metal-semiconductor contact (back-biased for normal injection *into* the semiconductor) the current entering the formula (9.2) is the reverse barrier current which is equal to the bias current [192].

9.4 Thermal Noise in Space-charge-limited Diodes [195]

We shall show below an example [195] of calculation of the thermal noise. One expects this type of noise to dominate in SCLC devices, where the shot noise is (at least in part) suppressed [196] by a mechanism discussed in the Appendix to Section 9.2.

The h.f. thermal noise of the square-law SCLC diode was calculated by van der Ziel [197] and Shumka [195] by use of the Langevin method, and a classical thermal-noise hypothesis.

According to the Langevin method the random fluctuations of the current *in the bulk* is described by the random drive function $H(x, t)$ as follows

$$J(t) = ep(x, t)\, v(x, t) + \varepsilon \frac{\partial E(x, t)}{\partial t} + H(x, t). \tag{9.43}$$

The calculations proceed in the same manner as described in Section 4.5 for the small-signal response, because $H(x, t)$ may be assumed very small. The unique difference is the fact that $J(t) - H(x, t)$ should be replaced for $J(t)$. The equations used are exactly the same as in Section 4.5. Each physical quantity should be replaced by a d.c. part plus a very small perturbation. The time t is eliminated by applying the Laplace transform. The electric field is*)

$$\mathcal{E}_1(T_x; s) = \frac{1 - \alpha v_0(T_x)}{v_0(T_x)} \int_0^{T_x} \frac{[\mathcal{J}_1(s) - \mathcal{H}(\eta; s)]\, v_0(\eta)}{1 - \alpha v_0(\eta)} [\exp s(\eta - T_x)]\, d\eta +$$

$$+ [\exp(-sT_x)] \left[\frac{1 - \alpha v_0(T_x)}{v_0(T_x)} \right] \frac{v_0(0)}{1 - \alpha v_0(0)} \mathcal{E}_1(0; s) \tag{9.44}$$

(a formula which is analogous to equation (4.50)).

The SCLC injection conditions require $E_0(0) = 0$ and $v_0(0) = 0$. The SCLC *square-law* diode is characterized by $N = 0$ and $\mu = ct$. The former condition requires $\alpha = \dfrac{eN}{J_0} = 0$ whereas the latter determines

$$J_0 T_x = \varepsilon E_0(x) = \frac{\varepsilon v_0(x)}{\mu} \tag{9.45}$$

or

$$v_0 = \frac{\mu J_0}{\varepsilon} T_x. \tag{9.46}$$

*) See the notations used in Section 4.5. $\mathcal{H}(x; s)$ or $\mathcal{H}(T_x; s)$ is the Laplace transform of $H(x, t)$.

Therefore, equation (9.44) becomes

$$\mathcal{E}_1(T_x; s) = -\frac{1}{v_0(T_x)}\left(\frac{\mu J_0}{\varepsilon}\right)\int_0^{T_x}\left[\mathcal{J}_1(s) - \mathcal{H}(\eta; s)\right] \eta \exp s(\eta - T_x)\, \mathrm{d}\eta. \tag{9.47}$$

At this point we shall make another assumption which does not, however, restrict the generality of the problem. We shall calculate *the short-circuit noise current*. Thus, the perturbation of the diode voltage is zero

$$\mathcal{V}_1(s) \equiv 0 \tag{9.48}$$

or

$$\int_0^L \mathcal{E}_1\, \mathrm{d}x = \int_0^{T_L} \mathcal{E}_1(T_x; s)\, v_0(x)\, \mathrm{d}T_x = 0. \tag{9.49}$$

From equations (9.47) and (9.49) one obtains

$$\mathcal{J}_1(s)\int_0^{T_L} \mathrm{d}T_x \int_0^{T_x} \eta \exp s(\eta - T_x)\mathrm{d}\eta = \int_0^{T_L} \mathrm{d}T_x \int_0^{T_x} \mathcal{H}(\eta; s)\, \eta \exp s(\eta - T_x)\, \mathrm{d}\eta. \tag{9.50}$$

The left-hand side (LHS) can be readily obtained by noting that

$$\frac{1}{\varepsilon}\int_0^{T_L} \mathrm{d}y\, [\exp(-sy)] \int_0^{y} v_0(\eta) \exp(s\eta)\, \mathrm{d}\eta = Z \tag{9.51}$$

is the impedance of the square-law diode (see Section 4.7 and equation (4.90) and thus

$$\text{LHS} = \mathcal{J}_1(s)\, \frac{\varepsilon^2}{\mu J_0}\, Z. \tag{9.52}$$

The integral in the right-hand side (RHS) of equation (9.49) can be integrated by parts and the result is

$$\text{RHS} = \frac{1}{s}\int_0^{T_L} T_x\, \mathcal{H}(T_x; s)\, [1 - \exp s(T_x - T_L)]\, \mathrm{d}T_x. \tag{9.53}$$

Therefore equations (9.50), (9.52) and (9.53) yield

$$\mathcal{J}_1(s)\, Z = \frac{\mu J_0}{s\varepsilon^2}\int_0^{T_L} T_x \mathcal{H}(T_x; s)\, [1 - \exp s(T_x - T_L)]\, \mathrm{d}T_x. \tag{9.54}$$

Here we shall replace $i\omega$ for s, $\underline{\mathbf{J}}_1(\omega)$ for $\mathcal{J}_1(s)$ and $\underline{\mathbf{H}}_1(x, \omega)$ for $\mathcal{H}(T_x; s)$:

$$\underline{\mathbf{J}}_1(\omega)\, Z(\omega) = \frac{\mu J_0}{i\omega\varepsilon^2} \int_0^{T_L} T_x\, \underline{\mathbf{H}}_1(T_x; \omega)\, [1 - \exp i\omega(T_x - T_L)]\, \mathrm{d}T_x \cdot \tag{9.55}$$

The thermal-noise hypothesis states that the spectral intensity of the random driving function is [195], [197]

$$<\underline{\mathbf{H}}_1(x, \omega)\, \underline{\mathbf{H}}_1^*(y, \omega)> = 4kTe\, \mu p_0(x)\, \delta(x - y) \tag{9.56}$$

where* denotes complex conjugate and δ is the Dirac function. Equation (9.55) expresses the fact that the thermal noise sources at the points x and y and a given instant t are uncorrelated. The spatial variable x may be replaced by T_x (due to one-to-one correspondence of these variables), i.e.

$$<\underline{\mathbf{H}}_1(T_x, \omega)\, \underline{\mathbf{H}}_1^*(T_x', \omega)> = 4kTe\mu\, \frac{p_0(T_x)}{v_0(T_x)}\, \delta(T_x - T'_x) \cdot \tag{9.57}$$

The spectral intensity for the short-circuit noise current (S_i) will be equal to [195], [197]

$$S_i = \frac{1}{|Z(\omega)|^2} \left(\frac{\mu J_0}{\omega\varepsilon^2}\right)^2 \int \int_0^{T_L} T_x^2 <\underline{\mathbf{H}}_1(T_x, \omega)\, \underline{\mathbf{H}}_1^*(T_x', \omega)> \times$$

$$\times [1 - \exp i\omega\, (T_x - T_L)]\, [1 - \exp - i\omega(T_x - T_L)]\, \mathrm{d}T_x\, \mathrm{d}\, T_x' = \frac{4kT \times 2R(\omega)}{|Z(\omega)|^2} \tag{9.58}$$

where $R\,(\omega)$ is the resistive part of the small-signal impedance (4.90)

$$R(\omega) = \frac{v_0(L)\, T_L^2}{\varepsilon} \frac{\theta - \sin\theta}{\theta^3}, \quad \theta = \omega T_L. \tag{9.59}$$

Equations (4.91) and (4.92) were also taken into account. Equation (9.59) may be also written

$$S_i = 4kT.2G(\omega). \tag{9.60}$$

$G(\omega) = R(\omega)|\, Z(\omega)|^{-2}$ is the parallel conductance (all computations were made for a unit-area device). The spectral intensity of the open-circuit voltage should be

$$S_v = 4kT.2R(\omega) \tag{9.61}$$

and can be found directly from equation (9.44) by putting $\mathcal{J}_1 = 0$ and calculating $\mathcal{V}_1$ [197]. Thus, we have

$$S_v = |Z(\omega)|^2 S_i. \tag{9.62}$$

The spectral intensity of the short-circuit current varies with frequency exactly as the parallel conductance $G(\omega)$ does (Figure 5.4, curve a = 0). By comparing equation (9.76) with the Nyquist formula, equation (9.1), we note that this equation can be, formally, applied if the conductance G in equation (9.1) is replaced by *twice* the device conductance. Strictly speaking, the Nyquist formula, which is derived under the assumption of thermodynamic equilibrium, cannot be applied as far as a non-zero direct current J_0 passes through the device [198]. On the other hand, the device is non-linear and it is not clear what kind of conductance have to be placed in equation (9.1). Webb and Wright [199] postulated that the differential (low-frequency) conductance G_i should occur in equation (9.1). Later, van der Ziel [200] argued that the correct resistance which should be used in the thermal-noise formula $S_v = 4\,kTR$ is the d.c. resistance V_0/J_0*). The second result [200] agrees with equation (9.60) because, for the SCLC square-law diode the d.c. resistance is twice the a.c. resistance and $S_i = S_v\ G_i^2 = 4kT\ (2/G_i)\ G_i^2 = 8kTG_i$, in agreement with the result of equation (9.60) derived here [195], [197].

A number of papers [202] — [207] reported experimental data on the so-called 'thermal-noise' discussed above. Hsu *et al.* [202] suggested that the equivalent noise resistance appearing in equation $S_v = 4\,kTR$ should be also replaced by V_0/J_0 for the SCLC diode with a characteristic of the form $J_0 = C\,V_0^{\alpha}$. More recently, Shumka [208] calculated the noise in SCLC solid-state diodes with field-dependent mobility and invalidated this statement†). The 'hypothesis of the d.c. resistance' [200] was also discussed by Nicolet [207]. He also reported [207] calculations which prove the result expressed in equation (9.61) for the low-frequency domain, by using the 'impedance-field method' put forward by Shockley, Copeland and James [210].

The same method [210] was used by Thim [137] to calculate the thermal noise in a bulk negative-mobility amplifier. Thim [137] analysed two cases characterized, respectively, by (a) ohmic injecting contact and highly non-uniform static electric field (Sections 5.5, 5.7 and 5.8); (b) non-ohmic injecting contact and almost uniform d.c. electric field (Section 6.6 and Ref. [154]). The above author [137] stated that the thermal noise level is lower in the latter case.

Finally, we note that a more rigourous theory of the transport (bulk) noise with applications to unipolar devices and especially SCLC diodes was developed by Sergiescu and his co-workers [198], [211] — [215]. These theoretical formulae were derived by using the basic microscopic and statistical properties of charge transport in solids. However, such a theory cannot yield simple analytical results involving either space-charge, or transit-time effects in solid-state devices.

---

*) A subsequent paper by van der Ziel and van Vliet [201] demonstrated that 'the noise source introduced in [197] is a *diffusion* noise source if the Einstein relation holds···' (see also Section III of the review paper [192]).

†) The 'thermal' noise in SCLC solid-state devices with hot carriers was discussed by Bougalis and van der Ziel [209].

5.9 Thermal Noise in Field-Effect Transistors

One of noise sources in field-effect transistors is the thermal noise of the conductive channel [192]. We shall consider below the simple model of the IGFET which was introduced in Section 8.1.

The low-frequency ($\omega \to 0$) behaviour will be discussed first. Therefore, the gradient of the particle current, which occurs due to the electric charge redistribution under the influence of the variation of the gate potential will be neglected. However, the channel current I_c will be a function of both x and t, due to the thermal noise source $H(x, t)$, i.e.

$$I_c(x, t) = g[V(x, t)] \frac{\partial V(x, t)}{\partial x} + H(x, t) \tag{9.63}$$

where, according to equation (8.10)

$$g(V) = \frac{\mu \varepsilon_i W}{h} [V_G - V_T - V]. \tag{9.64}$$

Because $H(x, t)$ is small, we may write

$$I_c(x, t) = I_{c,0} + I_{c,1}(x, t), \; |I_{c,1}| \ll I_{c,0} \tag{9.65}$$

$$V(x, t) = V_0(x) + V_1(x, t), \quad |V_1| \ll V_0. \tag{9.66}$$

By assuming that the gate is short-circuited to the source with respect to potential variations and by using the d.c. relation

$$I_{c,0} = g(V_0) \frac{\mathrm{d} V_0}{\mathrm{d}x} \tag{9.67}$$

one obtains

$$I_{c,1}(x, t) = \frac{\mathrm{d}}{\mathrm{d}x} [g(V_0) V_1(x, t)] + H(x, t). \tag{9.68}$$

We shall evaluate the short-circuit noise current, and therefore $V_1(0, t) = V_1(L, t) = 0$. By integrating equation (9.68) with respect to x from $x = 0$ to $x = L$, one obtains

$$I_{c,1}(x, t) = \frac{1}{L} \int_0^L H(x, t) \, \mathrm{d}x. \tag{9.69}$$

By taking the Fourier transform, equation (9.69) becomes

$$\underline{\mathbf{I}}_{c,1}(x, \omega) = \frac{1}{L} \int_0^L \underline{\mathbf{H}}_1(x, \omega) \, \mathrm{d}x. \tag{9.70}$$

By proceeding as in Section 9.4 one obtains the spectral intensity of the short-circuit noise current

$$S_{i,d} = \frac{1}{L^2} \int_0^L \int_0^L < \underline{\mathbf{H}}_1 (x, \omega) \underline{\mathbf{H}}_1^* (x', \omega) > \mathrm{d}x \, \mathrm{d}x' \, . \tag{9.71}$$

The thermal noise hypothesis yields the spatial cross-correlation spectrum of $H(x, t)$

$$< \underline{\mathbf{H}}_1 (x, \omega) \underline{\mathbf{H}}_1^* (x', \omega) > = 4 k T \mathrm{g}(x') \, \delta (x' - x) \tag{9.72}$$

(δ being the Dirac delta function). The final result of carrying out the integrations in equation (9.71) is

$$S_{i,d} = 4 k T g_{d,0} \frac{2}{3} \frac{1 - (1 - a)^{3/2}}{a} \tag{9.73}$$

where a is the normalized drain current, equation (8.27)

$$a = I_{D,0} / I_{D\,sat} \tag{9.74}$$

and $g_{d,0}$ is the low-current drain conductance, equation (8.42). This result was first obtained by Jordan and Jordan [216] (see also [188] and [217]). The thermal noise in the linear region ($a \to 0$) should be

$$\overline{i_d^2} = 4 k T g_{d,0} \Delta f \tag{9.75}$$

i.e., simply the Nyquist noise (9.1) of the channel resistance $g_{d,0}$. If the device is biased in the saturation region ($a = 1$ in equation (9.74)), the thermal noise is*)

$$\overline{i_d^2} = 4 k T g_{d,0} \Delta f \times \frac{2}{3} \, . \tag{9.76}$$

This noise depends upon the gate voltage (through $g_{d,0}$) but not upon the drain potential. For a given gate voltage, the short-circuit drain noise decreases monotonously with increasing drain voltage and remains constant in the saturation region. It can be easily shown that

$$\overline{i_d^2} = 4 k T g_{d,0} \Delta f \left(\frac{Q_{\mathrm{mob}}}{Q_{\mathrm{mob.\,sat.}}} \right) . \tag{9.77}$$

---

*) Note the fact that the 'noise conductance' $\left(\frac{2}{3} g_{d,0} \right)$ does not equal the differential conductance (which is zero in saturation), or the d.c. conductance $I_{D,0} / V_{D,0}$.

This formula shows that for a given gate to source voltage, the short-circuit noise is proportional to the total mobile charge stored in the channel Q_{mob} ($Q_{mob,\ sat}$ is the charge stored in saturation, both $g_{d,0}$ and $Q_{mob,\ sat}$ are fixed for a given gate bias) [218].

At higher frequencies, the drain noise i_d becomes frequency-dependent due to transit-time effects. Moreover, the noise voltage distribution along the channel

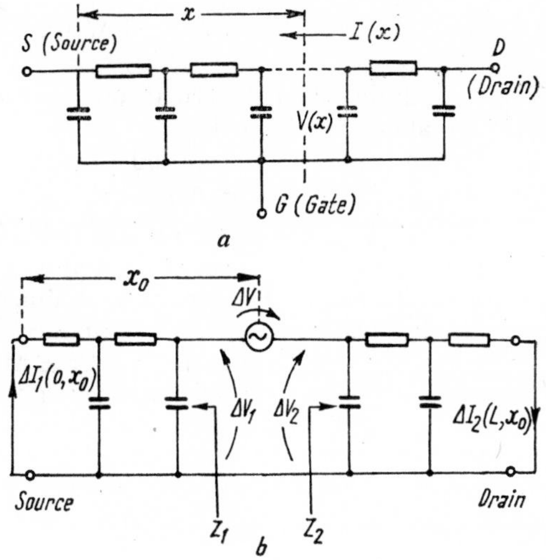

Figure 9.4. (a) Transmission line model of the IGFET; (b) method to calculate the thermal noise [188].

gives rise, by capacitive coupling, to a gate-noise current i_g. The gate noise is partially correlated with the drain noise. Therefore, we have to calculate $\overline{i_d^2}$, $\overline{i_g^2}$ and $\overline{i_g\, i_d^*}$ [192].

The calculation procedure is summarized below (see for example [188], [192]), Let us consider the transmission line model of the MOSFET. Figure 9.4a shows that the resistance of the channel and the distributed capacitance between the channel and the gate constitute a non-uniform, active transmission line. The equations describing the electrical behaviour of this line are given by equations (8.17) and (8.18) and were solved in Chapter 8.

We shall focus our attention on the thermal noise of the resistance of a small section Δx_0 (centred at x_0) of the channel. The corresponding thermal noise voltage generator ΔV drives two transmission lines (Figure 9.4b): one extends to the source and the other to the drain. These two transmission lines are driven, respectively, by ΔV_1 and ΔV_2, and are short circuited at their ends. The short-circuit currents at the source and the drain (due to *that* voltage generator at $x = x_0$) are ΔI_1 (0, x_0) and ΔI_2 (L, x_0), respectively. The voltages ΔV_1 and ΔV_2 are given by

$$\Delta V_1 = -\frac{z_1}{z_1 + z_2}\Delta V,\ \Delta V_2 = \frac{z_2}{z_1 + z_2}\Delta V \tag{9.78}$$

where z_1 and z_2 are the impedances of the transmission lines looking to the left and to the right-hand sides, respectively, from the point x_0. The part of the drain noise current due to ΔV is

$$\Delta i_d = \Delta I_2 (L, x_0). \tag{9.79}$$

The part of the gate noise current due to ΔV at $x = x_0$ is

$$\Delta i_g = -\Delta I_1 (0, x_0) + \Delta I_2 (L, x_0). \tag{9.80}$$

The total drain or gate thermal noise is the square sum of either equation (9.79) or (9.80) along the channel.

The values of $\Delta I_1 (0, x_0)$ and $\Delta I_2(L, x_0)$ can be found by using the solution of the wave (or transmission-line) equations, i.e. the solution of equations (8.19).

The Nyquist formula can be also used to express the value of $\overline{\Delta V^2}$. Such a calculation of $\overline{i_d^2}$, $\overline{i_g^2}$ and $\overline{i_g^* i_d}$ is very long and was performed by Shoji [188]*). These noise generators depend in a complicated manner upon both the operating frequency ω and the d.c. bias voltages. We reproduce here the results derived by Shoji [188] for moderately high frequencies and for a MOSFET biased in saturation. By expanding into the power series of frequency and retaining only the first term one obtains

$$\overline{i_d^2} = 4\,k\,T\,g_{d,0} \times \frac{2}{3} \tag{9.81}$$

(which is identically the low frequency result, equation (9.76), derived by using a different method)

$$\overline{i_g^2} = \frac{64}{135}\,\omega^2 k\,T \frac{C_{gc}^2}{g_{d,0}} \tag{9.82}$$

$$\overline{i_g^* i_d} = \frac{4}{9} i\,\omega\,k\,T\,C_{gc} \tag{9.83}$$

C_{gc} being the gate-to-channel capacitance defined in the preceding chapter.

The noise behaviour (and also the small-signal a.c. parameters) for moderately high frequencies can be also obtained by applying an approximate method of solving the wave equation. Thus, the solution is expressed as a power series of the operating frequency with unknown coefficients [180]. By collecting equal terms in $i\omega$ one obtains differential equations which can be solved successively to yield these unknown coefficients. Such a method is especially useful for studying the MOSFET with finite substrate conductivity (the substrate effect was ignored in the previous chapter and in this Section) because the basic differential equation cannot be solved exactly.

---

*) A similar calculation was performed by Klaassen [220] for the junction-gate FET.

The effect of the substrate upon the thermal noise in MOSFET was studied by Rao [221] and Rao and van der Ziel [222] (see also [223])*) .

It seems generally accepted that the mechanism of thermal noise in field-effect devices is well understood and verified experimentally [192]. Other sources of noise are also present and can dominate the device behaviour in certain frequency and temperature ranges [192].

Problems

9.1. Justify intuitively the effect of the relaxation mechanism upon the injection noise (Section 9.2).

9.2. Calculate the short-circuit noise factor F_0 for the negative-mobility Schottky-barrier diode, by considering d.c. neutrality. Discuss the frequency dependence of F_0.

9.3. Find the high-frequency thermal noise of the SCLC resistor (n^+nn^+ structure, Section 5.5).

Hint: Use equation (9.43) where $v_0(0) = 0$, $\mu = \text{const}$ but $\alpha \neq 0 (N > 0)$.

9.4. Derive the equations describing the electrical behaviour of the transmission line of Figure 9.4a and compare them with equations (8.17) and (8.18).

9.5. Derive equations (9.79)—(9.81) by using the van der Ziel and Ero [180] method, sketched at the end of Section 9.5.

---

*) For 'hot carrier' effects upon the low-frequency drain noise $\overline{i_d^2}$ see Klaassen [224] and Baechtold [225].

Part 3

Non-linear Theory: Transient and Large-signal Behaviour

10

First-order Non-linear Effects: V.H.F Detection and Frequency Multiplication

10.1 Introduction

The non-linear theory should describe such non-linear modes of operation as: the large signal behaviour of active devices in oscillating and amplifying circuits, the turn-on and turn-off of diodes and transistors, various non-linear applications as rectification, detection, mixing, modulation, and analogic multiplication.

In certain cases, a non-linear quasi-stationary analysis suffices. For example, the large-signal amplification properties of a transistor or the detection (rectification) capabilities of a semiconductor diode may be examined by using the steady-state characteristics. The results are valid only for relatively low frequencies. The upper limit in frequency may be approximately evaluated on the basis of the small-signal theory.

There are, however, cases when such a quasi-stationary analysis is insufficient or even useless. A *non-linear and non-stationary* theory is indispensable for studying the transit-time oscillators. A non-stationary analysis is also necessary for large-signal transient, for v.h.f. detection and multiplication.

We may expect that the non-linear and non-stationary theory raises difficult mathematical problems and this is really the case. Very often such an analysis cannot be done by using an exact analytical calculation. Therefore two alternatives have to be considered: either to made drastic simplifications concerning the basic hypotheses and other approximations, or to use numerical techniques and computer calculations. The first approach may lead to an over-simplified analysis which is not satisfactory for design purposes, or even does not retain the basic features of the problem. The disadvantage of the second alternative results from the fact that one loses from the generality of the analysis and sometimes, from the understanding of the physical phenomena.

Consequently, *a diversity of methods* should be used in non-linear problems. Some of them are illustrated in the last part of this book.

The present chapter deals with an approach which allows the study of non-linear transit-time effects. This is a technique of successive approximations, based upon the assumption of a relatively small, but finite, signal. The zero-order approximation yields the steady-state (d.c.) characteristic, the first-order approximation gives the small signal behaviour and the next approximation yields non-linear effects. This theory is used to study the high-frequency detection and frequency multiplication properties of SCLC solid-state diodes. Other applications will be discussed in the next chapter.

10.2 Statement of the Problem

Consider a one-dimensional semiconductor region (Figure 3.1b) between the injecting plane $x = 0$, called conventionally cathode, and the collecting plane, at $x = L$ (anode). The fixed space-charge $(-N)$ is assumed constant and uniformly distributed in space. The diffusion current is neglected. The starting equations (Section 4.2) are written down below

$$J(t) = ep(x, t)\, v(x, t) + \varepsilon \frac{\partial E(x, t)}{\partial t} \qquad (10.1)$$

$$\frac{\partial E(x, t)}{\partial x} = \frac{e[p(x, t) - N]}{\varepsilon} \qquad (10.2)$$

$$v = v(E) \qquad (10.3)$$

$$\Phi(x, t) = -\int E(x, t)\, dx. \qquad (10.4)$$

For calculation purposes let us assume the fact that $J = J(t)$ is given and the voltage drop

$$V(t) = \int_0^L E(x, t)\, dx \qquad (10.5)$$

should be calculated. Therefore, equations (10.1) — (10.3) yield a first-order equation with partial derivatives:

$$\left(\varepsilon \frac{\partial E}{\partial x} + eN\right) v(E) + \varepsilon \frac{\partial E}{\partial t} = J(t). \qquad (10.6)$$

This equation is quasi-linear (i.e. linear with respect to the derivatives). A boundary condition such as

$$E(0, t) = \text{given function of time} \qquad (10.7)$$

should determine the complete solution. Then, $V = V(t)$ will be found from equation (10.5).

The mathematical method used here for solving equation (10.6) is a technique of successive approximations, based upon the assumption that the variable part of the physical quantities (electric field etc.) is relatively small as compared to the d.c. part. This is an adaptation of Grinberg's analysis of the planar vacuum diode [226]. This approach is, in a certain sense, equivalent to the classical Benham-Müller-Llewellyn method (Section 4.3).

The principle of this method may be sketched as follows. *Each injected carrier is labelled with its emission moment* τ. This is necessary because when the applied signal changes fast as compared to the carrier transit-time, each injected carrier experiences a different field distribution during its travel towards the anode, i.e. it has a 'history' which is different from that of the carriers emitted before or after $t = \tau$. The reader has noticed, of course, that in the linear theory the charge carriers were not individualized, and that the transit-time was taken the same for all carriers and equal to the d.c. transit-time T_L.

Let us denote by $x_{t,\tau}$ the distance covered until the time t, by the hole leaving the cathode ($x = 0$) at the time τ. Of course, we have

$$x_{\tau,\tau} = 0. \tag{10.8}$$

The transit-time $t - \tau$ of the hole emitted at the moment τ and reaching the anode at the moment t is implicitly given by

$$x_{t,\tau} = L. \tag{10.9}$$

We also denote by $E_{t,\tau}$ the electric field acting at the time t upon the hole emitted at the time τ. Clearly, the condition (10.7) becomes

$$E_{\tau,\tau} = \text{given}. \tag{10.10}$$

The hole velocity $v(x, t)$ may be written

$$v(x, t) = \frac{\mathrm{d}x_{t,\tau}}{\mathrm{d}t} = v(E_{t,\tau}). \tag{10.11}$$

The integral (10.5) should be calculated at a given time t, and thus $\mathrm{d}x$ may be replaced by

$$\mathrm{d}x = \frac{\partial x_{t,\tau}}{\partial \tau}\,\mathrm{d}\tau. \tag{10.12}$$

Therefore equation (10.5) becomes

$$V(t) = \int_{\tau=t}^{\tau(L,t)} E_{t,\tau} \frac{\partial x_{t,\tau}}{\partial \tau}\,\mathrm{d}\tau. \tag{10.13}$$

Here $\tau(L, t)$ is given implicitly by equation (10.9).

We already pointed out that E (and therefore $V(t)$) will be calculated by assuming $J(t) =$ given. The basic equation is equation (10.6), which may be written successively

$$\varepsilon \frac{\partial E(x, t)}{\partial x} \frac{\mathrm{d}x}{\mathrm{d}t} + eNv(x, t) + \varepsilon \frac{\partial E(x, t)}{\partial t} = J(t) \tag{10.14}$$

$$\varepsilon \frac{\mathrm{d}E(x, t)}{\mathrm{d}t} + eNv\,[E(x, t)] = J(t) \tag{10.15}$$

$$\frac{\partial E_{t,\tau}}{\partial t} + \frac{eN}{\varepsilon} v(E_{t,\tau}) = \frac{1}{\varepsilon} J(t). \tag{10.16}$$

In fact, we made a change of independent variable; $E(x, t)$ was replaced by $E_{t,\tau}$. Due to this change the total derivative in equation (10.15) was replaced by a partial derivative. This is the conversion from Eulerian to Lagrangian variables, and is discussed in detail by Birdsall and Bridges in their monograph on transit-time effects [74].

10.3 Basic Non-linear Equations for SCLC Semiconductor Diodes

In this chapter we shall assume

$$v = \mu E, \;\; \mu = \text{const.} \tag{10.17}$$

i.e. low-field conditions. The case of field-dependent mobility is too complicated for an analytical non-linear theory to be developed. The saturated-velocity case will be examined in the next chapter. From equations (10.16) and (10.17) one obtains

$$\frac{\partial E_{t,\tau}}{\partial t} + \omega_r E_{t,\tau} = \frac{1}{\varepsilon} J(t), \tag{10.18}$$

where ω_r is given by

$$\omega_r = \frac{e\mu N}{\varepsilon} \tag{10.19}$$

and may be positive (majority carrier injection), negative or zero. It may be easily checked by derivation that

$$E_{t,\tau} = \frac{1}{\varepsilon} \int_{\tau}^{t} J(\xi) \,[\exp - \omega_r(t - \xi)]\, \mathrm{d}\xi + E_{\tau,\tau} \tag{10.20}$$

is the solution of equation (10.18) which satisfies the boundary condition, equation (10.10). Because in this chapter we intend to discuss only the SCL injection we shall now introduce the proper boundary condition

$$E_{\tau,\tau} \equiv 0. \tag{10.21}$$

The method remains valid for arbitrary boundary conditions. Equation (10.11) can be integrated by taking into account equations (10.17), (10.20) and (10.21). The result is

$$x_{t,\tau} = \frac{\mu}{\varepsilon\omega_r}\int_{\tau}^{t} J(\xi)\,[1 - \exp - \omega_r(t - \xi)]\,\mathrm{d}\xi. \tag{10.22}$$

By differentiating this equation we have

$$\frac{\partial x_{t,\tau}}{\partial\tau} = -\frac{\mu}{\varepsilon\omega_r}[1 - \exp - \omega_r(t - \tau)]\,J(\tau) \tag{10.23}$$

and introducing this in equation (10.13), one obtains [227]:

$$V(t) = \frac{\mu}{\varepsilon^2\,\omega_r}\int_{\tau=\tau(L,t)}^{\tau=t} [1 - \exp - \omega_r(t - y)]\,J(y)\,\mathrm{d}y\int_{y}^{t} J(\xi)\,[\exp\omega_r(t - \xi)]\,\mathrm{d}\xi. \tag{10.24}$$

Therefore, the voltage drop $V(t)$ may be calculated if $J = J(t)$ is given. Unfortunately, in practice $V(t)$ is given and $J(t)$ should be found and this is a much more difficult task. Another difficulty arises from the fact that $\tau = \tau(L, t)$ in the above equation is given implicitly by equation (10.22) for $x_{t,\tau} = L$. The device response can still be calculated by using a technique of successive approximations, as shown below.

To conclude this Section we note the fact that the change of independent variable ($x \to \tau$) made in equation (10.5) is valid only if $\tau = \tau(x, t)$ is a uniform function of x, i.e.

$$\frac{\partial x_{t,\tau}}{\partial\tau} \neq 0. \tag{10.25}$$

Because of equation (10.23), one finds

$$J(t) \neq 0, \text{ i.e. } J(t) > 0. \tag{10.26}$$

(class A operation).

The physical significance of the condition (10.25) is that at a given time instant t and at an arbitrary position x should not exist carriers emitted at different moments τ. In other words, a hole cannot be overtaken by holes emitted later.

10.4. Mathematical Technique

Due to the lack of space we shall discuss here only the special case $\omega_r = 0$ (N=0) [228], i.e. the behaviour of the square-law SCLC diode (Sections 12, 4.7 and 5.5). The general problem for $\omega_r \neq 0$ was also solved [229] — [231] and the results will be given here in a graphical form.

The basic equations of the problems (10.20), (10.22) and (10.24) become, respectively, ($\omega_r \to 0$)

$$E_{t,\tau} = \frac{1}{\varepsilon}\int_\tau^t J(\xi)\,\mathrm{d}\xi \tag{10.27}$$

$$x_{t,\tau} = \frac{\mu}{\varepsilon}\int_\tau^t (t-\xi)\,J(\xi)\,\mathrm{d}\xi \tag{10.28}$$

$$V(t) = \frac{\mu}{\varepsilon^2}\int_{\tau(L,t)}^t (t-y)J(y)\,\mathrm{d}y \int_y^t J(\xi)\,\mathrm{d}\xi. \tag{10.29}$$

Let us consider a periodical signal $V(t)$ of angular frequency ω, applied on the diode terminals. The diode current should be of the form [226].

$$J(t) = J_0\left[1 + \sum_{n=1}^{\infty} \nu_n \sin(n\omega t + \alpha_n)\right] \tag{10.30}$$

where J_0, ν_n, α_n are still unknown quantities. By introducing $J(t)$ in equation (10.28), one obtains

$$x_{t,\tau} = \frac{\mu J_0}{2\varepsilon}\frac{\alpha^2}{\omega^2}\left\{1 + 2\sum_{n=1}^{\infty} \nu_n[A(n\alpha)\sin(n\omega t + \alpha_n) - B(n\alpha)\cos(n\omega t + \alpha_n)]\right\} \tag{10.31}$$

where

$$\alpha = \omega(t-\tau) \tag{10.32}$$

and

$$A(z) = \frac{1}{z^2}\int_0^z y\cos y\,\mathrm{d}y, \quad B(z) = \frac{1}{z^2}\int_0^z y\sin y\,\mathrm{d}y. \tag{10.33}$$

Zero-order approximation. We put all ν_n equal to zero. For $x_{t,\tau} = L$ and $\alpha = \alpha_0$ in equation (10.31) one obtains

$$\alpha_0 = \omega\left(\frac{2\varepsilon\, L}{\mu J_0}\right)^{1/2} \tag{10.34}$$

and

$$\tau = \tau(L, t) = t - \left(\frac{2\varepsilon L}{\mu J_0}\right). \tag{10.35}$$

Then, equation (10.29) yields

$$V_0 = \frac{\mu J_0^2}{3\varepsilon^2}(t-\tau)^3 \tag{10.36}$$

or

$$J_0 = \frac{9}{8}\varepsilon\mu\frac{V_0^2}{L^3} \tag{10.37}$$

which is the well-known square law calculated for $\mu =$ const., $N = 0$ and SCLC conditions.

First-order approximation. In this stage of approximation we shall calculate the small-signal (linear) response. If the variable part of $V(t)$ is much smaler than the d.c. part, then $J(t)$ should be of the same form as $V(t)$ and all ν_n in equation (10.30) will be small, of the first-order of magnitude (d.c. quantities are zero-order).

By introducing in equation (10.29) $J(t)$ given by equation (10.30) and by neglecting the product of first-order quantities (which is second order in ν_n) one obtains

$$V(t) = \frac{\mu J_0^2}{\varepsilon^2}(I_a + I_b + I_c) \tag{10.38}$$

where

$$I_a = \int_{\tau(L,t)}^{t} (t-y)^2\,\mathrm{d}y \tag{10.39}$$

$$I_b = \int_{\tau(L,t)}^{t} (t-y)\sum_{n=1}^{\infty}\frac{\nu_n}{n\omega}[\cos\psi_n(y) - \cos\psi_n(t)] \tag{10.40}$$

$$I_c = \int_{\tau(a,t)}^{t} (t-y)^2\sum_{n=1}^{\infty}\nu_n \sin\psi_n(y)\,\mathrm{d}y \tag{10.41}$$

$$\psi_n(y) = n\omega y + \alpha_n. \tag{10.42}$$

The carrier transit-time $t - \tau(L, t)$ should be evaluated first. Equation (10.31) may be written

$$x_{t,\tau} = \frac{\mu J_0^2}{2\varepsilon}\,\frac{\alpha^2}{\omega^2}[1 + 2\zeta(\alpha)] \tag{10.43}$$

where

$$\zeta(\alpha) = \sum_{n=1}^{\infty}\nu_n[A(n\alpha)\sin\psi_n(t) - B(n\alpha)\cos\psi_n(t)] \tag{10.44}$$

is a first-order quantity. Hence $\zeta(\alpha)$ may be approximated as

$$\zeta(\alpha) \simeq \zeta(\alpha_0) \tag{10.45}$$

and the error occurring in evaluating equation (10.43) is of the second-order of magnitude,

$$L \simeq \frac{\mu J_0}{2\varepsilon} \frac{\alpha^2}{\omega^2} [1 + 2\zeta(\alpha_0)]. \tag{10.46}$$

α may be calculated from equations (10.34) and (10.46). The result is

$$\alpha \simeq \alpha_0 [1 - \zeta(\alpha_0)] \tag{10.47}$$

and hence

$$\tau(L, t) = t - \frac{\alpha_0}{\omega} [1 - \zeta(\alpha_0)]. \tag{10.48}$$

This expression of $\tau(L, t)$ will be used to calculate $V(t)$ in equation (10.38). Note the fact that as long as we evaluate the integrals (10.40) and (10.41), ζ may be neglected in equation (10.48). The final result is

$$V(t) = V_0 \left\{ 1 + \sum_{n=1}^{\infty} \nu_n [S_1(n\alpha_0) \cos \psi_n(t) + S_2(n\alpha_0) \sin \psi_n(t)] \right. \tag{10.49}$$

where

$$S_1(z) = \frac{3}{z^3} \left(1 - \cos z - \frac{z^2}{2} \right) \tag{10.50}$$

$$S_2(z) = \frac{3}{z^3} (z - \sin z) \tag{10.51}$$

and V_0 is given by equation (10.37). The d.c. component is not modified by the signal. The response is linear, i.e. it may be calculated separately for each harmonic, by using the small signal impedance $Z = Z(n\omega)$ where n is the order of the harmonic. The normalized impedance is

$$\left. \begin{aligned} \overline{Z} &= \frac{Z}{R_i} = 2 S_2(\theta) + i 2 S_1(\theta) \\ R_i &= Z|_{\omega \to 0} = \frac{1}{2} \frac{V_0}{J_0} \end{aligned} \right\} \tag{10.52}$$

(R_i is the incremental resistance)

$$\theta = \alpha_0 = \omega [t - \tau(L, t)] = \omega T_L^0 \tag{10.53}$$

is the d.c. transit-angle, and T_L^0 the d.c. hole transit-time. It can be easily shown that the impedance, equation (10.52), is identical with equation (4.90), derived by Shao and Wright [10] and other authors.

Second-order approximation. Let us consider an electrical circuit containing the SCLC diode, a voltage source

$$V_b(t) = V_{b0}(1 + \nu \sin \omega t) \tag{10.54}$$

and other (passive) circuit elements. We assume that

$$\nu \ll 1. \tag{10.55}$$

The current through the diode should be a periodical function $J = J(t)$ having the same period $2\pi/\omega$ as the voltage, equation (10.54). The current may be expressed by a Fourier series, with unknown coefficients. Moreover, in the s branch of the circuit flows a current which may be expressed as [226]

$$I^{(s)}(t) = I_0^{(s)} + \sum_{n=1}^{\infty} I_n^{(s)} \sin(n\omega t + \alpha_n^{(s)}). \tag{10.56}$$

$I_n^{(s)}$ and $\alpha_n^{(s)}$ are still unknown but may be considered as functions of ν. Because ν is a small quantity, $I^{(s)}(t)$ may be expressed as a power series of ν. Let us consider the circuit branch which contains the SCLC diode. The current through it is assumed known, and the voltage drop $V(t)$ can be calculated according to equations (10.28) and (10.29) and expressed as a power series in ν. We recall the fact that this development contains unknown coefficients. These coefficients will be determined by also calculating $V(t)$ (the voltage across the diode) from Kirchhoff's equations and identifying the coefficients of ν^k $(k = 0, 1, 2 \ldots)$ from the equality of the two expressions obtained for $V(t)$.

Two particular circuits will be considered:

(1) The voltage, equation (10.54), is applied directly on the diode terminals

$$V(t) \equiv V_b(t) = V_{b,0}(1 + \nu \sin \omega t) \tag{10.57}$$

(2) A series circuit contains the voltage-source, equation (10.54), the SCLC diode and a series resistor R_s, i.e.

$$R_s I + V(t) = V_b(t) \tag{10.58}$$

where $I = JA$ is the diode current (A = the electrode area).

As shown above, we have to calculate $V(t)$ as a power series of ν. This is a most difficult problem and requires a long and tedious calculation [228].

Let us write the diode current (periodical function of time)

$$J = J(t) = J_0\left[1 + \sum_{n=1}^{\infty} \nu_n \sin(n\omega t + \alpha_n)\right] = J_0[1 + \delta(t)] \tag{10.59}$$

where

$$\delta(t) = \sum_{n=1}^{\infty} \nu_n \sin(n\omega t + \alpha_n). \tag{10.60}$$

The expression (10.59) is identical with equation (10.30) but has a different significance. Here, the applied voltage has a *sinusoidal* variable part $\nu V_{b,0} \sin \omega t$. If $V_{b,0}$ is of zero-order of magnitude and $\nu V_{b,0}$ is first-order, then $J(t) - J_0$ should be, to the first-order of magnitude, also sinusoidal i.e. $J(t) - J_0 = \nu_1 \sin(\omega t + \alpha)$ or $\nu_1 \sim \nu$. If second and higher-order of magnitude quantities are taken into account, each ν_n will be expressed as a series with respect to the ν^k where k is an integer. By taking into account some properties of the trigonometric functions, one obtains

$$\nu_{2p+1} = \sum_{k=0}^{\infty} \alpha_k^{(2p+1)} \nu^{2k+1} = \alpha_0^{(2p+1)} \nu + \alpha_1^{(2p+1)} \nu^3 + \alpha_2^{(2p+1)} \nu^5 + \dots \tag{10.61}$$

$$\nu_{2p} = \sum_{k=1}^{\infty} \beta_k^{(2p)} \nu^{2k} = \beta_1^{(2p)} \nu^2 + \beta_2^{(2p)} \nu^4 + \dots \quad . \tag{10.62}$$

The d.c. component J_0 should be of the form

$$J_0 = J_0(\nu) = \bar{J}_0 \left[1 + \sum_{k=0}^{\infty} \rho_k \nu^{2(k+1)}\right] \tag{10.63}$$

where $\bar{J}_0$ is the d.c. component in the absence of signal ($\nu = 0$).

In the second-order approximation we shall neglect all third and higher-order quantities. For example, J_0 should be of the form

$$J_0 \simeq \bar{J}_0(1 + \rho_0 \nu^2). \tag{10.64}$$

The calculations presented in Appendix to this Section yield the following expression for the voltage across the diode terminals

$$V(t) = \bar{V}_0 \Big\{ 1 + \frac{1}{2} \nu^2 \rho_0 + \nu \alpha' (S_{11} \cos \psi_1 + S_{21} \sin \psi_1) -$$

$$- \frac{3}{4} \nu^2 (\alpha')^2 [(M_1 - A_1^2 - B_1^2) + (M_2 + A_1^2 - B_1^2) \cos 2\psi_1 +$$

$$+ (M_3 + 2A_1B_1) \sin 2\psi_1] + \nu_2(\varphi_1 \cos 2\psi_2 +$$

$$+ \varphi_2 \sin 2\psi_2) + \sum_{n=3}^{\infty} \nu_n [S_{1n} \cos \psi_n(t) + S_{2n} \sin \psi_n(t)] \Big\} \tag{10.65}$$

where, according to equation (10.62)

$$\nu_2 = \beta_1^{(2)} \nu^2 = \beta \nu^2 \tag{10.66}$$

and S_{ij}, A_1, B_1, M_1 etc. are defined in the Appendix.

10.5 Application to Detection with Square Law Diodes

On the basis of the above theory we shall calculate the detection characteristic of the square law SCLC diode ($\theta_r = 0$, $N = 0$) [228], [232].

Let us first evaluate the non-linear effects under the assumption of negligible series resistence ($R_s \to 0$ in equation 10.58), i.e. the voltage (10.54) appears directly on the diode terminals (equation 10.57).

Because ψ_1 is defined as $\omega t + \alpha_1$, we may write $\sin \omega t = \sin (\psi_1 - \alpha_1)$ and thus

$$V(t) = V_0 [1 + \nu \sin (\psi_1 - \alpha_1)]. \tag{10.67}$$

Because of equation (10.57), we shall identify equation (10.67) with equation (10.65)

$$V_0[1 + \nu (\cos \alpha_1 \sin \psi_1 - \sin \alpha_1 \cos \psi_1)] =$$

$$= \overline{V}_0 \Bigg\{ 1 + \nu \alpha' (S_{11} \cos \psi_1 + S_{21} \sin \psi_1) +$$

$$+ \nu^2 \left[\frac{\rho_0}{2} - \frac{3}{4} \alpha^2 (M_1 - A_1^2 - B_1^2) + \right.$$

$$+ \left(\beta \varphi_1 - \frac{3}{4} (\alpha')^2 [M_2 + A_1^2 - B_1^2] \right) \cos 2 \psi_1 +$$

$$+ \left. \left(\beta \varphi_2 - \frac{3}{4} (\alpha')^2 [M_3 + 2A_1 B_1] \right) \sin 2 \psi_1 \right] +$$

$$+ \sum_{n=3}^{\infty} \nu_n [S_{1n} \cos \psi_n + S_{2n} \sin \psi_n] \Bigg\}. \tag{10.68}$$

As shown in the previous Section, we should equalize the coefficients of ν^k ($k = 0,1,2, \ldots$). i.e. the terms of the same order of magnitude on both sides of the identity (10.68) should equal each other. On the other hand, relationship (10.68) holds for any time instant t, i.e. for an arbitrary $\psi_1 = \omega t + \alpha_1$. Thus, we should equalize separately the coefficients of $\cos \psi_1, \sin \psi_1, \cos 2\psi_1, \sin 2\psi_1, \ldots$ Therefore, one obtains

$$\bar{V}_0 = V_0 \tag{10.69}$$

$$\cos \alpha_1 = \alpha' S_{21}, \ \sin \alpha_1 = - \alpha' S_{11} \tag{10.70}$$

$$\frac{1}{2} \rho_0 = \frac{3}{4} (\alpha')^2 [M_1 - A_1^2 - B_1^2] \tag{10.71}$$

$$\beta \varphi_1 = \frac{3}{4} (\alpha')^2 [M_2 + A_1^2 - B_1^2] \tag{10.72}$$

$$\beta \varphi_2 = \frac{3}{4} (\alpha')^2 [M_3 + 2A_1 B_1] \tag{10.73}$$

$$\nu_n = 0 \text{ for } n = 3, 4, 5, \ldots . \tag{10.74}$$

From equation (10.69) one obtains the steady-state characteristic

$$\bar{J}_0 = \frac{9}{8} \varepsilon\mu \frac{\bar{V}_0^2}{L^3} . \tag{10.75}$$

Equations (10.70) yield

$$\alpha' = \frac{1}{\sqrt{S_{11}^2 + S_{21}^2}}, \ \tan \alpha_1 = - \frac{S_{11}}{S_{21}} \tag{10.76}$$

which corresponds to the small signal impedance for the fundamental frequency (see the preceding Section). Note the fact that in the second-order approximation, the a.c. impedance is not affected by the amplitude of the signal.

Equations (10.71) and (10.76) determine*)

$$\rho_0 = \frac{3}{2} \frac{M_1 - A_1^2 - B_1^2}{S_{11}^2 + S_{21}^2} \tag{10.77}$$

and if M_1, A_1, B_1 are replaced by their value from Appendix to Section 10.4, one obtains

$$\rho_0 = \frac{6}{[\bar{Z}(\theta)]^2} \frac{\theta^2 + 2(\cos \theta - 1)}{\theta^4} = \rho_0(\theta) \tag{10.78}$$

---

*) Equations (10.72) and (10.73) will be used to calculate the second harmonic (Section 10.7).

where

$$\bar{Z}(\theta) = 2S_{11} + i2S_{21} = 2S_1(\theta) + 2iS_2(\theta) \qquad (10.79)$$

is the small signal (normalized) impedance given by equations (10.50)—(10.54), and also calculated in Section 4.7[10].

According to equation (10.64), an increase ΔJ_0 of the d.c. component J_0 with respect to the steady-state value $J_0 = J_0|_{\nu=0}$ should occur. This will be given by

$$\Delta J_0 = J_0 - \bar{J}_0 = \bar{J}_0 \, \nu^2 \rho_0(\theta) = \frac{\bar{J}_0}{V_0^2} V_1^2 \rho_0(\theta) \qquad (10.80)$$

where

$$V_1 = \nu V_0 \qquad (10.81)$$

is the amplitude of the sinusoidal signal (see equation (10.57))

ΔJ_0 is called here the detected current. Because ΔJ_0 is proportional to the square of the applied voltage, the detection is named 'parabolic' or 'power' detection.

The d.c. characteristic, equation (10.75), may be written ($\bar{V}_0 = V_0$)

$$\bar{J}_0 = K\bar{V}_0^2, \; K = (9/8)\,\varepsilon\mu L^{-3} \qquad (10.82)$$

and ΔJ_0 becomes

$$\Delta J_0 = KV_1^2 \rho_0(\theta) = \Delta J_0(\theta). \qquad (10.83)$$

The low frequency detected current is

$$\Delta J_0(0) = KV_1^2 \rho_0(0) = \frac{1}{2} KV_1^2. \qquad (10.84)$$

This result can be found directly from the steady-state characteristic, equation (10.82). The reader will easily demonstrate that the low-frequency detection with a d.c. biased square-law diode is a parabolic detection not only at relatively small but also at large amplitudes of the applied signal (provided that the voltage $V(t)$ does not go beyond the limits of the square-law region, equation (10.82))*). However, the frequency-dependent detected current, equation (10.83), does not have the same property, because for large-signal operation, large variations ΔT of the carrier transit-time (see equation (10.4.12)) should occur, and the calculations made in Appendix to Section 10.4 are no longer valid.

The normalized detected current

$$\frac{\Delta J_0(\theta)}{\Delta J_0(0)} = 2\rho_0(\theta) = \frac{12}{Z^2(\theta)} \, \frac{\theta^2 + 2(\cos\theta - 1)}{\theta^4} \qquad (10.85)$$

---

* The square-law semiconductor diode can be an excellent parabolic detector, its principal advantage being the extremely large dynamic range [233]. A recent paper [234] demonstrates experimentally the high-level (up to several volts) parabolic detection with SCLC silicon diodes.

is a function of the transit angle $\theta = \omega T_L^0$ only and is plotted in Figure 10.1 (dashed curve) [232]. A series expansion for small θ yields the result obtained by Wright [235], who predicted a small increase of ΔJ_0 at frequencies comparable to the transit-time frequency ($\theta = 2\pi$) (see Problem 10.3). The theory developed here also predicts small oscillations of ΔJ_0 with increasing frequency [232] and an almost constant

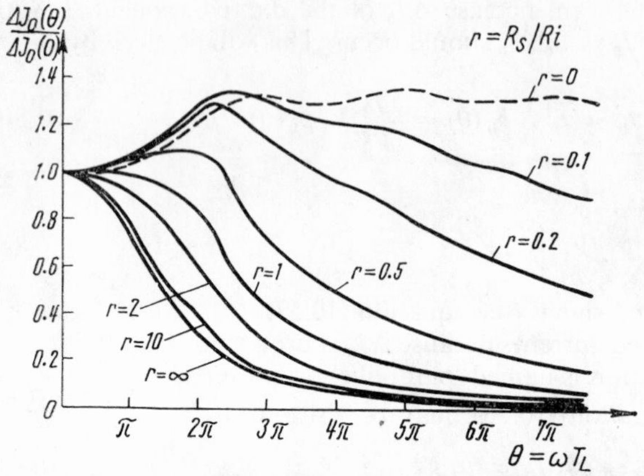

Figure 10.1. High-frequency detection characteristics of the square-law SCLC solid-state diode [232].

value at frequencies much higher than the transit-time frequency. The asymptotic value is [228]

$$\Delta J_0 \Big|_{\theta \to \infty} = \frac{4}{3} \Delta J_0(0) = \frac{2}{3} K V_1^2. \tag{10.86}$$

We shall consider a detection circuit which contains a resistance $R_s \neq 0$ in series with the SCLC diode. The detected current may be determined as follows. As a first approximation we consider that the signal V_{b1} is very small and calculate the steady-state current, thus determining the d.c. operating point and the incremental resistance R_i. The amplitude of the a.c. voltage on the diode terminals is

$$V_1 = V_{b,1} \left| \frac{Z(\theta)}{R_s + Z(\theta)} \right|. \tag{10.87}$$

As a second approximation, the detected current is calculated by replacing V_1 in the above equations ($V_{b,1}$ = constant). The result is [232]

$$\frac{\Delta J_0(\theta)}{\Delta J_0(0)} = \frac{12\,[\theta^2 + 2(\cos\theta - 1)]}{\theta^4} \left(\frac{r+1}{r+\bar{Z}(\theta)}\right)^2$$

where

$$r = R_s/R_i. \tag{10.88}$$

The normalized detection characteristics with r as a parameter are represented in Figure 10.1 [232]. The following features of these characteristic are important. If R_s is comparable to R_i, the detection effect will decrease appreciably at the transit-time frequency ($\theta = 2\pi$). Secondly, even a very small series resistance ($R_s \ll R_i$) will degradate the ultra-high frequency performances (see $r = 0.1$ curve on Figure 10.1). This is due to the diode capacitance which shunts the device at frequencies comparable to or higher than the transit-time frequency (Section 5.2). Thus, a smaller part of the applied voltage occurs across the diode and the detected current decreases.

10.6 Effect of Fixed Space Charge upon V.H.F. Detection with SCLC Solid-state Diodes

The detection characteristics of the $p^+\pi p^+$ SCLC silicon resistor ($N > 0,\ \theta_r > 0$) and the punch-through $p^+\nu p^+$ diode ($N < 0,\ \theta_r < 0$) were also derived in analytical form [229], [231], but are too complicated and are not given here. The normalized detection characteristics

$$\frac{\Delta J_0(\theta)}{\Delta J_0(0)} = \Lambda\,(\theta; \theta_r) \tag{10.89}$$

are represented in Figure 10.2 for majority carrier injection ($\theta_r > 0$) and in Figures 10.3 and 10.4 for minority carrier-injection ($\theta_r < 0$). The normalized transit-time

$$\theta_r = \omega_r T_L,\ \omega_r = \frac{e\mu N}{\varepsilon} \tag{10.90}$$

determines the steady-state bias, as shown by the normalized d.c. characteristics plotted, respectively, in Figure 3.21 ($\Delta \to \infty$) for θ_r positive (SCLC resistor) and in Figure 5.15 for θ_r negative (punch-through diode).

The low frequency detected current $\Delta J_0(0)$ (used as a normalization constant in Figures 10.2—10.5) changes with the d.c. bias (for a given device and constant signal amplitude) as shown in Figure 10.5. Note the fact that for the SCLC resistor, there exists a maximum of ΔJ_0. The change of sign which takes place for $\theta_r \simeq 0.57$ indicates an inflection point on the $J_0 - V_0$ characteristic of the punch-through diode: in the low current region ($|\theta_r| > 0.57$) the mean current decreases with the increasing signal amplitude.

Let us first discuss the detection characteristics for majority carrier injection (Figure 10.2). All characteristics tend to 'saturate' at frequencies much higher than the transit-time frequency ($\theta \gg 2\pi$). The asymptotic value of the detected current, $\Delta J_0(\infty)$ is plotted in Figure 10.5.

Figure 10.2 shows that if $\theta_r > 0.3$ (low-current region on the steady-state characteristic, $\Delta \to \infty$ in Figure 3.21), the detected current will first *decrease* with increasing frequency, whereas when biased in the square-law region ($\theta_r \to 0$), the detected current *increases* [232], [234]. This difference s unexpected (because the d.c. operating point $\theta_r = 0.3$ is practically in the square-law region) and might be of practical importance for an experimental verification.

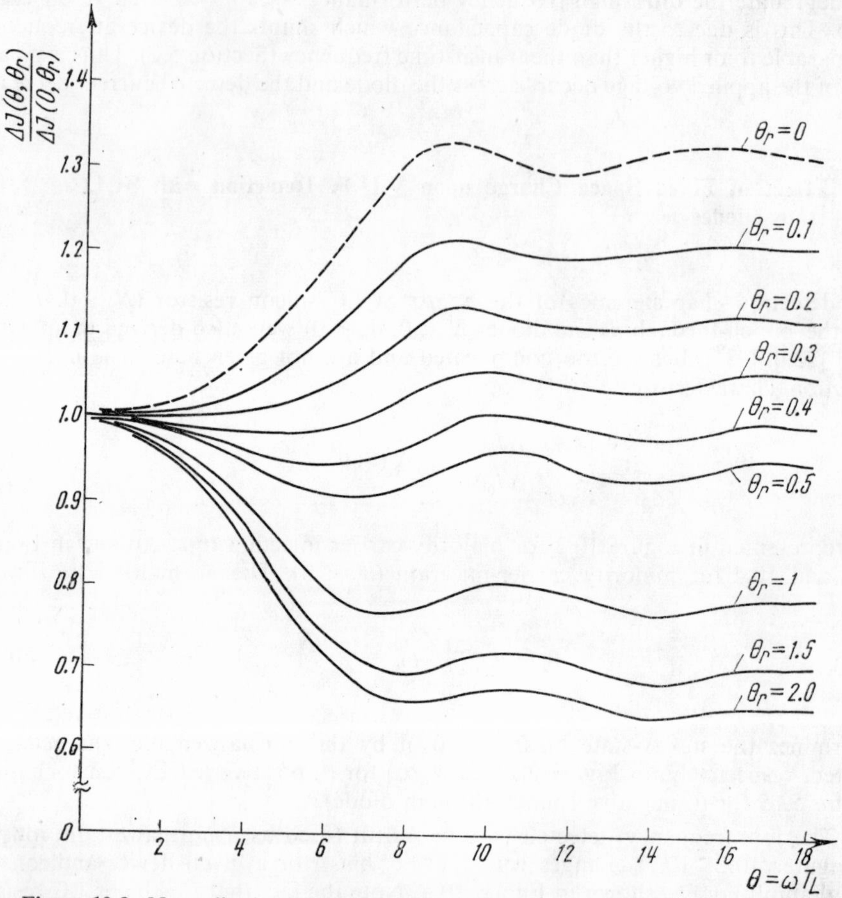

Figure 10.2. Normalized detection characteristics of the SCLC semiconductor resistor.

The variation of the detected current with frequency (Figure 10.2) is relatively small. The transit-time effects do not represent a serious limitation for v.h.f. detection. The frequency-dependence may be minimized by properly choosing the steady-state operating point. If the d.c. bias voltage is so that θ_r lies somewhere between 0.25 and 0.5 (Figure 10.2), the variation of ΔJ_0 with frequency will not exceed $\pm 10\%$, which is negligibly small for most applications [230].

The detection characteristics of the punch-through diode ($\theta_r < 0$) will be now considered. The asymptotic values $\Delta J_0(\infty)$ and $\Delta J_0(0)$ are plotted in Figure 10.5.

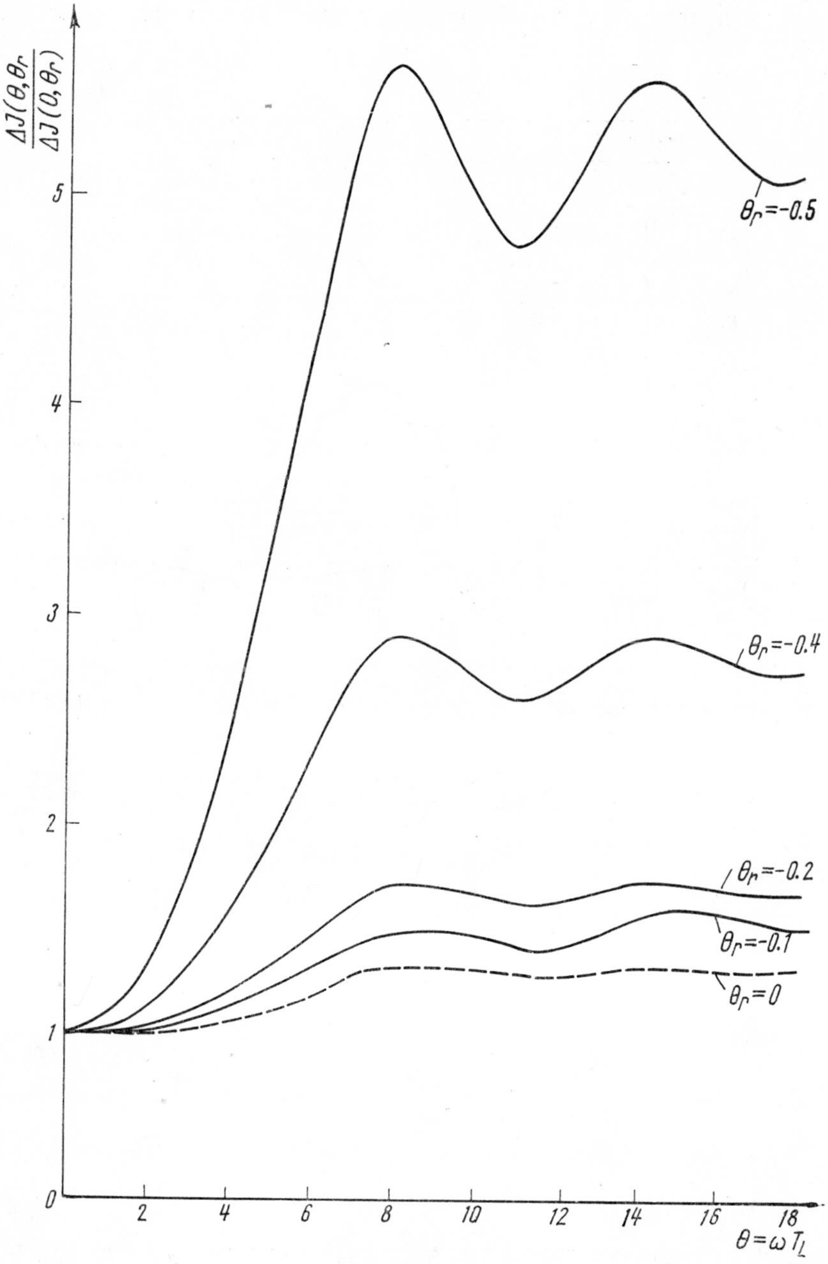

Figure 10.3. Normalized detection characteristics of the SCLC punch-through diode (I).

We distinguish three regions of the steady-state characteristic, according to Table 10.1.

Table 10.1

Region	θ_r	$\Delta J_0(0, \theta_r)$	$\Delta J_0(\infty, \theta_r)$
I	$-0.57 < \theta_r < 0$	> 0	> 0
II	$-0.94 < \theta_r < -0.57$	< 0	> 0
III	$\theta_r < -0.94$	< 0	< 0

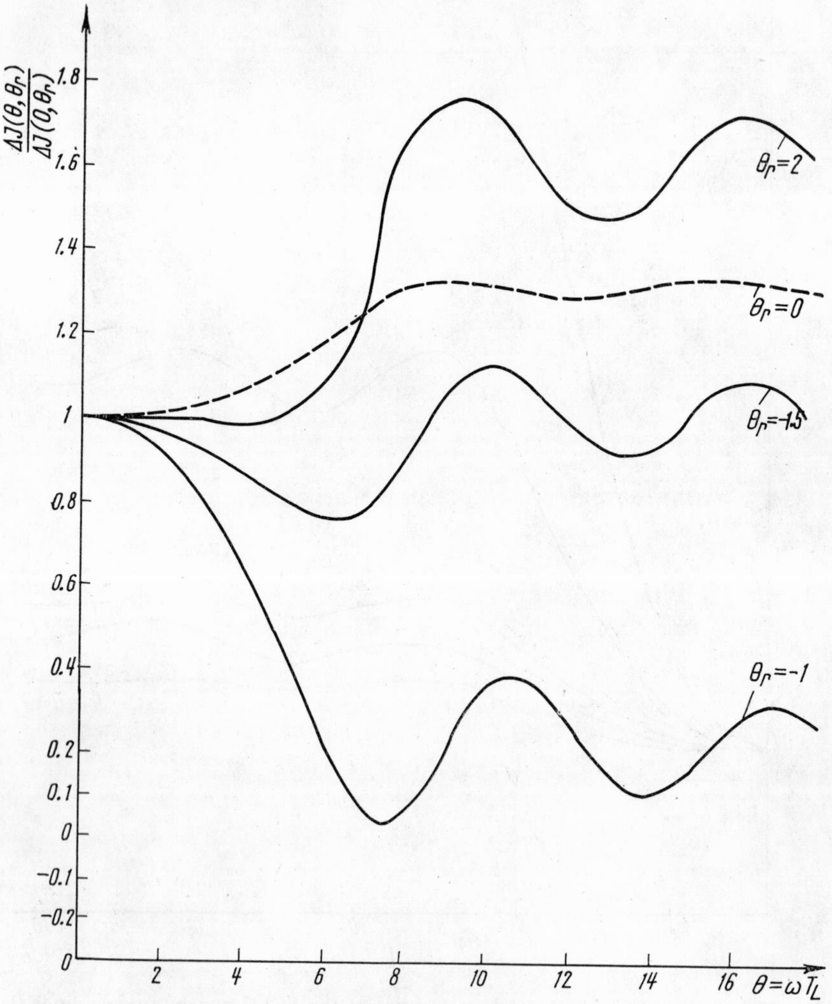

Figure 10.4. Normalized detection characteristics of the SCLC punch-through diode (II).

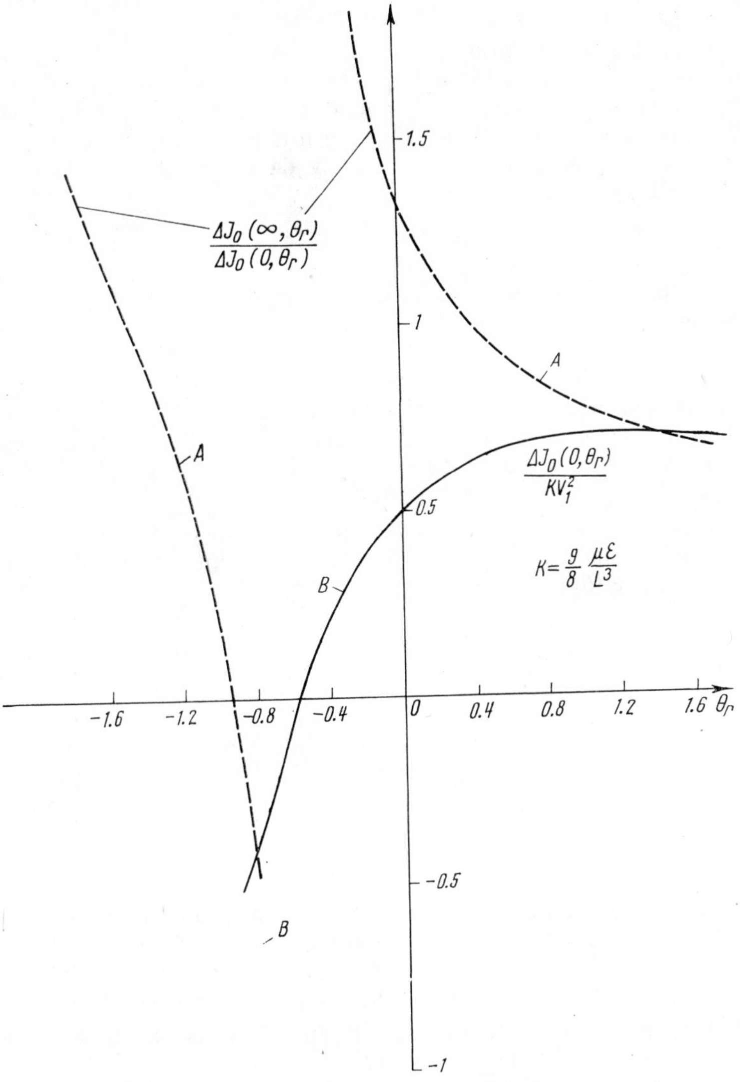

Figure 10.5. Bias dependence of the low-frequency detected current

Figure 10.3 represents characteristics obtained for ad.c. bias in the region I, whereas Figure 10.4 shows characteristics corresponding to region III. If the device is biased in the intermediate region II, the detection characteristics should cross the frequency axis, going from negative values of $\Lambda(\theta; \theta_r)$ (equation (10.89)) at low frequencies to positive values at very-high frequencies.

Therefore, the punch-through diode ($\theta_r < 0$) is not suitable for broad band detection, since very large variations of ΔJ_0 may occur around the transit-time frequency. However, the device could be used, at least in principle, for power detection at frequencies much higher than the transit-time frequency. The fact should be stressed that a series resistance may severely degradate these 'ideal' capabilities (Section 10.5) [229], [231]*)

10.7 Transit-time Effects upon Frequency Multiplication with SCLC Semiconductor Diodes

Let us first consider the square-law region of a SCLC diode ($\theta_r \to 0$). It can be shown that the current $J(t)$ calculated in the second-order approximation (Section 10.4) contains also a second harmonic (frequency 2ω). This is derived in the Appendix to this Section and plotted in Figure 10.6 ($\theta_r = 0$).

Similar calculations were also performed for $\theta_r \neq 0$. We are particularly interested in the amplitude A_2 of this harmonic and represent it normalized to its low frequency value ($\theta \to 0$), i.e.

$$\frac{A_2(\theta;\ \theta_r)}{A_2(0;\ \theta_r)} = G(\theta, \theta_r). \tag{10.91}$$

It can be shown (Problem 10.6) that $A_2(0; \theta_r)$ is identical with $\Delta J_0\,(0; \theta_r)$ (see Figure 10.5).

The h.f. multiplication characteristics for several steady-state operating points of the majority-carrier $p^+\pi p^+$ device are shown in Figure 10.6 and 10.7. Strong oscillations with increasing $\theta = \omega T^c{}_L$ do occur. One cycle corresponds to approximately π (compare with the small signal frequency characteristics represented in Figure 5.13). Note the fact that the second harmonic is *nearly rejected* at certain transit angles.

Figures 10.8 and 10.9 show $G\,(\theta; \theta_r)$ for the punch-through diode ($\theta_r < 0$). The oscillations of the characteristics from Figure 10.8 are even stronger than those found for majority-carrier injection. Figure 10.9 shows (for certain d.c. operating points) a 'step by step' increase with increasing frequency [229], [231].

---

*) We also reported in [236], [237] the effect of shallow traps (Section 5.10) upon the detection characteristics of the SCLC solid-state-diode, at moderately-high frequencies (the transit-time effect was neglected).

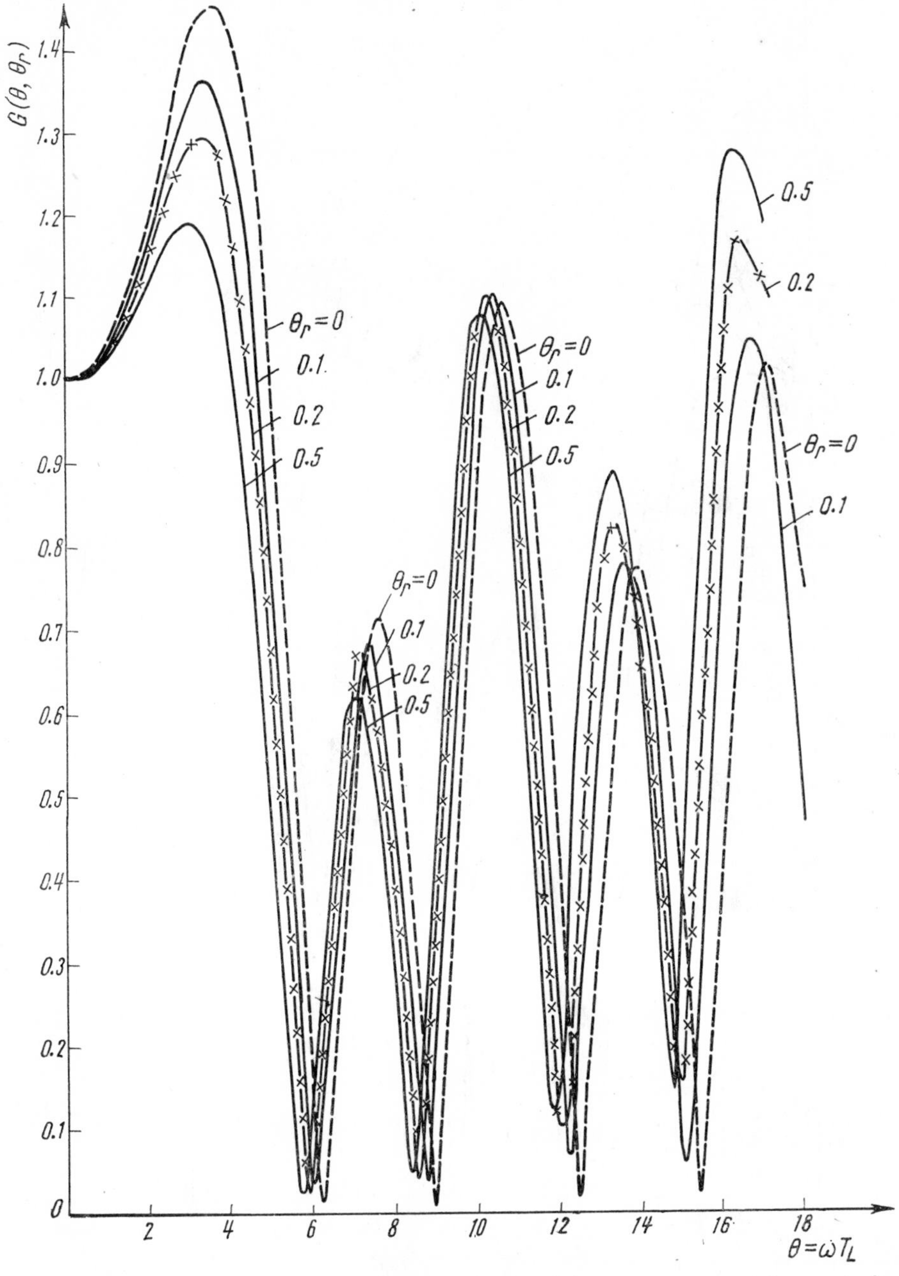

Figure 10.6. Frequency dependence of the amplitude of the current second harmonic for the SCLC resistor (I).

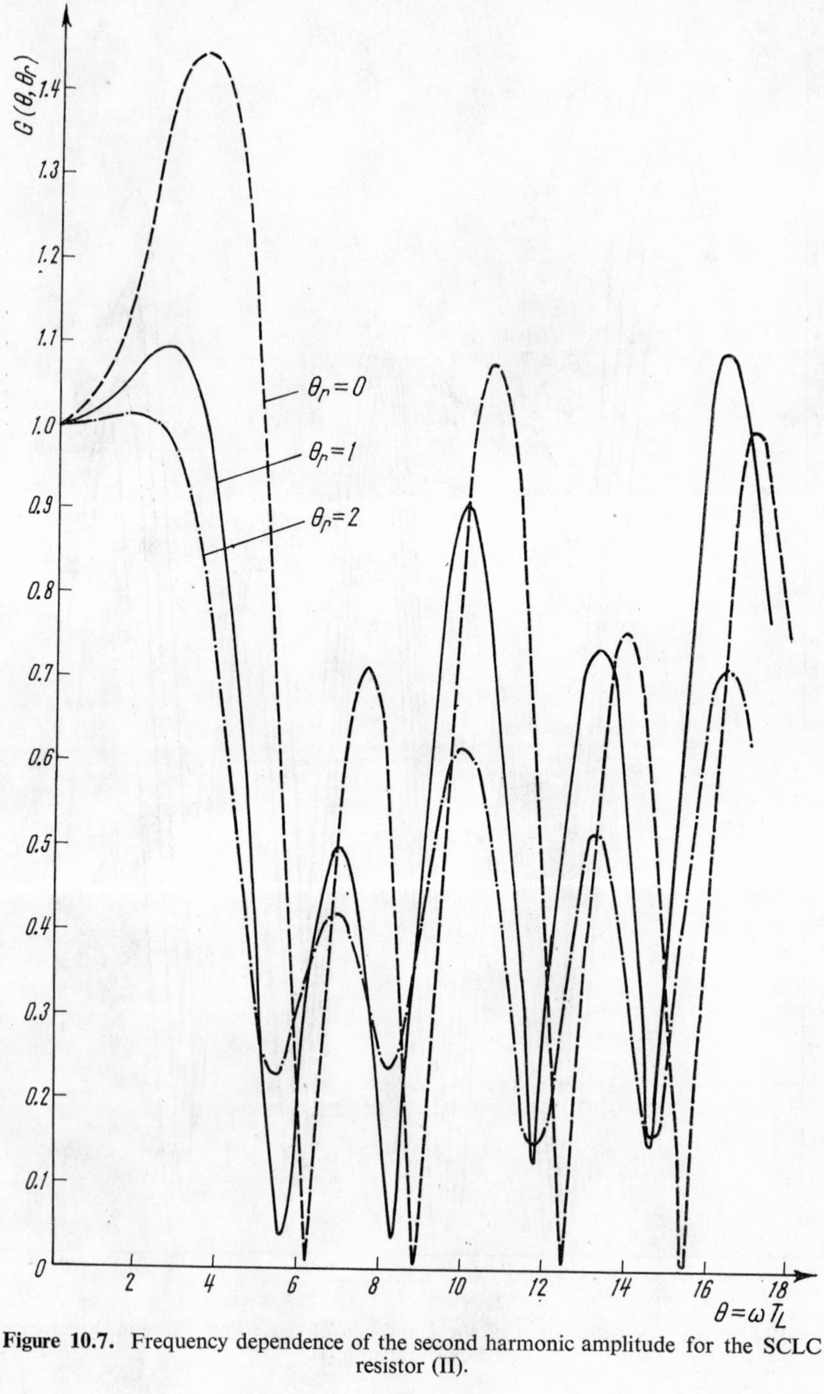

Figure 10.7. Frequency dependence of the second harmonic amplitude for the SCLC resistor (II).

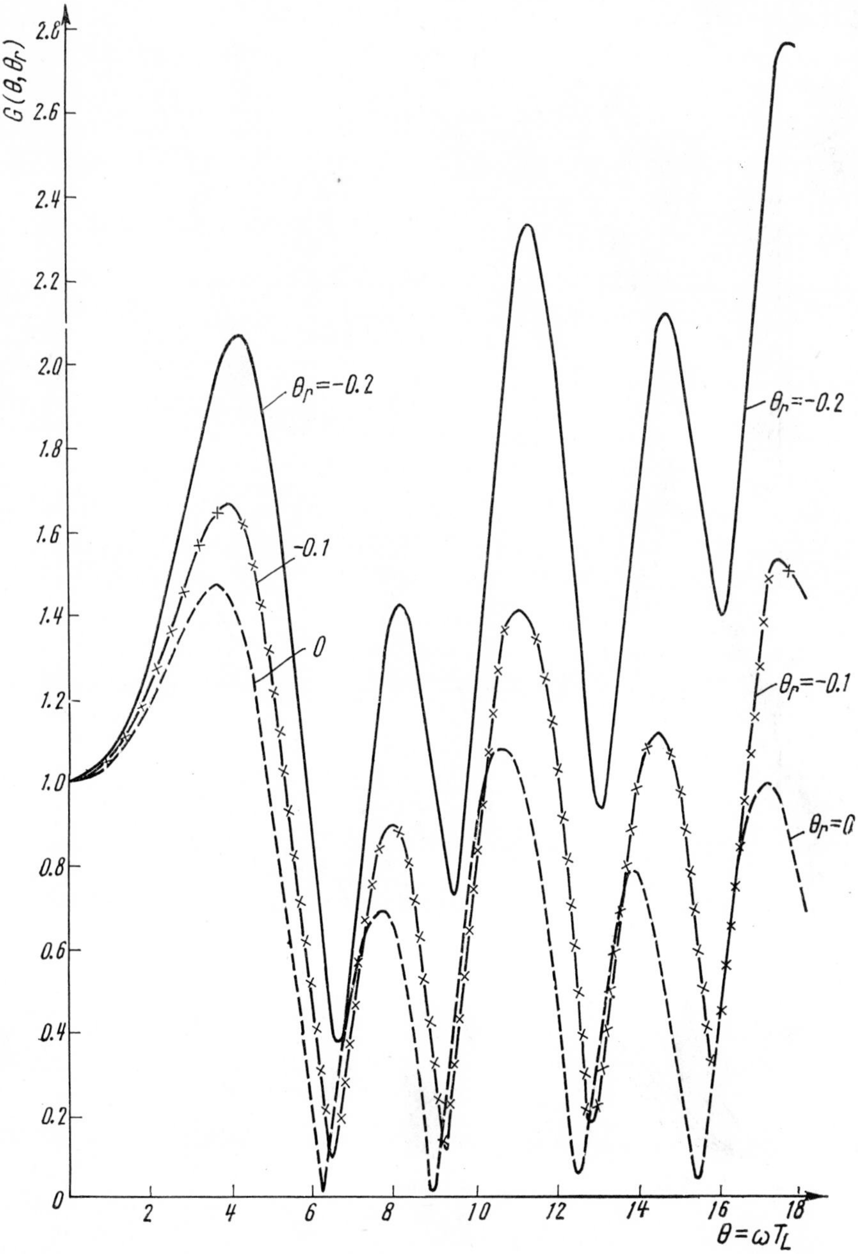

Figure 10.8. Frequency dependence of the second harmonic amplitude for the SCLC punch-through diode (I).

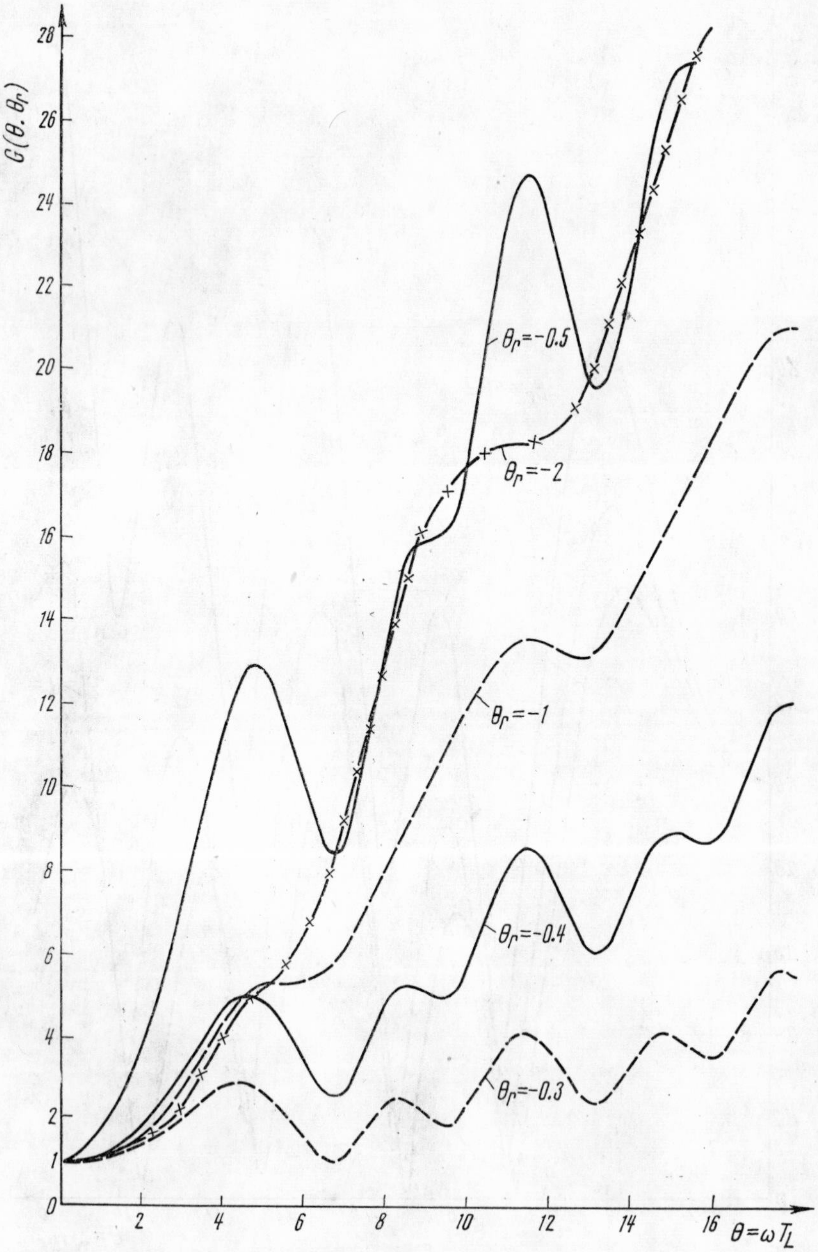

Figure 10.9. Frequency dependence of the second harmonic amplitude for the SCLC punch-through diode (II).

10.8 Conclusions

The analytical theory put forward in this paper is exact, in the limits of relatively-small signal amplitudes.

The third-order approximation will be even more complicated*). However, such an approach, based on successive approximations, may be performed *numerically*, by using a digital computer.

Concerning the physical model used for the above calculations, we note the fact that neglecting diffusion may lead to serious errors, especially in the case of the punch-through diode.

As shown by Kroemer [83], the position of the virtual cathode should be also modulated by the applied signal. This modulation is small for small signal operation and its effect can be neglected if the distance between the actual and the virtual cathode is very small as compared to the interelectrode distance. The question is whether in the second-order approximation when all second-order perturbations should be taken into account, this modulation can be neglected or not. Thus, the second-order (non-linear) theory may not prove satisfactory, even if the steady-state and small signal characteristics derived for the same model are verified experimentally.

Consequently, possible experimental measurements of the non-linear effects may prove interesting for both electronic circuits and device theory.

Problems

10.1. Verify that equations (10.50)–(10.53) lead to the same expression of the SCLC impedance as in equation (4.90).

10.2. Derive the frequency-dependence of the detected current ΔJ_0 by assuming that the alternating current is kept constant. Plot the above dependence for the SCLC square-law diode and suggest an intuitive explanation for the behaviour you found.

10.3. Consider the normalized detected current, equation (10.85). Find the series expansion of this quantity for low θ (see also [235]) and then determine the frequency range where equation (10.85) can be approximated by using only the first two terms of the above expansion.

10.4. By studying the high-frequency behaviour of the planar vacuum diode, Taub and Wax [238] indicated that by applying an a.c. signal of relatively small amplitude to this device, the active-power dissipation increases by a quantity which differs from the a.c. power consumed by the a.c. device resistance. The result of these authors

---

*) An example of such a calculation is shown in the next Chapter (fortunately, the model discussed there will be simpler than the present one). Such a calculation will give a correction to the a.c. impedance.

[238] apparently yields a correction to the results obtained by Llewellyn [73] and others. However, this correction added to the a.c. power is not an a.c. power, but a supplementary d.c. power (given by the d.c. bias source) corresponding to the variation ΔJ_0 of the bias current. Demonstrate a similar property for the SCLC solid-state diode [106].

Hint: Consider, first, the low-frequency behaviour.

10.5. Show, by using the parametric equations of the punch-through diode that an inflection point on the steady-state $J - V$ characteristic does indeed exist (see Table 10.1). Note that if the device is biased in a certain region of the $J - V$ characteristic, the d.c. power absorbed by the diode from the bias source *decreases* as the amplitude of the a.c. signal present in the circuit increases.

10.6. Demonstrate the fact that the bias dependence of the second-harmonic amplitude at very low-frequencies is exactly the same as the bias dependence of the low-frequency detected current.

10.7. Try to find general equations (which are similar to that given in Section 10.3) for the hot-carrier SCLC diode.

Hint: Assume zero fixed space-charge ($N = 0$) and consider a velocity field dependence of the form of equation (5.17).

11

Large-signal Behaviour of Saturated-velocity Transit-time Diode

11.1 Introduction: the Model

In this chapter we consider a one-dimensional unipolar semiconductor diode operating under high-field bias conditions such as the carrier velocity may be assumed) almost saturated (or limited) to a certain value

$$v = v_l = \text{const.} \tag{11.1}$$

The basic equations of Section 4.2 will be written

$$J(t) = ep(x, t)\, v_l + \varepsilon \frac{\partial E(x, t)}{\partial t} \tag{11.2}$$

$$\frac{\partial E(x, t)}{\partial x} = \frac{e}{\varepsilon} [p(x, t) - N(x)] \tag{11.3}$$

$$V(t) = -\int_0^L E(x, t)\, dx. \tag{11.4}$$

Here we denote by N the density of fixed space-charge (e.g. ionized impurities) which is assumed constant, $N \neq N(t)$ but might be non-uniform, $N = N(x)$.

A boundary condition is also necessary. This will be specified by taking into account the injection mechanism at the 'cathode' ($x = 0$). We shall give the particle current density at the cathode

$$j_c(t) = j(x, t)\,|_{x=0} = [ep(x, t)\, v_l]\,|_{x=0} = j(0, t) \tag{11.5}$$

as a function of the electric field $E(x, t)$ in the semiconductor just near the cathode, i.e.

$$j_c(t) = F(E_c), \qquad E_c = E_c(t) = E(0, t). \tag{11.6}$$

F is an instantaneous function of E_c, i.e. the time t does not appear explicitly in equation (11.6) (see also Section 6.1). A relation of the form of equation (11.6) is suitable for barrier-controlled injection, such as Schottky-emission or field emission in semiconductor (Section 3.3).

We stress the fact that the SCLC injection cannot be rigorously included in the above model. This is because the SCL current flows when the electric field vanishes in the vicinity of the emitting electrode: the electric field should be low in this region and the carrier velocity cannot reach its limiting velocity v_l.

On the other hand, neglecting diffusion in equation (11.2) seems to have no important effect upon the validity of our calculations, at least as far as N may be considered almost uniform. If N is highly non-uniform, the diffusion current is large.

11.2 Integration of Equations

The equations (11.2)—(11.4) can be, formally, integrated by using an approach which is similar to that indicated in Section 10.3.

From equations (11.2) and (11.3) we have

$$J(t) = eNv_l + \varepsilon \frac{\partial E}{\partial x} v_l + \varepsilon \frac{\partial E}{\partial t} = eNv_l + \varepsilon \frac{\mathrm{d}E}{\mathrm{d}t} \tag{11.7}$$

($v_l = \mathrm{d}x/\mathrm{d}t$). The total derivative in equation (11.7) means that we contemplate the change of the electric field by moving together with the charge carrier. Let $E_{t,\tau}$ be the electric field acting at the instant t upon the hole emitted at the moment τ. At the time t, this carrier will be at the distance $x = x(t, \tau)$ from the cathode. Since the carrier drifts at its limiting velocity v_l we have

$$x(t, \tau) = (t - \tau)\, v_l. \tag{11.8}$$

By replacing $\partial E_{t,\tau}/\partial t$ for $\mathrm{d}E(x, t)/\mathrm{d}t$ in equation (11.7) and integrating, we have

$$E_{t,\tau} = \frac{1}{\varepsilon} \int_{\tau}^{t} [J(t) - eNv_l]\, \mathrm{d}t + E_{\tau,\tau} \tag{11.9}$$

where

$$E_{\tau,\tau} = E_c(\tau) = E(0, \tau) \tag{11.10}$$

is the cathode field. Then, equation (11.9) becomes

$$E_{t,\tau} = \frac{1}{\varepsilon} \int_{\tau}^{t} J(\xi)\, \mathrm{d}\xi - \frac{e}{\varepsilon} \int_{0}^{x(t,\tau)} N(x)\, \mathrm{d}x + E_{\tau,\tau}. \tag{11.11}$$

From equations (11.4), (11.8) and (11.11) one obtains

$$V(t) = -\int_{\tau}^{t(L,\tau)} E_{t,\tau} \frac{\partial x}{\partial \tau} \, \mathrm{d}\tau = v_l \int_{\tau}^{t(L,\tau)} E_{t,\tau} \, \mathrm{d}\tau \tag{11.12}$$

and

$$V(t) = \frac{v_l}{\varepsilon} \int_{t-T_L}^{t} \mathrm{d}\tau \int_{\tau}^{t} J(\xi) \, \mathrm{d}\xi - \frac{e v_l}{\varepsilon} \int_{t-T_L}^{t} \mathrm{d}\tau \int_{0}^{x(t,\tau)} N(x) \, \mathrm{d}x + v_l \int_{\tau}^{t} E_{\tau,\tau} \, \mathrm{d}\tau \tag{11.13}$$

where

$$T_L = \frac{L}{v_l} \tag{11.14}$$

is the transit-time. The total current density $J(t)$ will be written at $x = 0$, as follows:

$$J(t) = j_c(t) + \varepsilon \frac{\mathrm{d}E_c}{\mathrm{d}t} = F(E_c) + \varepsilon \frac{\mathrm{d}E_c}{\mathrm{d}t}. \tag{11.15}$$

By replacing $J(t)$ in equation (11.13) ($E_c \equiv E_{\tau,\tau}$) one finds

$$V(t) = \frac{v_l}{\varepsilon} \int_{t-T_L}^{t} \mathrm{d}\tau \int_{\tau}^{t} F[E_c(\xi)] \, \mathrm{d}\xi + E_c(t) \cdot L + V_P \tag{11.16}$$

where

$$V_P = -\frac{e v_l}{\varepsilon} \int_{t-T_L}^{t} \mathrm{d}\tau \int_{0}^{x} N(x) \, \mathrm{d}x = -\frac{e}{\varepsilon} \int_{0}^{L} \mathrm{d}x \int_{0}^{x} N(\eta) \, \mathrm{d}\eta \tag{11.17}$$

is a constant, time-independent voltage. If the doping is uniform ($N(\eta) \rightarrow N =$ const.), then

$$V_P = -\frac{eNL^2}{2\varepsilon}. \tag{11.18}$$

From equation (11.16), there follows that the only effect of doping*) (uniform or not) is to modify the constant voltage V_P. We have already indicated in Section 6.5 that the d.c. and small-signal a.c. behaviour of the saturated-velocity diode do not

*) The dielectric relaxation mechanism does not operate because the differential mobility is zero (saturated velocity) and the *differential* relaxation times is infinite. In other words, due to the fact that the carrier velocity is constant and identical for all charge carriers, the repulsion forces between these carriers cannot determine a tendency of disappearance of the bulk space-charge.

depend upon N, except through an additional quantity in the expression of the steady-state voltage. Equation (11.16) indicates that the same situation occurs for an arbitrary dynamic behaviour (as far as the velocity may be assumed saturated). Moreover, the same property holds for nonuniformly doped semiconductor samples (but only as long as diffusion may be neglected).

V_P of equation (11.17) is negative for majority carrier and positive for minority-carrier injection. In the second case V_P is just the punch-through voltage V_{PT} which must be exceeded in order to allow a current to flow (Section 1.2 etc.).

The complete solution for the dynamic behaviour is given by equations (11.15) and (11.16) ,where F, defined by equation (11.6) should be specified also. If $J = J(t)$ is given, the differential equation (11.15) will give $E_c(t)$, and $V(t)$ will be obtained from equation (11.16). In most cases, however, we have to calculate $J(t)$ for a given $V(t)$. Assume, for example, that $V(t)$ is given as a periodical function of t. $V(t)$ will be expanded in a Fourier series with known coefficients. $J(t)$ and $E_c(t)$ can be also expressed as a Fourier series, but the coefficients of these expansions are still unknown. They can be determined, at least in principle, by using equations (11.15) and (11.16). Unfortunately, $F = F(E_c)$ is a complicate (transcendental) function and this fact leads to extremely involved relations between the Fourier coefficients of different order*).

Therefore, we should calculate the device response by using a method of successive approximations. Let us assume an applied voltage of the form

$$V(t) = V_0 (1 + \nu \sin \omega t). \tag{11.19}$$

The cathode field $E_c(t)$ should be periodical (period $2\pi/\omega$) and may be written as

$$E_c(t) = E_{c,0} [1 + \sum_{n=1}^{\infty} \nu_n \sin \psi_n] \tag{11.20}$$

where

$$\psi_n = n\omega t + \alpha_n. \tag{11.21}$$

In the series expansion (11.20). ν_n and α_n are unknown and should be determined from equation (11.16) where $F = F(E_c)$ is assumed given. Once $E_c(t)$ found, $J(t)$ follows from equation (11.15).

Zero-Order Approximation. The applied signal is zero ($\nu = 0$). The cathode field will be constant, $E_c = E_{c,0}$ ($\nu_n = 0$, $n = 1,2,3,\ldots$). By eliminating E_c between equations (11.15) and (11.16), one obtains the steady-state characteristic.

First-Order Approximation. ν is very small in equation (11.19) and E_c is approximately equal to

$$E_c(t) \simeq E_{c,0} [1 + \nu_1 \sin \psi_1] \tag{11.22}$$

*) The starting equations (11.15) and (11.16) are, however, relatively simple and, perhaps, a simulation on an analog computer is suitable.

where ν_1 is also very small (we assume a sinusoidal response to a sinusoidal perturbation). Then we use a two-term Taylor expansion

$$F(E_c) \simeq F(E_{c,0}) + \left.\frac{\mathrm{d}F}{\mathrm{d}E_c}\right|_{E=E_{c,0}} \times E_{c,0}\, \nu_1 \sin \psi_1 \tag{11.23}$$

in equation (11.16), and thus determine ν_1 and ψ_1. According to equations (11.15), (11.22) and (11.23)

$$J(t) = F(E_{c,0}) + \left.\frac{\mathrm{d}F}{\mathrm{d}E_c}\right|_{E=E_{c,0}} \times E_{c,0}\, \nu_1 \sin \psi_1 + \varepsilon\omega E_{c,0}\, \nu_1 \cos \psi_1 . \tag{11.24}$$

The sum of the second and of the third terms in the right-hand-side of the above equation is the small-signal alternating current. An a.c. impedance can be thus calculated.

Second- and Higher-order Approximations. In the second order approximation we shall assume

$$E_c(t) \simeq E_{c,0}\,[1 + \nu_1 \sin \psi_1 + \nu_2 \sin \psi_2] \tag{11.25}$$

where ν_1 is of the first order of magnitude and ν_2 is of the second order of magnitude) After introducing $E_c(t)$ in equations we shall separate the terms of different order. of magnitude on each side of the equality and we shall equalize their coefficients. The procedure is similar for higher-order approximations. A similar technique was developed in the previous chapter. The basic integral equation (11.16) is simpler for the saturated-velocity diode and we can go, without serious difficulties, beyond the second-order approximation: this will yield a correction to the alternating-current impedance. We shall find that the a.c. impedance depends upon the amplitude of the applied signal. This theory can be applied to moderate signal levels. Moreover, the small-signal condition can be now accurately defined.

11.3 Analysis of Schottky-barrier Emission Saturated-velocity Diode

As an example we shall study below the saturated-velocity diode with Schottky-barrier emission cathode. The d.c. and small signal properties of this device were already discussed in Section 6.5. According to equation (6.31), the boundary condition equation (11.6) should be written

$$j_c(t) = J_R \exp \sqrt{\frac{E_c\,(t)}{E_R}} . \tag{11.26}$$

The zero-order approximation ($\nu = 0$) gives immediately the steady-state $J - V$ characteristic, equation (6.35).

In the first-order approximation, we search a solution of the form of equation (11.22) (where ν_1 is very small) for the integral equation (11.16); $V(t)$ is of the form of equation (11.19) (ν very small). According to equations (11.6) and (11.26), $F(E_c)$ in equation (11.16) becomes successively [229]

$$F[E_c(\xi)] = J_R \exp \sqrt{\frac{E_c(\xi)}{E_R}} = J_R \exp \left\{ \frac{E_{c,0}}{E_R} [1 + \nu_1 \sin \psi_1(\xi)] \right\}^{1/2} \simeq$$

$$\simeq J_R \exp \left\{ \left(\frac{E_{c,0}}{E_R} \right)^{1/2} \left[1 + \frac{\nu_1}{2} \sin \psi_1(\xi) \right] \right\} = J_R \exp \sqrt{\frac{E_{c,0}}{E_R}} \times \exp \left[1 + \frac{\nu_1}{2} \sin \psi_1(\xi) \right] =$$

$$= F(E_{c,0}) \left[1 + \frac{\nu_1}{2} \sqrt{\frac{E_{c,0}}{E_R}} \sin \psi_1(\xi) \right]. \tag{11.27}$$

After calculations which are similar to those developed in Section 10.4, one obtains exactly the small-signal impedance (6.28), where Γ is given by equation (6.32).

The next step [229] is the second-order approximation. One assumes a solution of the form of equation (11.25) where ν_1 is the first-order and ν_2 the second-order of magnitude. By replacing equation (11.25) in equation (11.16) we have to calculate $F(E_c(\xi)] = J_R \exp \sqrt{E_c(\xi)/E_R}$. By expanding the radical in series and by neglecting the third and the higher-order terms we have

$$\left(\frac{E_c(\xi)}{E_R} \right)^{1/2} = \left(\frac{E_{c,0}}{E_R} \right)^{1/2} [1 + \nu_1 \sin \psi_1(\xi) + \nu_2 \sin \psi_2(\xi)]^{1/2} \simeq$$

$$\simeq \left(\frac{E_{c,0}}{E_R} \right)^{1/2} \left[1 + \frac{\nu_1}{2} \sin \psi_1 + \frac{\nu_2}{2} \sin \psi_2 - \frac{\nu_1^2}{8} \sin^2 \psi_1 \right]. \tag{11.28}$$

Therefore one obtains, to second-order,

$$F[E_c(\xi)] = J_R \left[\exp \left(\frac{E_{c,0}}{E_R} \right)^{1/2} \right] \times \left\{ 1 + \left(\frac{E_{c,0}}{E_R} \right)^{1/2} \left[\frac{\nu_1}{2} \sin \psi(\xi) + \frac{\nu_2}{2} \sin \psi_2(\xi) - \right.\right.$$

$$\left.\left. - \frac{\nu_1^2}{8} \sin^2 \psi_1(\xi) \right] \right\}. \tag{11.29}$$

Then equations (11.25) and (11.29) will be replaced in equations (11.15) and (11.16). We shall identify the coefficients of $\sin \psi_1$, $\cos \psi_1$, $\sin \psi_2$, $\cos \psi_2$ etc. obtaining thus a correction to the d.c. current calculated in the absence of the signal, as well as a second harmonic. The fundamental (frequency ω) is not modified by this second-order calculation, i.e. the a.c. impedance remains equal to the small-signal impedance and does not depend upon the signal amplitude.

The third-order approximation will be carried out by using a similar procedure [229]. The steady-state component and the second harmonic retain the same expression as in the second-order approximation. The new results are: the third harmonic and a new value for the fundamental. The alternating-current impedance should be

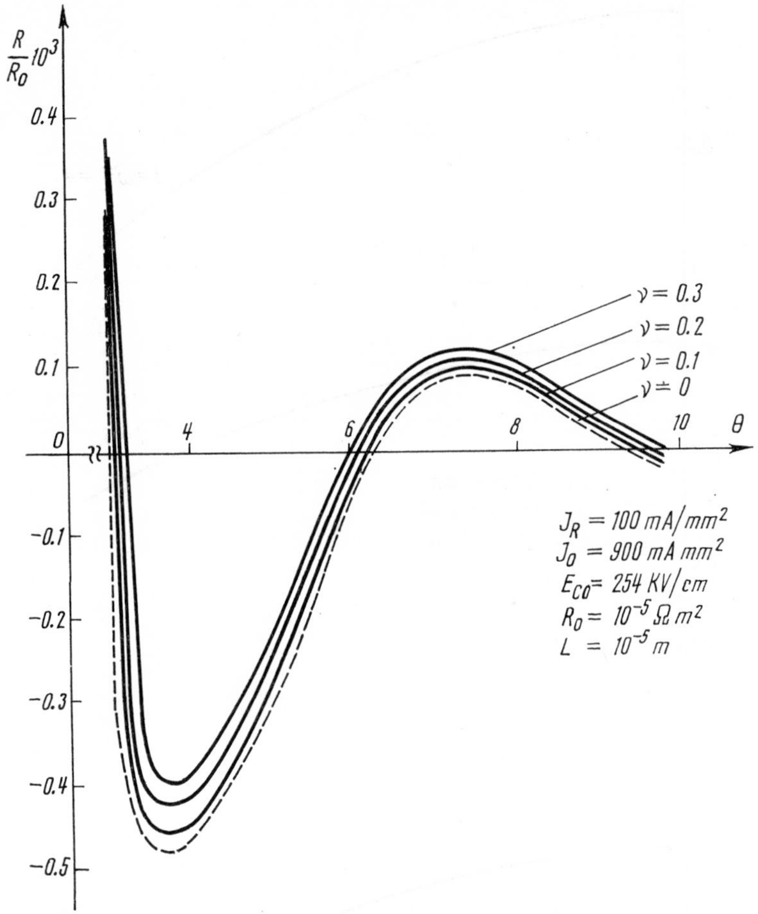

Figure 11.1. Frequency dependence of the series resistance of a MSM structure (ν is the normalized signal amplitude).

calculated again. This impedance depends upon the signal level through the parameter ν defined by equation (11.19), i.e.

$$\bar{Z} = \frac{Z}{R_0} = \bar{Z}\left(\theta; \Gamma, \frac{J_0}{J_R}, \nu\right) \tag{11.30}$$

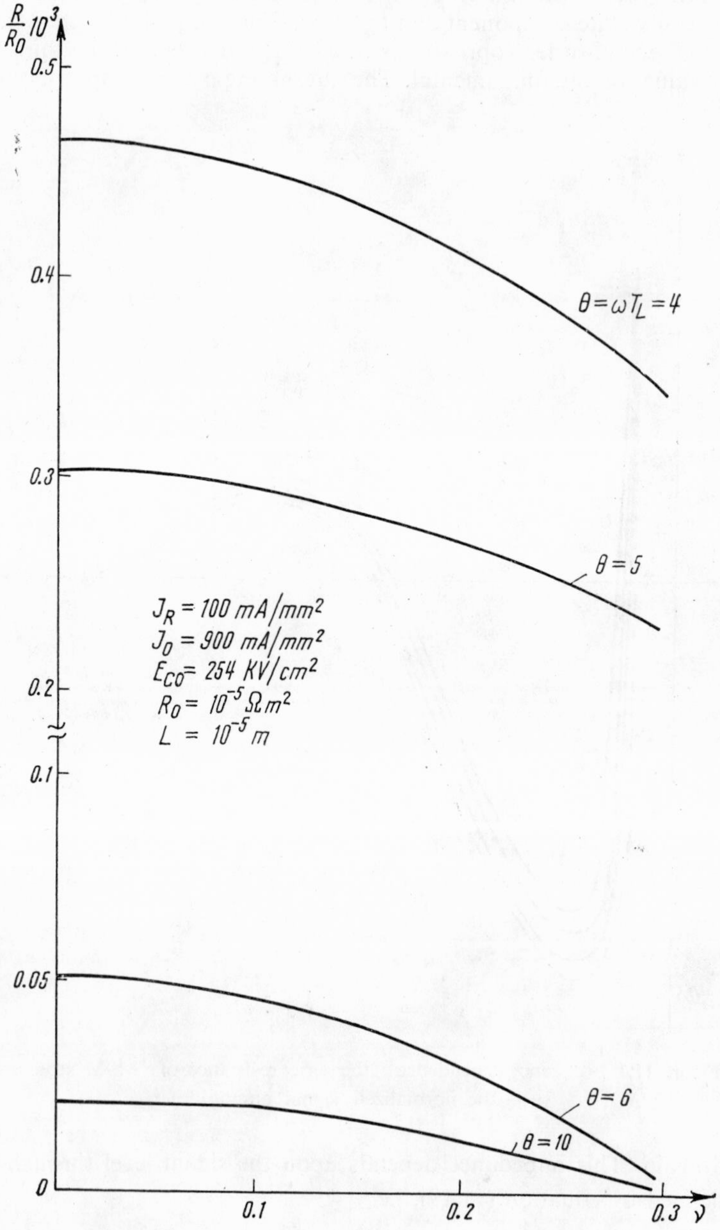

Figure 11.2. Dependence of the series resistance of a MSM diode upon the normalized signal amplitude, for a few transit angles.

where

$$R_0 = \frac{L^2}{\varepsilon v_l} \tag{11.31}$$

is a characteristic resistance.

As an example, Figure 11.1 shows the frequency dependence of the series resistance calculated for a M-Si-M structure, 10 μm thick, at room temperature ($R_0 \simeq 10^{-5}\ \Omega\text{m}^2$). The Richardson (saturation) current density is J_R=100 mA mm^{-2}.

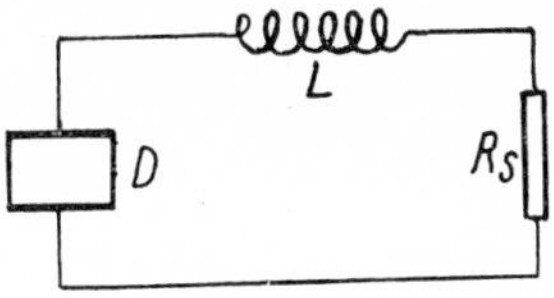

Figure 11.3. Series oscillating circuit indicating the MSM diode (D) and the load resistance Rs (the load proper plus the series parasitic positive resistance).

The bias current is J_0=900 mA mm^{-2} (E_c=254 kV cm^{-1}, whereas E_R=52.8 kV cm^{-1}). The frequency characteristics of Figure 11.1 also depend upon the normalized signal amplitude ν (see equation (11.19)). The above calculations are not valid for large ν, say above $\nu = 0.3$. Figure 11.1 shows that for $\nu < 0.1$ the small signal calculations ($\nu = 0$) predict the maximum absolute value of the negative resistance with less than 5% error. Figure 11.2 shows the dependence of the series (negative) resistance upon ν, for several frequencies ($\theta = 4,5,6,10$). Such graphs can be used to find the oscillation amplitude in a series circuit like that represented in Figure 11.3. The oscillations start if the device is properly tuned and if the absolute value of the device resistance is larger than the total series resistance. The final oscillations established in such a circuit have a frequency determined by the resonance of the device capacitance and the external inductance. The oscillation amplitude can be found by putting the external resistance equal to the absolute value of the device (negative) resistance. Once θ and $|R(\omega)|$ specified, Figure 11.2 yields the ratio ν, and thus the oscillation amplitude.

We stress the fact that the *decrease* of $|R(\omega)|$ with increasing ν is an essential feature. It can be easily proved that the condition

$$\frac{\partial}{\partial \nu} |R(\omega)| < 0, \text{ for } R(\omega) < 0 \tag{11.32}$$

is necessary for the dynamic stability of oscillations *).

11.4 Approximate Analysis of Oscillations in a $p^+ np^+$ (or $p^+ n \pi p^+$) Structure Biased below Punch-through [240]

Let us consider a punch-through diode, like that discussed in Section 1.2, with the only difference that the electric field at the punch-through is sufficiently high to deter-

*) Assume a perturbation which increases temporarily the oscillation amplitude. Because $|R(\omega)|$ decreases and becomes lower than the series positive resistance, the losses will overcome the power generation and the amplitude will tend to decrease, etc.

mine the carrier velocity saturation over the greatest part of the central region. As the punch-through voltage V_{PT} is exceeded, the current rises steeply. At larger voltages the current becomes space-charge limited. The SCL current is given by equation (6.36) and may be written

$$J_0 = \frac{2\varepsilon v_1}{L^2}(V_0 - V_{PT}), \quad V_{PT} = \frac{e N_D L^2}{2\varepsilon}. \tag{11.33}$$

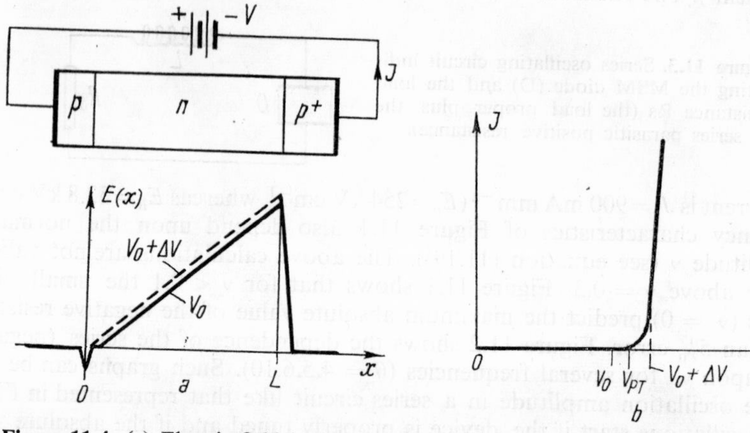

Figure 11.4. (a) Electric field inside a p^+np^+ structure biased at punch-through, (b) $J-V$ characteristic of the punch-through device.

Wright [88], Rüegg [240], Sheorey *et al.* [241] discussed the operation of the punch-through diode biased outside the SCLC region. Wright [88] considered the small signal operation for a saturated-velocity punch-through diode biased just at punch-through. He assumed a finite a.c. conductivity of the emitting plane; this conductivity is very sensitive at the direct current passing through the device [88]. We recall the fact that Shockley's negative resistance [14], [16] (see also Section 5.6) was calculated under SCL current (and low-field) conditions, i.e. infinite conductivity of the injecting plane (and constant mobility).

Rüegg [240] considered a similar p^+np^+ structure but biased slightly *below* punch-through. The applied alternative voltage drives the device into punch-through, and within a short time interval a considerable amount of mobile charge is injected into the drift (n) region. This is illustrated in Figure 11.4. Figure 11.4a shows the electric field inside a p^+np^+ structure for an applied voltage V_0 slightly lower than V_{PT} (solid line), and also for a voltage $V_0 + \Delta V > V_{PT}$ (dashed line). Figure 11.4b shows the pronounced non-linearity of the steady-state $J-V$ characteristic [240]. Because of this non-linearity, in a properly biased device the current can pass through the device as very short and large pulses (spikes of current). The injected charge requires a finite time to travel through the device. If the device operates at a sufficiently high frequency, such as the signal period becomes comparable to the carrier transit-time, a *negative-resistance* effect can be obtained due of the delay of the current with respect to the voltage. Rüegg [240] indicated two major defficiences of the structure of Figure 11.4a. First, because the injected carriers initially drift through a low-field region, the transit-time is large and the operation frequency is relatively low (one or a few GHz). Secondly, the large fields at the collecting junction can lead to avalanche breakdown. This restricts the a.c. field swing and, thus, the power output [240].

A practical punch-through structure disclosed by Rüegg [240] is represented in Figure 11.5. A fourth layer has been added between the base (n) and the collector

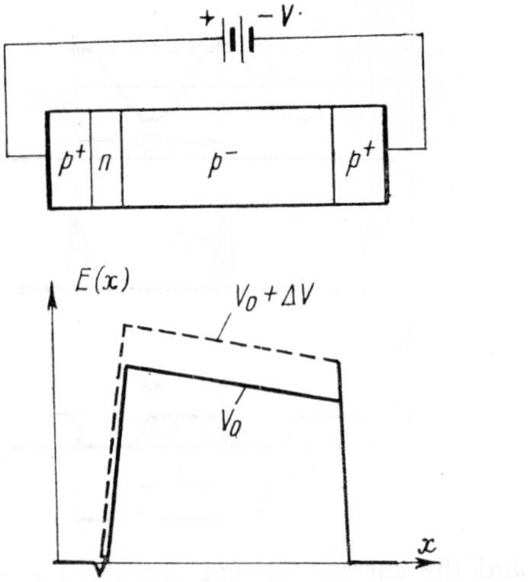

Figure 11.5. Electric field inside an improved punch-through structure [240].

(p^+). This new region is called the drift region: it is thick as compared to the base and lightly doped. Figure 11.5 shows the electric field configuration inside the device for a voltage V_0 slightly below V_{PT} (solid curve), and for $V_0 + \Delta V$ just above V_{PT} (dashed curve). The charge carriers injected when the applied voltage exceeds V_{PT} will travel at their limiting velocity in the drift space and in part of the base n, whereas the maximum field can be maintained well below the value required by the carrier multiplication process.

We shall discuss below the high-frequency non-linear behaviour of a p^+np^+ or $p^+np^-p^+$ structure. The simplified method of approach is essentially that used by Rüegg [240].

The total current density, equation (11.2) is the sum of the particle current and the displacement current. By integrating equation (11.2) with respect to x, from $x = 0$ (the emitting plane) to $x = L$ (the collecting plane), at a given instant t, one obtains

$$J(t) = \frac{1}{L}\int_0^L e\,p\,(x, t)\,v_l\,\mathrm{d}x + \frac{\varepsilon}{L}\frac{\partial}{\partial t}\int_0^L E(x, t)\,\mathrm{d}x \tag{11.34}$$

or

$$J(t) = \frac{Q_{\mathrm{mob}}(t)}{T_L} + C_0\frac{\mathrm{d}}{\mathrm{d}t}V(t), \quad C_0 = \frac{\varepsilon}{L} \tag{11.35}$$

where Q_{mob} is the total mobile charge (per unit of electrode area) contained between $x = 0$ and $x = L$ at a given moment t. $V(t)$ is the applied voltage, C_0 is the geometrical capacitance of the inter-electrode space and $T_L = L/v_l$ is the carrier transit-time. The total current (11.35) is the sum of the so-called induced current

$$j_{ind}(t) = \frac{1}{L}\int_0^L j(x, t)\,dx = \frac{Q_{mob}(t)}{T_L}, \quad (v = v_l) \tag{11.36}$$

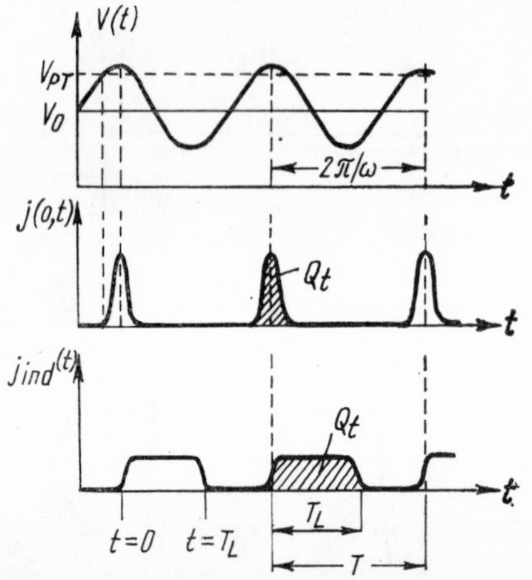

Figure 11.6. (a) Applied voltage waveform, (b) particle current injected into the drift region, (c) current induced by injected carriers moving with saturated velocity.

and the current flowing through the geometrical capacitance C_0 (which is equal to the depletion layer capacitance at $V = V_{PT}$).

Figure 11.6 shows typical waveforms for the applied voltage $V(t)$, the injected particle current $j(0, t)$ and the induced current $j_{ind}(t)$ [240]. The alternating signal $V_1 \sin \omega t$ superimposed on $V_0 < V_{PT}$ drives the device beyond punch-through for a *very short* time-interval during each cycle. When $V(t) > V_{PT}$ a spike of the injected particle current occurs. In this short 'injection interval', the mobile charge Q_{mob} inside the device increases rapidly. In a subsequent time interval, $V(t)$ falls below V_{PT} and Q_{mob} remains constant during the transit of the mobile charge towards the collector. During this 'transport interval', the current induced by the mobile charge travelling at the carrier limiting velocity is approximately a rectangular pulse, provided that the injection interval is extremely short. Figure 11.6 indicates a situation when the transport interval, which is roughly equal to the carrier transit-time T_L, is shorter than the period of $V(t)$. The whole mobile charge injected during one spike of the injected particle current leaves the inter-electrode space before a new cycle begins.

The induced current, equation (11.36), can be thus approximated by

$$\left.\begin{aligned} j_{ind}(t) &= \frac{Q_{mob}(t)}{T_L} = \frac{Q_t}{T_L}, \quad 0 < t < T_L \\ j_{ind}(t) &= 0, \qquad T_L < t < T \end{aligned}\right\} \tag{11.37}$$

where Q_t is the total charge (per unit area) injected during one cycle and $T = 2\pi/\omega$ is the signal period. The current j_{ind} can be decomposed in its sine-wave and cosine-wave component by means of Fourier analysis. We shall consider below only the component corresponding to the fundamental frequency ω [240]. The amplitude of the cosine component, which is phase with the voltage, is

$$J_{1R} = \frac{\omega Q_t}{\pi} \frac{\sin\theta}{\theta} = 2J_0 \frac{\sin\theta}{\theta} \tag{11.38}$$

where $\theta = \omega T_L$ is the transit-angle and J_0 the direct current. The amplitude of the sine-wave component, which lags the voltage by 90°, is equal to

$$J_{1I} = -\frac{\omega Q_t}{\pi} \frac{1-\cos\theta}{\theta}. \tag{11.39}$$

We note that the current flowing through the geometrical capacitance C_0 should be added to the imaginary component, equation (11.39).

The total injected charge Q_t should be now determined. The fundamental assumption of the calculation which will be developed below, is the fact that 'at the moment of punch-through the injection of charges from the emitter into the depletion region is limited by the space charge which these injected carriers set up' [240]. In other words, the electric field at the emitting plane is always zero. Because

$$\mathrm{d}Q_{mob} = \varepsilon\,\mathrm{d}E \tag{11.40}$$

where $\mathrm{d}E$ is the change experienced in the bulk electric field due to the charge injection at the emitter, we have

$$\mathrm{d}E = \frac{\mathrm{d}Q_{mob}}{\varepsilon} = \frac{j(0,t)\,\mathrm{d}t}{\varepsilon} \tag{11.41}$$

where $j(0, t) = j_c(t)$ is the injected particle current. The punch-through occurs at $t = -t_p$ (Figure 11.6). The total field change due to the injected space charge at time t is

$$\Delta E(t) = \frac{1}{\varepsilon}\int_{t_p}^{t} j(0,\xi)\,\mathrm{d}\xi. \tag{11.42}$$

Because the injection of the total charge Q_t takes place in a time which is short as compared to the carrier transit-time (Figure 11.6), the field change, equation (11.42), occurs in the vicinity of the emitter, over a distance which is short as compared to L. Therefore, the change in voltage is

$$\Delta V(t) = \int_0^t \Delta E(t)\,\mathrm{d}x \simeq \Delta E(t).L = \frac{L}{\varepsilon}\int_{t_p}^{t} j(0,\xi)\,\mathrm{d}\xi. \tag{11.43}$$

This voltage change is supplied by the a.c. voltage in excess of the punch through voltage (Figure 11.6)

$$V_{PT} = V_0 + V_1 \cos \omega t_p \tag{11.44}$$

and thus

$$\Delta V(t) = \frac{L}{\varepsilon} \int_{t_p}^{t} j(0, \xi) \, d\xi = V_1 (\cos \omega t - \cos \omega t_p) \, . \tag{11.45}$$

The particle current $j(0, t)$ can be found by differentiating equation (11.45). We have

$$\left. \begin{aligned} j(0, t) &= -\frac{\varepsilon \omega}{L} V_1 \sin \omega t, \quad t_p < t < 0 \\ j(0, t) &= 0, \qquad\qquad\qquad 0 < t < T - t_p \end{aligned} \right\} \tag{11.46}$$

and the total charge injected during one cycle should be

$$Q_t = \int_{t_p}^{0} j(0, \xi) \, d\xi = C_0 V_1 (1 - \cos \theta_p) \tag{11.47}$$

where

$$\theta_p = \omega t_p, \ \cos \theta_p = \frac{V_{PT} - V_0}{V_1} \tag{11.48}$$

is the punch-through angle [240]. We assume, as we have previously mentioned, $\theta_p \ll \theta$.

By using the above results it is possible to calculate the conductance G and the susceptance B of the device. These are, respectively,

$$G = \frac{\omega C_0}{\pi} \frac{\sin \theta}{\theta} (1 - \cos \theta_p) \tag{11.49}$$

$$B = \omega C_0 \left[1 - \frac{1 - \cos \theta}{\pi \theta} (1 - \cos \theta_p) \right] \tag{11.50}$$

and depend upon frequency directly and through the transit-angle $\theta = \omega T_L$. The parallel conductance may be also written

$$G = \left(\frac{1 - \cos \theta_p}{\pi} \right) \frac{\sin \theta}{R_0} \, , \ R_0 = \frac{L^2}{\varepsilon v_l} \, . \tag{11.51}$$

G is negative for $\pi < \theta < 2\pi$ and has a maximum absolute value for

$$\theta = \omega T_L = \frac{3\pi}{2} \tag{11.52}$$

(see also Section 6.5). θ_p is small and G increases as θ_p^2. G is inversely proportional to the square of the device thickness L. However, the device thickness is determined by the operating frequency ω_0, according to the optimum transit-angle, equation (11.52)

$$L_{(0)} = \frac{3\pi v_l}{2\omega_0} . \tag{11.53}$$

The Q of the device calculated at the optimum frequency is [240]

$$Q_{(0)} = \frac{B_{(0)}}{G_{(0)}} \simeq -\frac{3\pi^2}{2} \simeq -15 \tag{11.54}$$

for a punch-through angle $\theta_p = \pi/2$ *).

11.5 Conclusion

The operation mode discussed in the previous section is characterized by the fact that the injected particle current is zero for a part of the signal period, i.e. the charge injection takes place in pulses and the mobile charge propagates through the device in bunches. The approximate analysis presented before [240] assumes injection in sharp pulses (short time intervals as compared to the transit-time). A more exact analysis should also take into account a correct description of the injection mechanism. We have already indicated in Section 5.6 that the SCLC boundary condition is not applicable just above the punch-through because the potential barrier at the emitting plane is still high and the conductivity is low.

A more detailed description of the injection mechanism in the large signal operation of the punch-through diode was given by Sheorey, Lundström and Ash [241]. These authors assumed that the injected current depends exponentially upon the barrier height. The variation of the barrier height was evaluated by taking into account both the modulation due to the applied signal and the effect of the injected space charge (which takes the form of a sheet of injected carriers moving at their limiting velocity). The height and the location of the potential maximum in the absence of the signal was evaluated as follows. The voltage drop occurring between

*) In fact, the above calculations are not rigorously valid for such a large punch-through angle. Rüegg [240] also evaluated the device efficiency and indicated some design criteria.

the barrier and the collector was calculated by neglecting any mobile space charge. The Poisson equation was solved in the vicinity of the emitter thus yielding another relation between the barrier height and the position of the potential maximum. The latter is of the order of a few Debye lengths. Numerical calculations reported in [241] indicated the fact that the modulation of the barrier height*) does not modify appreciably the position of the barrier: the variation of this position is small as compared to the drift length L.

Sheorey *et al.* [241] also calculated the alternating current and the efficiency. Recently, Harth and Claassen [156] reported numerical data and experimental results on the large-signal operation of a punch-through diode.

Finally, we note that the problem of the large-signal behaviour of a saturated-velocity diode was discussed by Semichon and Constant [242]. Their analysis was somewhat more general than that of Section 11.1 because a time-dependent boundary condition was considered: therefore non-instantaneous emission mechanisms, such as emission by avalanche generation process can be discussed.

Problems

11.1. Show that if the equations of Section 11.2 are applied, formally, to the SCL current-diode with saturated carrier velocity, the high-frequency large-signal behaviour of this device will result to be linear.

11.2. Consider the model of Section 11.1 for a barrier-injection controlled diode. Show that if the applied voltage-signal is sinusoidal and large, the variation of the cathode field cannot be assumed sinusoidal and the non-linear effects *cannot* be calculated in this way.

11.3. Use the second-order approximation to calculate the detected current ΔJ_0 for the Schottky-barrier emission saturated-velocity diode. Compare the frequency and bias dependence of ΔJ_0 with the results found in the previous chapter for the SCLC diode.

---

*) It was shown [241] that although the applied a.c. voltage is sinusoidal, the modulation of the barrier height is not sinusoidal. This is an effect of the injected space charge. See also Problem 11.2.

12

Large-signal Transient Behaviour

12.1 Introduction

In this chapter we shall discuss the response of unipolar semiconductor devices to a large change in the applied signal.

A typical example is a unipolar diode subjected to a voltage step

$$V(t) = 0, \ t < 0; \ V(t) = V_0, \ t \geqslant 0. \tag{12.1}$$

This is an idealized signal, of course. The rise time of this voltage step should be finite in a real circuit, and should depend upon the diode and the external circuitry. Therefore, we assume the fact that the time constants involved in the transient behaviour of the semiconductor device itself are much longer than the above-mentioned rise time. We are interested in calculating the response $J = J(t)$ (the initial spike of current required to charge the diode capacitance is substracted).

Quantitatively, we expect the following behaviour. For $t \to \infty$ the current $J(t)$ should approach the steady-state value $J_0 = J_0(V_0)$, which can be calculated from the steady-state characteristic. The current $J(\infty)$ is carried by the stationary distribution of injected mobile charge *in transit* between electrodes. Such a distribution can be practically established in a time of the order of the carrier transit-time. The current should start from a finite value $J(0) \neq 0$, because in the initial moment, the current at the anode is equal to the displacement current and the latter is finite due to the change of the anode field accompanying the voltage variation. The detailed behaviour of $J(t)$ cannot be predicted by a quasi-stationary analysis, and the reasons for this may be clearly understood by following the mathematical analysis presented below.

12.2 Basic Equations

We shall consider first a unidimensional semiconductor (or insulator) diode. The following equations can be written down for the conduction space between the cathode plane $x = 0$ and the anode plane $x = L$:

$$J = epv + \varepsilon \frac{\partial E}{\partial t} \ \text{(Maxwell)} \tag{12.2}$$

$$\frac{\partial E}{\partial x} = \frac{e(p - N)}{\varepsilon} = \text{const.} \tag{12.3}$$

$$v = v(E) \tag{12.4}$$

$$V = \int_0^L E \, \mathrm{d}x \tag{12.5}$$

a suitable boundary condition (12.6)

Here, diffusion is neglected and the fixed space-charge is assumed uniform*).

The hole density can be eliminated between equations (12.2), (12.3) and $J(t)$ is given by

$$J(t) = \left(\varepsilon \frac{\partial E}{\partial x} + eN \right) v(E) + \varepsilon \frac{\partial E}{\partial t}. \tag{12.7}$$

We shall integrate this equation over the diode length, thus obtaining

$$J(t)L = \varepsilon \int_0^L v(E) \frac{\partial E}{\partial x} \mathrm{d}x + eN \int_0^L v(E) \, \mathrm{d}x + \varepsilon \frac{\mathrm{d}V}{\mathrm{d}t}. \tag{12.8}$$

Now we particularize introducing

$$v = \mu E, \ \mu = \text{const.} \tag{12.9}$$

and $J(t)$ may be written successively

$$J(t) = \frac{\varepsilon\mu}{L} \int_0^L E \frac{\partial E}{\partial x} \mathrm{d}x + \frac{eN\mu}{L} V(t) + \frac{\varepsilon}{L} \frac{\mathrm{d}V}{\mathrm{d}t} = \frac{\varepsilon\mu}{2L} \left[E^2(L, t) - E^2(0, t) \right] + $$

$$+ \frac{V(t)}{R_0} + C_0 \frac{\mathrm{d}V}{\mathrm{d}t} \tag{12.10}$$

where

$$R_0 = \frac{L}{eN\mu}, \ C_0 = \frac{\varepsilon}{L} \tag{12.11}$$

(notations introduced before and repeated here for convenience).

---

*) Carrier trapping is not included in this model, but reference to the trapping effects in insulator diodes will be made later.

Assume, then, the fact that $J(t)$ is injected under SCL conditions (the cathode is an infinite supply of charge carriers). Thus equation (12.6) becomes

$$E|_{x=0} = E(0, t) \equiv 0 \tag{12.12}$$

and equation (12.11) may be written

$$J(t) = \frac{\varepsilon\mu}{2L} E^2(L, t) + \frac{V(t)}{R_0} + C_0 \frac{\mathrm{d}V}{\mathrm{d}t} \cdot \tag{12.13}$$

Let us denote by t_1 the moment when the front of the *injected* space-charge reaches the anode $x = L$. For $t < t_1$, the total current is simply the displacement current at the anode

$$J(t) \doteq \varepsilon \frac{\mathrm{d}E(L, t)}{\mathrm{d}t}, \quad t < t_1 \tag{12.14}$$

and equations (12.13) and (12.14) yield a differential equation in

$$E(L, t) = E_L, \tag{12.15}$$

namely

$$\varepsilon \frac{\mathrm{d}E_L}{\mathrm{d}t} - \frac{\varepsilon\mu}{2L} E_L^2 = \frac{V(t)}{R_0} + C_0 \frac{\mathrm{d}V(t)}{\mathrm{d}t} \tag{12.16}$$

where $V = V(t)$ is given.

12.3 Turn-on of SCLC Insulator Diodes

The above equations will be used to study the current flowing through an SCLC insulator diode ($N = 0$, $\mu = \text{const.}$) subjected to the voltage step, equation (12.1). This problem was first studied by Many and Rakavy [243].

For $0 < t < t_1$, equation (12.16) should be used and because $R_0 \to \infty$ ($N=0$) and $V(t) = \text{const.}$, it takes the simple form

$$\frac{\mathrm{d}E_L}{\mathrm{d}t} = \frac{\mu}{2L} E_L^2 \tag{12.17}$$

and can be integrated by separation of variables. We shall introduce the normalized variables

$$\bar{E}_L = \frac{E_L}{V_0/L}, \quad \bar{t} = \frac{t}{L^2/\mu V_0} \cdot \tag{12.18}$$

The result of integration is

$$\bar{t} = -\frac{2}{\bar{E}_L} + \text{const.} \tag{12.19}$$

At $t = 0$ the electric field inside the crystal is uniform, because of the absence of the injected space-charge

$$E(x, 0) = \frac{V_0}{L} \tag{12.20}$$

and thus the boundary condition for equation (12.19) is

$$\bar{E}_L(0) = 1$$

and the solution may be written

$$\bar{E}_L = \left(1 - \frac{\bar{t}}{2}\right)^{-1}, \; 0 < t < t_1 \tag{12.21}$$

In the same time interval $0 < t < t_1$, the normalized current*)

$$\bar{J} = \frac{J(t)}{\varepsilon\mu V_0^2/L^3} \tag{12.22}$$

should be given by equation (12.14). By using equation (12.21) one obtains [243]

$$\bar{J} = \bar{J}(\bar{t}) = \frac{1}{2}\left(1 - \frac{\bar{t}}{2}\right)^{-2}, \; 0 < t < t_1. \tag{12.23}$$

The time t_1, may be determined as follows. The space between the moving front of the injected space-charge and the anode does not contain any space charge and the electric field is uniform and equal to $E_L = E_L(t)$, as shown in Figure 12.1a. Thus, the carriers at the leading edge of the injected charge move at a velocity

$$v = \mu E_L . \tag{12.24}$$

The transit-time of *these* carriers is t_1 and can be determined from

$$L = \int_0^L \mathrm{d}x = \int_0^{t_1} v \, \mathrm{d}t = \mu \int_0^{t_1} E_L \, \mathrm{d}t \tag{12.25}$$

*) The notation J differs from that introduced in Chapter 3 and is used in this chapter only.

where $E_L(t)$ is given by equation (12.21). The normalized *initial* transit-time t_1 results

$$\bar{t}_1 = 2(1 - e^{-1/2}) \simeq 0.787 \qquad (12.26)$$

and is *smaller* than the transit-time of a unique electron (no space-charge) in the same structure ($L^2/\mu V_0$, see equation (3.81)). The stationary behaviour will be, however, established after a period of time longer-than t_1.

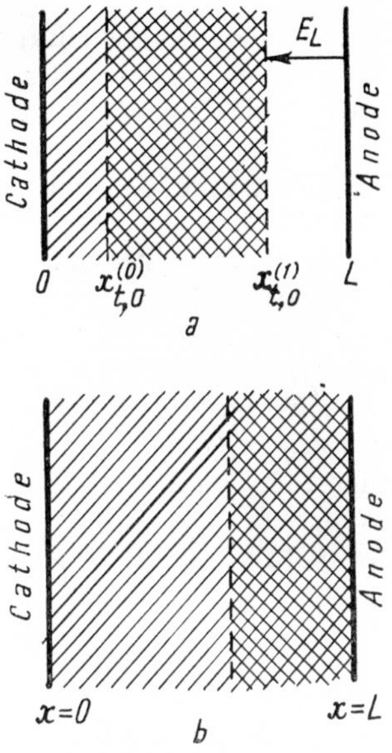

Figure 12.1. Distribution of the injected space-charge at a time moment $t < t_1$ (a) and $t_1 < t < t_2$ (b). The double hatched region is occupied by charge carriers injected instantaneously at $t = 0$.

We shall analyse a subsequent time interval $t_1 < t < t_2$, where t_2 will be defined as follows.

Under SCLC conditions we accept that at $t = 0$ a *finite* mobile charge is injected at the cathode. This is necessary for providing the SCLC boundary condition, equation (12.12). At $t = 0$, the electric field in the bulk is uniform and equal to V_0/L, as shown by equation (12.20). The electric field at $x = 0^+$ and $t=0$ (or, better, $t = 0^+$, i.e. just after the application of the voltage step) should be also equal to V_0/L because the space-charge cannot propagate itself instantaneously and does not yet exist at $x > 0$. On the other hand, on the cathode side of the $x = 0$ plane (i.e. at $x = 0^-$) the electric field should be zero, according to the SCLC condition. It is this *discontinuity* of the electric field at $x = 0$ for $t = 0$ that forces us to accept that a *finite* charge is injected instantaneously at $t = 0$: this charge founds itself in the $x = 0$ plane and has an infinite density*) (which,

*) It may be readily shown that neglecting diffusion in these circumstances is a somewhat arbitrary assumption. This problem will be discussed later.

by Poisson equation, explains the discontinuity of the electric field). The mobile charge (per unit area) injected at $t = 0$ is denoted by Q_0 and is given by

$$\frac{Q_0}{\varepsilon} = E(0^+, 0) - E(0^-, 0) = \frac{V_0}{L} \tag{12.27}$$

or

$$Q_0 = C_0 V_0, \; C_0 = \frac{\varepsilon}{L}. \tag{12.28}$$

For $t > 0$, the mobile charge Q_0 propagating towards the cathode should disperse because of electrostatic repulsion. This dispersion is 'mathematically' possible because the holes injected at $t = 0$ in an infinitely thin sheet, start moving at different velocities (we must accept that in this infinite ly thin sheet there is a field gradient and, consequently, a velocity gradient). These 'initial' carriers continues to travell under the influence of a different electric field intensity, and, in other words, have a different 'history'. They form a sheet of finite and variable thickness which travels in front of the injected space-charge, as shown in Figure 12.1. The moment when the trailing edge of this 'initial' sheet reaches the anode is denoted by t_2. Thus, for $t_1 < t < t_2$, the particle current at the anode is carried by holes injected instantaneously at $t = 0$. For $t > t_2$ the same current will be carried by holes injected at $t > 0$.

In order to calculate $J(t)$ for $t > t_1$, it is necessary to identify (or to label) the charge carriers which carry the particle current and to determine the electric field variation. In Chapter 10, we labelled the mobile carriers after their emission moment τ. The electric field $E_{t,\tau}$ which acts at the moment t upon the hole emitted at the moment τ, is given by ($N = 0$ and $\theta_r \to 0$ in equation (10.20))

$$E_{t,\tau} = \frac{1}{\varepsilon} \int_\tau^t J(\xi)\, \mathrm{d}\xi, \; \tau > 0 \tag{12.29}$$

and this hole travelled a distance equation (10.22)

$$x_{t,\tau} = \frac{\mu}{\varepsilon} \int_\tau^t t - \xi) J(\xi)\, \mathrm{d}\xi, \tau > 0. \tag{12.30}$$

However, the emission moment τ is not sufficient to identify the carriers from the charge sheet emitted at $t = 0$. We shall introduce a new parameter, φ_0, which is the electric field experienced at $t = 0$ by a charge carrier from the initial sheet. As shown above, φ_0 varies between 0 (on the cathode side of the emitting plane) and V_0/L (on the anode side of the same plane). The electric field acting at the moment $t \geqslant 0$ on a charge-carrier emitted at $t = 0$ under the action of a particular 'initial' field φ_0, is given by

$$E_{t,0}^{(\overline{\varphi}_0)} = \frac{1}{\varepsilon} \int_0^t J(\xi)\, \mathrm{d}\xi + \varphi_0, \; \overline{\varphi}_0 = \frac{\varphi_0}{V_0/L} \tag{12.31}$$

(note the fact that $E_{0,0}^{(\overline{\varphi}_0)} = \varphi_0$). In the same moment, the particular carrier considered has travelled a distance

$$x_{t,0}^{(\overline{\varphi}_0)} = \frac{\mu}{\varepsilon}\int_0^t \mathrm{d}y \int_0^y J(\xi)\,\mathrm{d}\xi + \mu\varphi_0 t \tag{12.32}$$

which was obtained by integrating $v = \mu E$, i.e.

$$\frac{\mathrm{d}x_{t,0}^{(\overline{\varphi}_0)}}{\mathrm{d}} = \mu E_{t,0}^{(\overline{\varphi}_0)} \; . \tag{12.33}$$

The position of the leading and the trailing edges of the charge layer emitted at $t = 0$ (Figure 12.1) are, respectively, $x_{t,0}^{(1)}$ and $x_{t,0}^{(0)}$. Equation (12.32) yields

$$x_{t,0}^{(1)} - x_{t,0}^{(0)} = \mu \frac{V_0}{L} t \tag{12.34}$$

and the thickness of this layer increases linearly in time. At $t = t_1$, by definition $x_{t_1,0}^{(1)} = L$ and due to (12.26) $x_{t,0}^{(0)} = L - 0.787\,L \approx 0.2L$ i.e. the space-charge layer emitted at $t = 0$ extends over almost 80% of the inter-electrode space.

From equations (12.31) and (12.32) it follows that, at a given moment t the electric field $E_{t,0}^{(\overline{\varphi}_0)}$ inside the initial layer depends linearly upon the distance $x_{t,0}^{(\overline{\varphi}_0)}$. Hence, according to the Poisson equation (12.3) it follows that the hole density should be uniform inside the 'initial' layer. Now, because the total charge contained in this layer is constant and equal to Q_0 (see equation (12.28)) and because the layer thickness decreases linearly in time according to equation (12.34), the uniform charge density should decrease hyperbolically (i.e. inversely proportional) with the time t. It may be easily found that the charge density in this layer is given by

$$p_{t,0} = \frac{\varepsilon}{e\mu t}$$

and note that the dielectric relaxation time inside this layer is exactly equal to the time t, which points out that the charge dispersion is due to the electrostatic repulsion.

The charge distribution inside the conduction space, for $t < t_1$ is shown in Figure 12.2, reproduced after Many and Rakavy [243]. The abrupt front of the injected space charge is unrealistic, and is a consequence of neglecting diffusion.

The transient current may be now calculated for $t > t_1$. The current at the anode has both a particle and a displacement component and is given by

$$J(t) = \frac{\varepsilon\mu}{2L} E_L^2(t) \tag{12.35}$$

which is a result of equation (12.13) where $R_0 \to \infty$ ($N = 0$) and $V(t) = \text{const.}$ for $t > 0$. Here, for $t_1 < t < t_2$ E_L is $E_{t,0}^{(\overline{\varphi}_0)}$ where φ_0 is determined by

$$x_{t,0}^{(\overline{\varphi}_0)} = L. \tag{12.36}$$

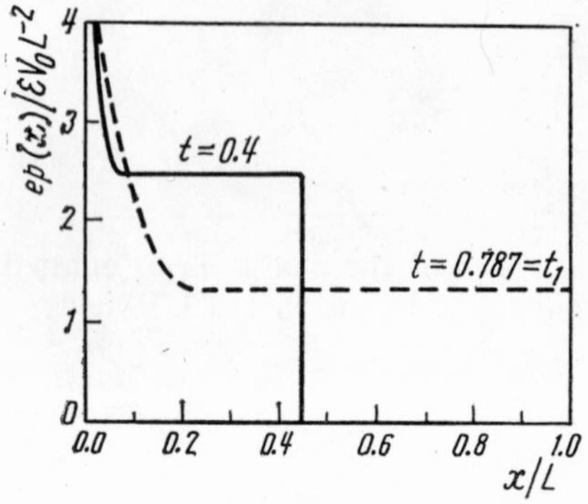

Figure 12.2. Space-charge distribution inside the crystal at two instants following the onset of injection [243].

By combining equations (12.31), (12.32) and (12.36), one obtains

$$E_L(t) = \frac{1}{\varepsilon}\int_0^t J(\xi)\,\mathrm{d}\xi + \frac{L - \dfrac{\mu}{\varepsilon}\displaystyle\int_0^t \mathrm{d}y \int_0^y J(\xi)\,\mathrm{d}\xi}{\mu t} \tag{12.37}$$

where J will be replaced by its value given by equation (12.35). Changing to normalized variables in equation (12.18), one finds

$$\overline{t}\left(\overline{E}_L - \frac{1}{2}\int_0^{\overline{t}} \overline{E}_L^2\,\mathrm{d}\xi\right) = \left(1 - \frac{1}{2}\int_0^t \mathrm{d}y \int_0^y \overline{E}_L^2\,\mathrm{d}\xi\right) \tag{12.38}$$

and by differentiating with respect to $\overline{t}$

$$\frac{\mathrm{d}\overline{E}_L}{\mathrm{d}\overline{t}} = \frac{\overline{E}_L^2}{2} - \frac{\overline{E}_L}{\overline{t}}. \tag{12.39}$$

The same differential equation was obtained by Many and Rakavy [243] by using a different approach. These authors solved this equation by making a change of dependent variable

$$\overline{E}_L = -\frac{2\Omega(\overline{t})}{\displaystyle\int \Omega(\overline{t})\,\mathrm{d}\overline{t}} \tag{12.40}$$

by which equation (12.39) becomes

$$\frac{\mathrm{d}\Omega}{\mathrm{d}\bar{t}} + \frac{\Omega}{\bar{t}} = 0 \tag{12.41}$$

and one obtains

$$\bar{E}_L(\bar{t}) = \left[\left(\beta - \frac{1}{2}\ln\bar{t}\right)\bar{t}\right]^{-1}, \; \bar{t}_1 < \bar{t} < \bar{t}_2 \tag{12.42}$$

where β is an integration constant. This constant will be determined requiring the continuity of $\bar{E}_L(\bar{t})$ at $\bar{t} = \bar{t}_1$. The final result is [243]

$$\bar{E}_L(\bar{t}) = \frac{e^{1/2}\bar{t}_1}{\bar{t}}\left[1 - (e^{1/2} - 1)\ln\left(\frac{\bar{t}}{\bar{t}_1}\right)\right]^{-1}, \; \bar{t}_1 \leqslant \bar{t} \leqslant \bar{t}_2 \tag{12.43}$$

$$J(t) = \frac{1}{2}e\left(\frac{\bar{t}_1}{\bar{t}}\right)^2\left[1 - (e^{1/2} - 1)\ln\left(\frac{\bar{t}}{\bar{t}_1}\right)\right]^{-2}, \; \bar{t}_1 \leqslant \bar{t} \leqslant \bar{t}_2\,. \tag{12.44}$$

The time t_2 is defined as the moment when the trailing edge of the charge-layer emitted at $t = 0$ reaches the anode. Thus

$$x^{(0)}_{t_2,0} = L. \tag{12.45}$$

The calculations are relatively long but do not set any problem. Many and Rakavy [243] have found that

$$\bar{t}_2 = 1.915\,\bar{t}_1 = 1.51. \tag{12.46}$$

These authors outlined the fact that

$$\bar{J}(\bar{t}_2) = 1.11 \tag{12.47}$$

whereas the asymptotic (steady-state) value is

$$\bar{J}(\infty) = 1.125 \tag{12.48}$$

(compare equations (4.85) and (12.22)). Therefore one may conclude that the steady-state behaviour is practically reached after a time of the order*)

$$t \gtrsim 2\frac{L^2}{\mu V_0} \quad (\bar{t} \gtrsim 2). \tag{12.49}$$

---

*) The solution for $\bar{t} > \bar{t}_2$ could be found numerically [243], starting from the equations developed in this section.

The result of the above calculations is shown in Figure 12.3. By the Gauss theorem we have

$$\varepsilon E_L(t) = Q_{\mathrm{mob}}(t) \tag{12.50}$$

where Q_{mob} is the total mobile charge (per unit of electrode area). Equations (12.36) and (12.50) yield

$$J(t) = \frac{\mu}{2\varepsilon L} Q_{\mathrm{mob}}^2(t) \tag{12.51}$$

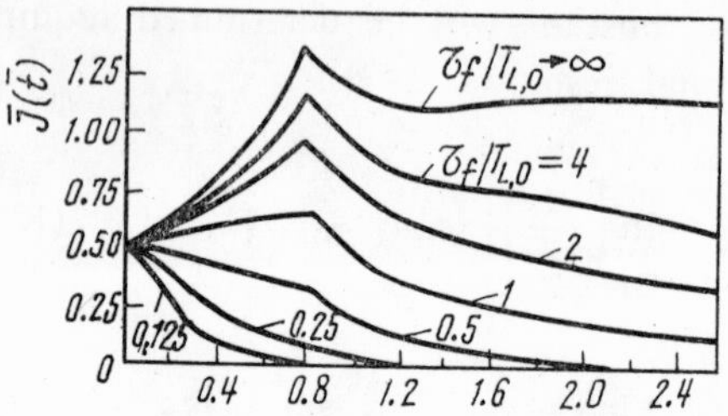

Figure 12.3. Turn-on transient SCL current of an insulator diode (the mobility is assumed constant). The curve $(\tau_f/T_{L,0})\to\infty$ corresponds to the absence of trapping [243].

and thus the total current is directly related to the total mobile charge which may be found between the electrodes at the same time moment t. In particular the initial ($t = 0$) and final ($t \to \infty$) values of $J(t)$ may be easily calculated.

At the time t_1, when the front of the injected space-charge reaches the anode, $J(t)$ (Figure 12.3) reaches a maximum and therefore $Q_{\mathrm{mob}}(t)$ also has a maximum. At the beginning of this transient régime, there are more favourable conditions for the injection of the mobile charge. At $t = t_1$ the mobile charge exceeds the value required to sustain the steady-state current. Then ($t > t_1$) this large injected charge is evacuated at the anode more rapidly than a fresh charge is injected at the cathode (because the charge already present in the crystal opposes to injection) and $Q_{\mathrm{mob}}(t)$ decreases below the steady-state value, etc. Many and Rakavy [243] appreciated that 'it is probable that the current exhibits an oscillatory behaviour, but the amplitude is so strongly damped that only the first cycle is of significance' [243]. Such an oscillatory behaviour was calculated for the *small-signal* transient behaviour and is shown in Figure 5.7. [110]. The oscillations predicted by Many and Rakavy [243] were theoretically demonstrated by Schilling and Schachter [244]. These authors studied [244] the a.c. small-signal stability of the short-circuited diode. They calculated the zeros of the a.c. impedance $Z = Z(s)$ where $s = \sigma + i\omega$ and Z is given by equation (4.90) ($N = 0$, $\mu =$ const., SCLC injection) i.e. the zeros of

$$\frac{1}{p^3}\{p^2 + 2p + 2[1 - \exp(p)]\} = 0, \; p = sT_L \tag{12.52}$$

and concluded that possible perturbations should lead to decreasing and rapidly damped oscillations.

Finally, the effect of trapping will be briefly discussed with reference to Figure 12.3 [243]. Here τ_f and τ_t are free-electron (hole) lifetime and the mean time spent by a charge-carrier in traps, respectively (see Section 3.13). The curves shown in Figure 12.3 are calculated for $\tau_f/\tau_t \to 0$, which denotes a severe trapping effect, and $J(t)$ should tend asymptotically towards zero, or to a negligible small-current as compared to the reference current (which is almost equal to the SCL current in the absence of trapping). The parameter on these curves is the ratio between the lifetime of a free carrier τ_f and the space-charge-free carrier transit-time, equation (3.81) $T_{L,0} = L^2/\mu V_0$. If the trapping is 'slow' as compared to the charge transport ($\tau_f \gg T_{L,0}$), the initial shape of the current transient is maintained. The transit-time of the leading front (indicated by the maximum of $J(t)$) is slightly longer, due to the charge trapped near $x = 0$ which increases the field in the cathode region, and thus decreases (because $\int_0^L E\,\mathrm{d}x = V_0 = \text{const.}$) the anode field under which the leading front moves. If the trapping is fast ($\tau_f \leqslant T_{L,0}$) then the transient behaviour is completely modified, as shown by Figure 12.3.

The above results are of importance in measurements on insulators containing a very large amount of structural defects and thus conducting a very small steady-state SCLC flow. If the transit-time is made sufficiently short, such as the trapping may be considered slow, then the carrier mobility may be determined from SCLC transient measurements.

12.4 Transient Behaviour of SCLC Semiconductor Resistors

The device to be considered in this Section is a $p^+\pi p^+$ (or $n^+\nu n^+$) SCLC resistor. The method of analysis is similar (see also equations 10.22 and 10.24 in Chapter 10, where θ_r is positive) but the calculations are, of course, more complicated. For the initial interval, $t < t_1$ the total current is equal to the displacement current at the anode and from equations (12.13) and (12.14) (where R_0 = finite because $N \neq 0$ and $V(t)$ is given by equation (12.1)) one obtains

$$\varepsilon \frac{\mathrm{d}^2 E_L}{\mathrm{d}t} = \frac{\varepsilon\mu}{2L} E_L^2 + \frac{V_0}{R_0}. \tag{12.53}$$

This equation may be written in normalized variables (12.18)

$$\frac{\mathrm{d}\overline{E}_L}{\mathrm{d}t} - \frac{\overline{E}_L^2}{2} = \theta_r' \tag{12.54}$$

where

$$\theta_r' = \frac{T_{L,0}}{\tau_r} = \frac{L^2/\mu V_0}{\tau_r}; \tag{12.55}$$

τ_r being the dielectric relaxation time

$$\tau_r = \frac{\varepsilon}{eN\mu}. \qquad (12.56)$$

The variables may be separated in equation (12.54) and the solution may be expressed in terms of elementary functions. Many and Rakavy [243] indicated that $J(t)$ may be expressed analytically for $t < t_1$. For $t > t_1$ no analytical solution seems to exist [243].

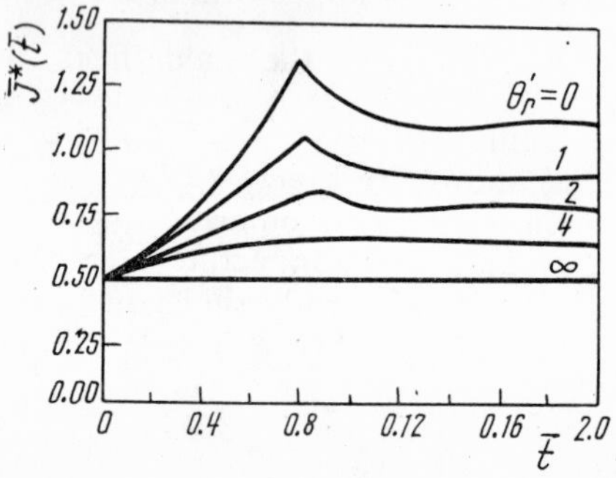

Figure 12.4. The time dependence of SCL excess current density $J^*(t)$ in conducting crystals. The curve $\theta_r' = 0$ corresponds to zero conductivity at thermal equilibrium [243].

We reproduce here, after Many and Rakavy [243] the time-dependence of the current

$$J^*(t) = J(t) - e\,N\mu\,(V_0/L) \qquad (12.57)$$

(the ohmic current was substracted, $N > 0$) for various numerical values of the normalized parameter $\theta_r' = T_{L,0}/\tau_r$. These curves are shown in Figure 12.4. The space-charge d.c. transit-time T_L does not differ very much from $T_{L,0}$ calculated in the absence of the space-charge and thus θ_r' is approximately equal to θ_r defined by equation (3.69). We note the fact (Figure 12.4) that if the device is switched-on in the square law region of the d.c. characteristic (Figure 3.21, $\Delta \to \infty$), the transient behaviour will be almost identical with the response of the ideal SCLC insulator diode ($\theta_r' \to 0$, see the previous Section).

12.5 The Effect of Diffusion Current

We have already indicated that the theory which neglects diffusion is inconsistent because it predicts an abrupt leading front, which leads to an infinitely large diffusion current.

Following Schilling and Schachter [245], we give below some approximate conditions required for neglecting diffusion. The current-continuity equation (12.2) should be modified as follows:

$$J = e\mu pE - eD\frac{\partial p}{\partial x} + \varepsilon\frac{\partial E}{\partial t} = J(t) \tag{12.58}$$

where the second term in the right-hand part is the diffusion current. Following the same procedure as in Section 12.3 (SCL injection, $N = 0$, $\mu = \text{const.}$, a voltage step of the form (12.11)), we obtain instead of equation (12.35)

$$J(t) = \frac{\mu\varepsilon}{2L}E_L^2 - \frac{eD}{L}\int_{x=0}^{x=L}\frac{\partial p}{\partial x}\mathrm{d}x \tag{12.59}$$

(where p is a function of both space and time). The diffusion current can be neglected if

$$\frac{1}{2}\mu\varepsilon E_L^2 \gg \left| eD\int_0^L \frac{\partial p}{\partial x}\mathrm{d}x \right| . \tag{12.60}$$

This condition will be rewritten by using the results derived in the absence of diffusion (Section 12.3). The carrier distribution is represented in Figure 12.2. We shall take into account only the *impulse* of diffusion current at the front of charge (diffusion will be neglected in the cathode region). One obtains [245]

$$\left.\frac{\partial p}{\partial x}\right|_{\text{front}} = -p_{\text{front}}(t)\,\delta(x_{\text{front}}) \tag{12.61}$$

where δ is the Dirac function, $x_{\text{front}} = x_{t,0}^{(1)}$ (Section 12.3) and $p_{\text{front}} = p_{t,0}$ is given by equation (12.34)*⁾. The electric field E_L in (12.60) will be replaced according to equations (12.18) and (12.21). The final result is

$$\frac{1}{2}\frac{V_0^2}{L^2}\left[\frac{1}{1 - t/2T_{L,0}}\right]^2 \gg \frac{\varepsilon D}{\mu t} \tag{12.62}$$

or, by using Einstein relation $\mu = \dfrac{e}{kT}D$

$$\frac{eV_0}{kT}\left[\frac{1}{1 - t/2T_{L,0}}\right] \gg \frac{2T_{L,0}}{t}. \tag{12.63}$$

---

*⁾ At the front the charge density changes abruptly from $p_{t,0}$ to zero (Figure 12.2).

Clearly, this condition cannot be satisfied in the vicinity of $t = 0$. Let t_{cr} be the time t which transforms equation (12.63) into an equality. One finds

$$t_{cr} = 2\,T_{L,0}\,\frac{eV_0}{kT} \tag{12.64}$$

and because of equations (12.18) and (12.26)

$$t_{cr}/t_1 = 2.54\,kT/eV_0. \tag{12.65}$$

According to these calculations, if V_0 is much greater than kT/e, the diffusion current may be neglected during most of the initial period of the transient, t_1. We shall also calculate the distance d_{cr} travelled by the injected space-charge in the time t_{cr}

$$\mathrm{d}_{cr} = \int_0^{t_{cr}} v_{\text{front}}\,\mathrm{d}x = \mu \int_0^{t_{cr}} E_L\,\mathrm{d}t = 2\mu\,T_{L,0}\,\frac{V_0}{L}\ln\left[1 - \frac{t_{cr}}{2T_{L,0}}\right]^{-1} \tag{12.66}$$

and expanding the logarithm ($t_{cr} \ll T_{L,0}$)

$$d_{cr} \simeq 2L\,\frac{e\,V_0}{kT} \tag{12.67}$$

i.e. d_{cr} may be neglected in comparison with L if $V_0 \gg \dfrac{kT}{e}$.

The above calculations are similar to the classical approach to the problem of SCL current in solids, due to Mott and Gunney [1]. The approximate character of these calculations was pointed out during the demonstration. There is, however, an implicit and capital approximation, which we wish to discuss here. This is the problem of the boundary conditions which should be used in a theory which includes diffusion. In this discussion we follow to a paper by Lampert and Schilling [246].

Consider further the model of the SCLC insulator diode discussed in Section 12.3. We expect that the above approximate theory will yield correct results at sufficiently long time intervals after the initial moment. The description of current-flow properties in the immediate neighbourhood of the cathode and at the early time of the transient response will be incorrect. We have assumed from the beginning of this chapter that the resistance in series with the device may be made arbitrarily small and the diode capacitance is charged instantaneously at the potential difference, V_0. Then, the correct boundary condition at the time $t = 0^+$ is

$$p\,(x, 0) = p_1(x), \;\; E(x, 0) = E_0\,(x) + \frac{V_0}{L} \tag{12.68}$$

where $p_0(x)$ and $E_0(x)$ correspond to the mobile carrier 'cloud' existing in insulator at $t < 0$ (see Problem 12.4). The total current for $t \geqslant 0$ is given by equation (12.58).

Integrating over the interelectrode distance and taking into account $V = \text{const.}$, one obtains

$$JL = e\mu \int_0^L pE\,\mathrm{d}x - eD \int_0^L \frac{\partial p}{\partial x}\,\mathrm{d}x. \tag{12.69}$$

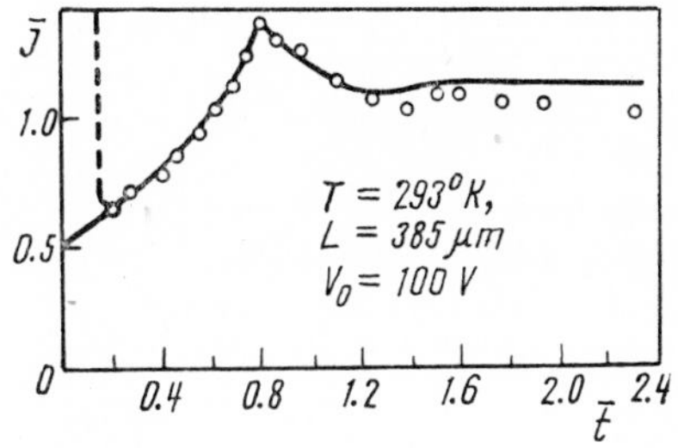

Figure 12.5. Comparison of theoretical transient of SCLC according to Many and Rakavy [243] (solid line) with experimental results of Lemke and Müller [249] (dots).

We wish to calculate the current at a given time instant, namely $t = 0^+$. By using the conditions (12.68) and taking into account the fact that the net current at $t = 0^-$ is zero, we obtain [246]

$$J(0^+) = \frac{e\mu}{L^2} V_0 \int_0^L p_0(x)\,\mathrm{d}x = \frac{Q_0'}{T_{L,0}}. \tag{12.70}$$

Here Q_0' is the mobile charge existing *inside* the insulator, at $t < 0$ (or $t = 0^-$) and its value is determined by the real conditions existing in the device, namely at the cathode interface. We quote from Lampert and Schilling [246]: Equation (12.70) 'tells us that the initial current, following the instantaneous capacitive surge, is determined by the Mott-Gurney electron (hole) cloud in the insulator and indeed gives a measure of this cloud'

First experimental results on transient large-signal behaviour were reported by Mark and Helfrich [247] on anthracene and by Many *et al.* [248] on single crystals of iodine. The SCL injection was provided by strong irradiation. The theory was also verified by Lemke and Müller [249] for $p^+\pi p^+$ silicon devices. Figure 12.5 shows the good agreement with theory at time moments comparable to the carrier transit-time. The decay at later time moments is due to the traps contained in crystal. The initial very large overshoot is attributed to the current charging the diode capacitance, according to Lampert and Schilling [245]. There is, however, some evidence that the diffusion current influences the decay of the initial large current [249].

12.6 Electrode-limited Transient-current in Insulators [250]

The results derived in the precedent Sections of this chapter are based upon the assumption on an ohmic injecting electrode at which the electric field is always zero. Practically, this condition is realised if the injecting electrode is a sufficiently large reservoir of carriers and can be considered a practically infinite one from the point of view of the current which it must supply at all times and voltages considered.

As shown above, in the experiments on transient currents in insulators, this infinite reservoir of carriers is often provided by generating carriers under the influence of radiation (illumination). There are, however, other situations which may be encountered in practice and will be considered below [250].

In the first case, the light-pulse is sufficiently intense to provide SCL injection conditions until at the time t_A, when it collapses.

In the second case, the light (steady-state) illumination is relatively weak, such that the zero-field at cathode is never established. Alternatively, the pulse duration is assumed to be short compared to the transit time, so that the entire mobile charge is injected into the bulk practically instantaneously.

One assumes throughout this Section μ = const., $N=0$ and negligible diffusion.

Case 1. The cathode field is zero for $t < t_A$ and let t_A be shorter than the time t_1 when the front of injected charge reaches the anode

$$E(0, t) = 0, \qquad t < t_A \leqslant t_1. \tag{12.71}$$

After the time t_A, and until the time t_1, the total charge contained in the crystal is constant and is given by

$$Q_A = Q_{\text{mob}}(t_A) = \varepsilon E_L(t_A). \tag{12.72}$$

This quantity may be readily calculated, because for $0 \leqslant t \leqslant t_A$ the situation is identical with that studied in Section 12.3 and the variation of the electric field is known from equations (12.18) and (12.21). Thus one obtains

$$Q_A = \frac{\varepsilon V_0}{L} \, \frac{1}{1 - t_A/2T_{L,0}}, \quad T_{L,0} = \frac{L^2}{\mu V_0}. \tag{12.73}$$

By the Gauss theorem we have

$$E(L, t) - E(0, t) = \frac{Q_A}{\varepsilon}, \; t_A \leqslant t \leqslant t_1. \tag{12.74}$$

Another equation which should be used for $t > t_A$ is given by equation (12.10), where $R_0 \to \infty$ $(N \to 0)$ and $V = V_0$ = const. Thus, we have

$$J(t) = \frac{\varepsilon\mu}{2L}\left[E^2(L, t) - E^2(0, t)\right]. \tag{12.75}$$

Finally, for $t < t_1$ the anode current has not a particle component and equation (12.14) is also valid. By using equation (12.74) we obtain

$$J(t) = \varepsilon \, \frac{\mathrm{d}E(L, t)}{\mathrm{d}t} = \varepsilon \, \frac{\mathrm{d}E(0, t)}{\mathrm{d}t}, \; t_A \leqslant t \leqslant t_1. \tag{12.76}$$

By combining the above equations, a differential equation in $E(L, t) = E_L(t)$ can be obtained, namely ($t_A \leqslant t \leqslant t_1$).

$$\frac{dE_L}{dt} - \frac{\mu}{L}\frac{Q_A}{\varepsilon} E_L + \frac{\mu}{2L}\frac{Q_A^2}{\varepsilon^2} = 0. \tag{12.77}$$

By using the initial condition, equation (12.72), at $t = t_A$, one finds [250]

$$E_L(t) = \frac{V_0/2L}{1 - t_A/2T_{L,0}}\left[1 + \exp\left(\frac{t - t_A}{T_{L,0} - t_A/2}\right)\right] \tag{12.78}$$

and then, from equations (12.76) and (12.78) the normalized current

$$\bar{J}(t) = \frac{J(t)}{\left(\frac{\mu\varepsilon V_0^2}{L^3}\right)} = \frac{1}{2}\left(1 - \frac{t_A}{2T_{L,0}}\right)^{-2} \exp\left(\frac{t - t_A}{T_{L,0} - t_A/2}\right). \tag{12.79}$$

The last two relations are valid only in the time interval $t_A \leqslant t \leqslant t_1$.

We stress the fact that the time t_1 is not yet known. It should be calculated from

$$L = \mu\int_0^{t_A} E(L, t)\,dt + \mu\int_{t_A}^{t_1} E(L, t)\,dt \tag{12.80}$$

where $E(L, t)$ is given by equation (12.21) for the calculation of the first integral and by equation (12.28) for the second integral. The calculations made by Weisz *et al.* [250] indicated that the transit-time t_1 is very close to that calculated in Section 12.3 under permanent space-charge limitation, namely $t_1 \simeq 0.8\,T_{L,0}$.

We shall now consider the situation $t > t_1$. The total current is

$$J(t) = \varepsilon\frac{dE(0, t)}{dt} = \varepsilon\frac{dE(L, t)}{dt} + e\mu\, p(L, t)\, E(L, t) \tag{12.81}$$

and may be calculated analytically for the time interval $t_1 < t < t_2$. Here t_2 is the transit-time of the trailing edge of the space-charge layer injected instantaneously at $t = 0$.

It may be shown that the space-charge density in the above-mentioned layer is again given by equation (12.34) (the demonstration is identical). Thus, for $t_1 \leqslant t \leqslant t_2$

$$p(L, t) = \frac{\varepsilon}{e\mu t}. \tag{12.82}$$

The total mobile charge $Q_{\mathrm{mob}}(t)$ is no longer constant because now the charge-carriers leave the crystal at the anode. We have

$$E(L, t) - E(0, t) = \varepsilon Q_{\mathrm{mob}}(t). \tag{12.83}$$

Equation (12.75) may be written

$$J(t) = \frac{\mu}{2L} Q_{\mathrm{mob}}(t) \left[2E(L,t) - \frac{Q_{\mathrm{mob}}(t)}{\varepsilon} \right]. \tag{12.84}$$

From equations (12.81) — (12.83) one obtains

$$E(L, t) = -\frac{t}{\varepsilon} \frac{\mathrm{d}Q_{\mathrm{mob}}}{\mathrm{d}t} \tag{12.85}$$

$$J(t) = \varepsilon \frac{\mathrm{d}E(L, t)}{\mathrm{d}t} - \frac{\mathrm{d}Q_{\mathrm{mob}}}{\mathrm{d}t}. \tag{12.86}$$

By eliminating $J(t)$ and $E(L, t)$, the last three equations yield a differential equation in $Q_{\mathrm{mob}}(t) = Q$, namely

$$\frac{\mu t}{\varepsilon L} Q \frac{\mathrm{d}Q}{\mathrm{d}t} + \frac{\mu}{2\varepsilon L} Q^2 = 2 \frac{\mathrm{d}Q}{\mathrm{d}t} + t \frac{\mathrm{d}^2 Q}{\mathrm{d}t^2}. \tag{12.87}$$

This equation may be integrated once and the result is

$$\frac{\mu t}{2\varepsilon L} Q^2 = Q + t \frac{\mathrm{d}Q}{\mathrm{d}t} + \text{const.} \tag{12.88}$$

The solution of this equation was expressed by Weisz *et al.* [250] in terms of Bessel functions. We reproduce here from the original paper [250] the results in graphical form — these are the first four curves from the top in Figure 12.6. It may be seen that if the pulse duration t_A is comparable to the transit-time t_1, the initial 'cusp' of the transient SCLC current ($t_A \to \infty$) is only slightly modified. Of course, $J(\infty)$ for $t_A =$ finite, should be zero because no more carriers are injected after $t = t_A$.

Case 2. Here we assume that the light-pulse duration is short compared to the transit time, and the charge Q_p injected by the light is smaller than $Q_0 = C_0 V_0$ (equation 12.28)

$$Q_p < Q_0 = C_0 V_C. \tag{12.89}$$

Under these conditions the entire charge Q_0 may be injected into the sample instantaneously at $t = 0$ and this is the case discussed here. The electric field at $x = 0$ is never equal to zero [250].

We shall study first the time interval $0 < t < t_1$. Equations (12.75) and (12.76) are again valid (here $t_A = 0$) and we can also write ($t < t_1$)

$$E(L, t) - E(0, t) = Q_p/\varepsilon. \tag{12.90}$$

By eliminating Q_p and $J(t)$ one obtains ($t < t_1$)

$$\frac{\mathrm{d}E(L, t)}{\mathrm{d}t} - \frac{\mu Q_p}{\varepsilon L} E(L,t) + \frac{\mu Q_p^2}{2L\varepsilon^2} = 0. \tag{12.91}$$

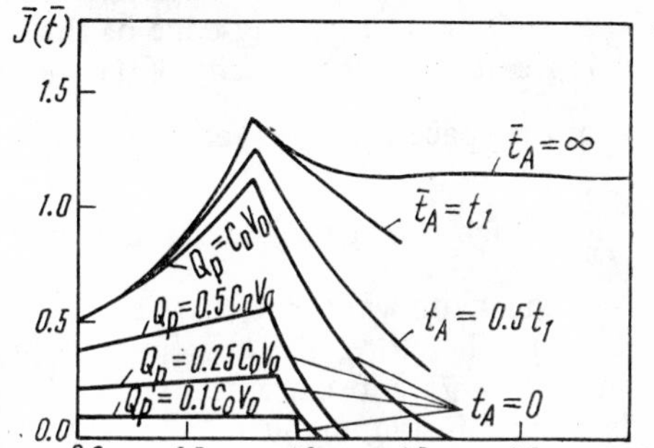

Figure 12.6. Current versus time curves due to injected charge in insulators. t_A is the time during which the field at the injected electrode vanishes. Also shown are current plots for $t_A = 0$, but with $Q_p < C_0V_0$ [250].

The solution should satisfy the initial condition

$$E(L, 0) = V_0/L. \tag{12.92}$$

Therefore, one obtains

$$E(L, t) = \left(\frac{V_0}{L} - \frac{Q_p}{2\varepsilon}\right) \exp\,(\mu Q_p t/L\varepsilon) + Q_p/2\varepsilon, \quad t < t_1 \tag{12.93}$$

$$J(t) = \varepsilon \frac{\mathrm{d}E(L, t)}{\mathrm{d}t} = \frac{\mu Q_p}{L}\left(\frac{V}{L} - \frac{Q_p}{2\varepsilon}\right) \exp \frac{\mu Q_p t}{\varepsilon L}, \quad t < t_1. \tag{12.94}$$

The transit-time t_1 will be determined in the usual manner

$$L = \int_0^{t_1} \mu E\ \mathrm{d}t = \frac{\mu\, Q_p\, t_1}{2\varepsilon} + \frac{\varepsilon L}{Q_p}\left[\frac{V_0}{L} - \frac{Q_p}{2\varepsilon}\right] \exp\left[\frac{\mu Q_p\, t_1}{L\varepsilon} - 1\right] \tag{12.95}$$

but an analytical solution is not possible. We note the fact [250] that if

$$Q_p \ll C_0 V_0 \tag{12.96}$$

then

$$J(t) \approx \frac{\mu Q_p}{L}\frac{V_0}{L} = \frac{Q_p}{T_{L,0}} = \text{const.} \tag{12.97}$$

and

$$t_1 \simeq T_{L0} \tag{12.98}$$

and this is exactly what we expect in the case of a very small injected-charge pulse.

The charge injected instantaneously at $t = 0$ will gradually disperse while moving towards the anode. At $t = 0$, the electric field at $x = 0^-$ (on the cathode side) should be $V_0/L - Q_p/\varepsilon$. The leading edge of the injected space-charge travels at a velocity which is greater than the velocity of the trailing edge by a quantity $\mu(Q_p/\varepsilon)$. Hence, the width of the space charge layer is $w = \mu t Q_p/\varepsilon$. The charge density is $ep = \frac{Q_p}{w} = \frac{\varepsilon}{\mu t}\cdot$ (which is exactly the result found in Section 12.3). Note the fact that at $t = T_{L,0}$, the width w is $(Q_p/C_0V_0)L$ and is much smaller than L if the condition (12.96) is satisfied. Therefore, the current "induced" by a small injected charge should vanish at $t_1 = T_{L,0}$, because the space charge layer ($w \ll L$) will disappear at the anode almost instantaneously.

The current transient for $t > t_1$ and $Q_p =$ finite, can be calculated by the same method as in the first part of this Section. The results obtained by Weisz *et al.* [250] are reproduced in Figure 12.6 and denoted by $t_A = 0$. Similar results were obtained by Papadakis [251], [252].

Finally, we note the fact that the results discussed until now are important for two categories of problems: first the measurement of carrier properties in high resistivity materials and secondly (see this Section), the drift of carriers and the resulting current in solid-state radiation detectors. The latter problem represents the subject of a monograph [253].

12.7 Transient Response of Insulated-gate Field-effect Transistors

We consider below the simple model of Section 8.1 for the IGFET. The dynamic behaviour should be studied by using either equations (8.17) and (8.18) or equation (8.19). These differential equations are non-linear and an exact analytical equation cannot be found. The problem was solved by numerical methods by O'Reilly [254] (he also considered the effect of a single trapping level). Das [255] calculated the device response by approximating the model, which is that of a *non-linear* RC distributed line, by an "effective" linear RC distributed line. This equivalence is supported by experimental data. "The effective capacitance should be such that the total charge storage in the non-linear and linear cases are exactly the same when subjected to identical boundary conditions" [255].

We reproduce below the results reported by Burns [173]. Let us consider the large-signal transient response $I_D = I_D(t)$ of an n-channel enhancement transistor ($V_T > 0$) to a gate-voltage step of value $+V_G$ (Figure 12.7). The drain voltage is assumed to be V_D

$$V_D > V_{D\,sat} = V_G - V_T \tag{12.99}$$

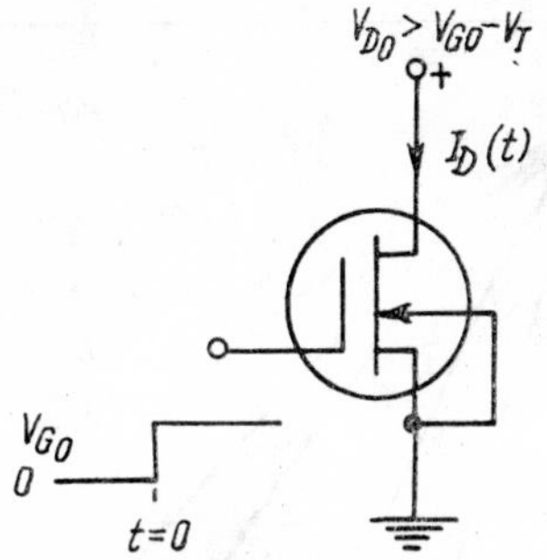

Figure 12.7. An IGFET subjected to a gate-to-source voltage step at $t = 0$.

i.e. the transistor will be switched on in saturation. By using equations (8.16), (8.19), (8.27), (8.29) and (8.31), the basic differential equation (8.19) may be written in normalized form

$$\frac{\partial}{\partial z}\left(\mathrm{v}\frac{\partial \mathrm{v}}{\partial z}\right) = \frac{\partial \mathrm{v}}{\partial t'} \tag{12.100}$$

where

$$z = 1 - \frac{x}{L} \tag{12.101}$$

$$t' = t/T_0,\ T_0 = L^2/\mu(V_G - V_T) \tag{12.102}$$

$$\mathrm{v} = \frac{\psi}{V_G - V_T} = 1 - \frac{V(x, t)}{V_G - V_T} = \mathrm{v}(z, t'). \tag{12.103}$$

Burns [173] indicated that this equation corresponds to 'nonstationary one-dimensional diffusion where the diffusion coefficient is directly proportional to the concentration'. The solution $\mathrm{v}(z, t')$ satisfies the following boundary conditions

$$\mathrm{v}(1, t') = 1 \text{ (source potential} = 0) \tag{12.104}$$

$$\mathrm{v}(0, t') = 0 \text{ (drain potential} = V_{D\,sat}) \tag{12.105}$$

$$\mathrm{v}(z, 0) = 0 \text{ (initial mobile charge} = 0). \tag{12.106}$$

Equation (12.100) indicates that the rate of change of the potential should be low at low voltages and is equal to zero at $t=0$ ($t'=0$, v $=0$). 'This behaviour is entirely different from that exhibited in a linear RC line, where the voltage is never identically zero at any finite distance and time. The solution will, therefore, be in the form of a traveling wave front where, for any given time, the voltage will be identically zero for distances greater than a certain value' [173].

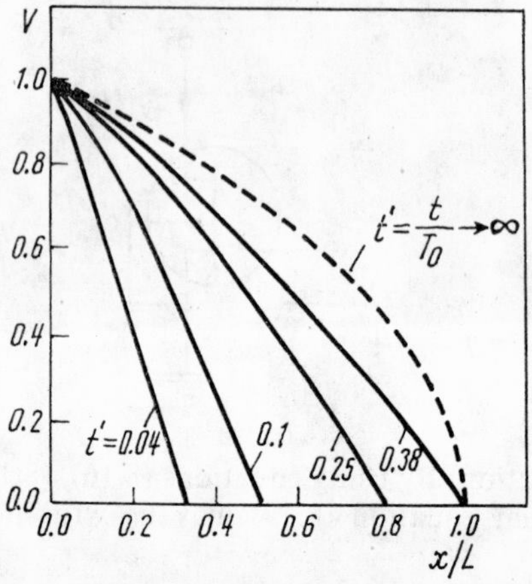

Figure 12.8. Voltage distribution in the channel as a function of time [173].

The solution is composed from two parts coresponding respectively to:

(a) An initial time interval starting at $t=0$ (the moment when the gate-voltage step is applied) and ending at t_1, when the front of the injected space-charge reaches the drain.

(b) The time after t_1, when the voltage distribution changes gradually to its steady-state value.

For $0 < t < t_1$, the differential equation (12.100) can be transformed into an ordinary differential equation by a proper change of variable and then integrated numerically [173]. Figure 12.8 shows the voltage distribution in the channel at various time moments before t_1. The transit time of the leading front is [173]

$$t_1 = 0.38\ T_0\ (t_1' = 0.38). \tag{12.107}$$

Figure 12.8 also shows the potential distribution in the channel for $t' \to \infty$ (steady-state distribution).

The analysis for $t > t_1$ can be carried out [173], approximately, by using a small-signal calculation. The reason for this is the fact that the potential distribution for $t = t_1$ is close to the steady-state distribution (Figure 12.8). The calculations are similar to that appearing in Chapter 8. The solution can be expressed in terms of Bessel functions and it was evaluated by using a computer [173].

The final results are shown in Figure 12.9, where the drain-current is plotted versus the normalized time t'. It can be seen that for the drain current to reach its final value, it is necessary, in addition to the *delay time* t_1, a *rise time* t_r which is

$$t_r = 0.3\ T_0. \tag{12.108}$$

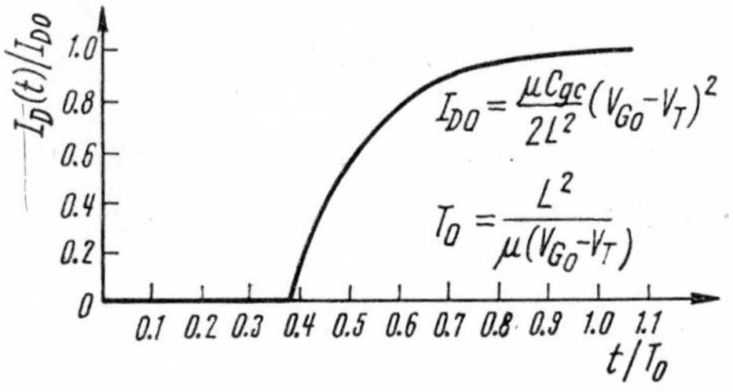

Figure 12.9. Large-signal drain-current response to a step of the gate-to-source voltage applied at $t = 0$ [173].

Burns [173] concludes that 'as a consequence of the distributed nature of the MOS transistor channel, the drain current does not follow an applied voltage step, but rather exhibits a time delay as well as a finite rise time...'*).

The existence of a finite time delay and rise time in the large signal turn-on behaviour was first determined theoretically [173] and then verified experimentally on special constructed devices [173]. The same devices (Figures 8.6) were used for small-signal measurements (Figures 8.7 and 8.8). Because the substrate is insulating (sapphire), the parasitic capacitances between the transistor proper and the substrate are eliminated. In present-day devices and integrated circuits, there are these capacitances which limit the switching speed [256]. The present theory [173] indicates the ultimate limit (the intrinsic limit) of the device switching performance.

Problems

12.1. Find the transient response of the square-law insulator (semiconductor diode) with SCL current to a ramp of voltage $V(t) = \alpha t$ [257].

Hint: The basic differential equation is again equation (12.13) where $R_0 \to \infty$ ($N = 0$). Before the arrival of the front of the injected charge, the current density is given by

$$\bar{J}\ (\bar{t}) = \left(\cos \frac{\bar{t}}{\sqrt{2}}\right)^{-2}, \tag{P.12.1}$$

where t and J are normalized as in equations (12.18) and (12.22), except the fact that V_0 is now $V_0 = V(T_0) = \alpha T_0$, etc.

---

*) The total turn-on times is of the order of T_0. This time constant is (for a given V_G) the product of the low-current drain resistance and the gate-channel capacitance. The same quantity, plays a key rôle in the small-signal theory (see Section 8.2).

12.2. Consider the response of a unipolar semiconductor diode with a 'saturated' contact (Section 3.3, the particle current cannot exceed the saturation current J_R) to a voltage step, equation (12.1). One assumes majority carrier injection and constant mobility. If J_Ω is the ohmic current and J_{SCLC} the steady-state SCL current, then J_R will satisfy the following equation

$$J_\Omega(V_0) < J_R < J_{SCLC}(V_0). \tag{P.12.2}$$

Derive the equations and discuss the response for the initial period of the transient.

12.3. Discuss the transient response to a voltage step of the Schottky-barrier-injection saturated-velocity diode (Section 11.1).

12.4. Show that the electron atmosphere in insulator in the vicinity of a metal-insulator contact, in thermal equilibrium is described by

$$n(x) = N_0\left(\frac{x}{x_0} + 1\right)^{-2}, \tag{P.12.3}$$

where

$$x_0 = \left(\frac{2\,\varepsilon kT}{e^2 N_0}\right)^{1/2} \tag{P.12.4}$$

and N_0 is the electron density inside the insulator at the metal-insulator interface. Derive a condition which provides that the total charge in an insulator diode (one contact injecting and the other rectifying separated by the distance L), when no current flows, is much smaller than the total charge accumulated in the device when a potential difference $+V_0$ is applied between the diode terminals (see also [246]).

12.5. Show that in the turn-on process in the IGFET (Figure 12.7), the front of the injected charge reaches the drain after a time interval which is considerably shorter than the steady-state transit-time, and explain intuitively this result.

13

High-field Domains in Negative-mobility Semiconductors

13.1 Introduction

We have already shown in Section 1.4 that a medium with negative differential conductivity is potentially unstable: a non-uniformity of the electric field due to a local space-charge will tend to grow, instead of decay by the normal dielectric relaxation process.

In certain conditions, this growing field non-uniformity takes the form of a *high-field domain*. This domain consists in an accumulation layer and an adjacent depletion layer and moves through the semiconductor together with the mobile carriers which form the above accumulation layer. The growing domain may eventually reach a stable configuration during its transit. In this chapter we shall concentrate upon the propagation of stable high-field domains, but the domain dynamics, their formation, and disappearance will be also discussed.

The behaviour of negative-mobility (or transferred-electron) devices may be quite diverse, depending upon the device geometry, doping profile, homogeneity of the semiconductor, properties of the electrical contacts, and the external circuit. Other modes of operation than that characterized by the cyclic propagation of high-field domains, will be discussed in the next chapter. However, a *complete* description of all these processes is beyond the scope of this book.

Most of the properties of the transferred-electron devices depend directly upon the relationship between the (average) carrier velocity and the intensity of the (local) electric field. This dependence is sometimes named the *static* $v - E$ characteristic.

Figure 13.1 shows, comparatively, the $v - E$ dependence for electrons in two semiconductor materials used to construct negative-mobility devices, namely GaAs (Figure 13.1a) and InP (Figure 13.1b). Despite the similarities there are important differences between them [258]. The high-field region for InP is characterized by a positive differential mobility, whereas in GaAs the electron velocity saturates or even continues to decrease at very high field intensities [259]. We anticipate that the existence of certain modes of operation depends upon the existence of the increasing high-field region of the $v - E$ characteristic.

We note from Figure 13.1 that InP requires a higher threshold field E_M for a negative-mobility to occur. However, InP has the advantage of a higher peak-to-valley ratio (v_M/v_m) and a higher absolute value of the differential mobility*).

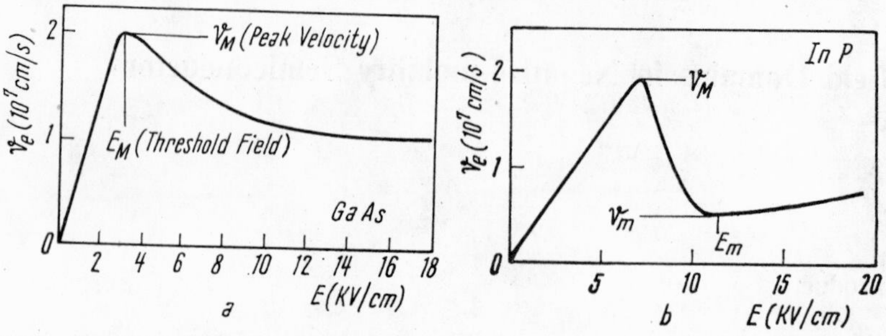

Figure 13.1. Velocity-field dependence for electrons in (a) GaAs and (b) In P [258].

13.2 Formation of a High-Field Domain

Consider a one-dimensional model, a diode consisting in a semiconductor slab of thickness L, between two plane-parallel contacts. A *constant* potential difference V is applied between these contacts.

The qualitative analysis of Section 1.4 has indicated that if the doping-length product is sufficiently high and the bias-field V/L is below the threshold field E_M, then the device will exhibit normal ohmic conduction (the bulk space-charge is negligible). However, if the bias field is increased above the threshold field, field non-uniformities may grow up and propagate through the semiconductor. This possibility was outlined in Section 1.4.

A mathematical description of these phenomena requires the current continuity equation (where the displacement current should be taken into account)

$$J = env(E) - e\frac{\partial}{\partial x}\{D(E)n\} + \varepsilon\frac{\partial E}{\partial t} \tag{13.1}$$

as well as Poisson's equation

$$\frac{\partial E}{\partial x} = \frac{e}{\varepsilon}(n - N_D). \tag{13.2}$$

Here electron conduction instead of hole conduction is considered. However, the electrons are treated as positively-charged particles†) and the equations are exactly the same as in Section 2.5.

---

*) The small-signal analysis of Chapter 6 clearly indicates that high v_M/v_m and large $|\mu_d|$ are advantageous.

†) The ionized donors should be considered as negatively charged and should yield a fixed space-charge density ($-N_D$).

Figure 13.2 shows qualitatively the growth of a field non-uniformity characterized by a local increase of the electric field, whereas in the remaining of the semiconductor bulk the electric field is uniform. Such a non-uniformity may be initiated, for example, by a notch in the doping profile: an increase of the electric field is required to provide the current continuity in a region of lower conductivity. The shape of the field non-uniformity is characteristic for a dipole configuration: an accumulation

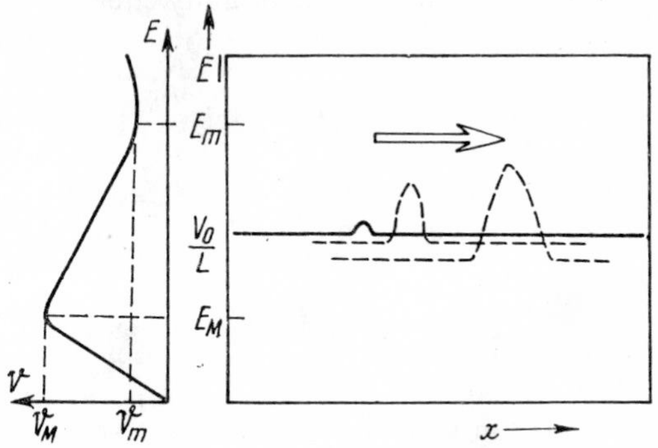

Figure 13.2. High-field domain formation in a negative-mobility semiconductor.

layer followed (towards positive x) by a depletion layer. Outside the dipole the semiconductor is assumed neutral*). As the time goes, the moving dipole increases through the accumulation of new charge carriers towards its trailing edge; this is due to the fact that in a medium characterized by a negative differential mobility the higher the field the lower the electron velocity.

However, because the voltage drop on the dipole increases and the total bias voltage is held constant, the field in the neutral regions must decrease. As soon as the external field (outside the domain) falls below the threshold field intensity, the dipole domain should tend towards a stable configuration.

Let us denote by $E_{N,1}$ and $E_{N,2}$ the electric field in the neutral regions on the cathode side (small x) and anode side, respectively. We shall prove that $E_{N,1}$ and $E_{N,2}$ should be equal. By requiring the current continuity one obtains [260]

$$J(t) = eN_D\, v(E_{N,1}) + \varepsilon \frac{\partial E_{N,1}}{\partial t} = eN_D v(E_{N,2}) + \varepsilon \frac{\partial E_{N,2}}{\partial t} \tag{13.3}$$

or

$$\varepsilon \frac{\partial}{\partial t}[E_{N,1} - E_{N,2}] = eN_D[v(E_{N,2}) - v(E_{N,1})]. \tag{13.4}$$

*) For the simplicity of analysis we consider that only one field non-uniformity exists in the bulk.

The perturbation occurs, say, at $t = 0$ and at this moment $E_{N,1} = E_{N,2}$, $v(E_{N,1}) = v(E_{N,2})$ and thus $\partial(E_{N,1} - E_{N,2})/\partial t = 0$. Therefore, at any time moment both sides of equation (13.4) should be zero and we have the following identity:

$$E_{N,1} = E_{N,2} = E_N \tag{13.5}$$

where E_N is the so-called external field (with respect to the domain)*).

By writing again the continuity equation, this time outside and inside the dipole, we have [260]

$$J(t) = eN_D v(E_N) + \varepsilon \frac{\partial E_N}{\partial t} = env(E) - \varepsilon \frac{\partial}{\partial x}\{D(E)\cdot n\} + \varepsilon \frac{\partial E}{\partial t} \tag{13.6}$$

and, by using equation (13.2)

$$\frac{\partial}{\partial t}(E - E_N) = \frac{eN_D}{\varepsilon}[v(E_N) - v(E)] + \frac{\partial}{\partial x}(Dn) - v(E)\frac{\partial E}{\partial x}. \tag{13.7}$$

This equation will be integrated with respect to x, from a plane x_1 situated on the left of the domain (Figure 13.2) until a plane x_2 which remains always on the right of the domain. This integration takes place at a certain time instant t and the integrals of the last two terms in equation (13.7) will be zero. The result is

$$\frac{d}{dt}\int_{x_1}^{x_2}(E - E_N)\,dx = \frac{eN_D}{\varepsilon}\int_{x_1}^{x_2}[v(E_N) - v(E)]\,dx. \tag{13.8}$$

The integral in the left-hand side is the *excess* voltage drop V_d on the dipole domain

$$V_d = \int_{x_1}^{x_2}(E - E_N)\,dx \tag{13.9}$$

and thus

$$\frac{dV_d}{dt} = \frac{eN_D}{\varepsilon}\int_{x_1}^{x_2}[v(E_N) - v(E)]\,dx. \tag{13.10}$$

The above relationship is useful for describing the domain dynamics. Let us find how fast the domain grows *initially*. Because the perturbation produced by the domain can be assumed small, the electron velocity may be approximated by a two-term series expansion

$$v(E) \simeq v(E_N) + \left(\frac{dv}{dE}\right)_{E=E_N} \times (E - E_N) \tag{13.11}$$

*) Because of equation (13.5), the total electric charge of the domain should be zero, i.e. the domain satisfies a necessary condition of being a dipole.

and by replacing $v(E)$ in equation (13.10) one obtains

$$\frac{\mathrm{d}V_d}{\mathrm{d}t} = -\frac{V_d}{\tau_d} \tag{13.12}$$

where τ_d is the differential relaxation time

$$\tau_d = \frac{\varepsilon}{eN_D\left(\dfrac{\mathrm{d}v}{\mathrm{d}E}\right)_{E=E_N}}. \tag{13.13}$$

Because $(\mathrm{d}v/\mathrm{d}E) < 0$ and $\tau_d < 0$, the excess voltage V_d will increase exponentially with the time constant $|\tau_d|$ (see also Section 1.4).

However, as the domain increases above a certain limit the rate of increase of V_d should decrease and, finally, should vanish. By simple inspection of equation (13.10) we find that as far as the electric field is so that $(\mathrm{d}v/\mathrm{d}E) < 0$ everywhere in the device, we have $v(E) \leqslant v(E_N)$ because $E \geqslant E_N$ and $\mathrm{d}V_d/\mathrm{d}t > 0$ (V_d increases and the domain itself grows). If E_N falls below the threshold field the integral in the right-hand side of equation (13.10) may vanish or take negative values. The moving domain can be *stable* only if $\mathrm{d}V_d/\mathrm{d}t = 0$, which requires

$$\int_{x_1}^{x_2} [v(E_N) - v(E)]\,\mathrm{d}x = 0. \tag{13.14}$$

The distribution of the electric field and the electric charge in a stable dipole domain will be examined in the next Section.

13.3 Stable High-field Domains

We shall consider below a stable non-uniformity which moves towards the anode with the velocity v_D. If x' is the coordinate of a system which moves together with this non-uniformity at the same velocity, in the positive x direction, then

$$x' = x - v_D t \tag{13.15}$$

and we should carry out the following replacement in equations (13.1) and (13.2)

$$\frac{\partial E}{\partial x} \to \frac{\mathrm{d}E}{\mathrm{d}x'}, \quad \frac{\partial (Dn)}{\partial x} \to \frac{\mathrm{d}(Dn)}{\mathrm{d}x'}, \quad \frac{\mathrm{d}E}{\mathrm{d}t} = -v_D\frac{\mathrm{d}E}{\mathrm{d}x} \tag{13.16}$$

to obtain

$$J = env(E) - e\frac{\mathrm{d}(Dn)}{\mathrm{d}x'} - \varepsilon v_D \frac{\mathrm{d}E}{\mathrm{d}x'} \tag{13.17}$$

$$\frac{\mathrm{d}E}{\mathrm{d}x'} = \frac{e}{\varepsilon}(n - N_D)\,. \tag{13.18}$$

The spatial variable x' can be eliminated between the above two equations and the result is

$$J = env(E) - \frac{e^2}{\varepsilon}(n - N_D)\frac{\mathrm{d}(Dn)}{\mathrm{d}E} - ev_D(n - N_D). \tag{13.19}$$

On the other hand, we have

$$J = e\,N_D v_N \tag{13.20}$$

where

$$v_N = v(E_N) \tag{13.21}$$

is the 'external' velocity. Equations (13.19) and (13.20) yield [260], [261]

$$\frac{e(n - N_D)}{\varepsilon}\frac{\mathrm{d}(Dn)}{\mathrm{d}E} = n[v(E) - v_D] - N_D(v_N - v_D) \tag{13.22}$$

and this may be considered the basic phenomenological equation describing the behaviour of stable non-uniformities. These non-uniformities are: high-field domains, low-field domains, accumulation layers, depletion layers. This equation is quite general: note the fact that we have made no assumption upon the shape of the velocity-field characteristic, upon the applied voltage, etc. The unique assumption is the superposition of drift and diffusion currents in equation (13.1) *).

Equation (13.22) should give, at least in principle, $n = n(E)$ and this dependence will be then used in conjunction with (13.18) to obtain the configuration of the domain, i.e.

$$n = n(x')\,,\; E = E(x'). \tag{13.23}$$

However, equation (13.22) is a non-linear differential equation †) and in most cases an analytical dependence, equation (13.23), cannot be found.

An usual assumption in the elementary theory of high-field domains is [260], [261]

$$D = \text{const.} \tag{13.24}$$

---

*) This problem was discussed in Chapter 2.

†) $v = v(E)$ and $D = D(E)$ are assumed given.

i.e. the diffusion constant is considered independent of field. This hypothesis is arbitrary. However, it has the merit of simplicity and allows a resonable understanding of the properties of stable layers.

It may be checked by differentiation that for $D = \text{const.}$ equation (13.22) admits the following formal solution

$$\frac{n}{N_D} - \ln \frac{n}{N_D} - 1 = \frac{\varepsilon}{e N_D D} \int_{E_N}^{E} \left\{ [v(E) - v_N] - \frac{N_D}{n} [v_N - v_D] \right\} \mathrm{d}E \cdot \quad (13.25)$$

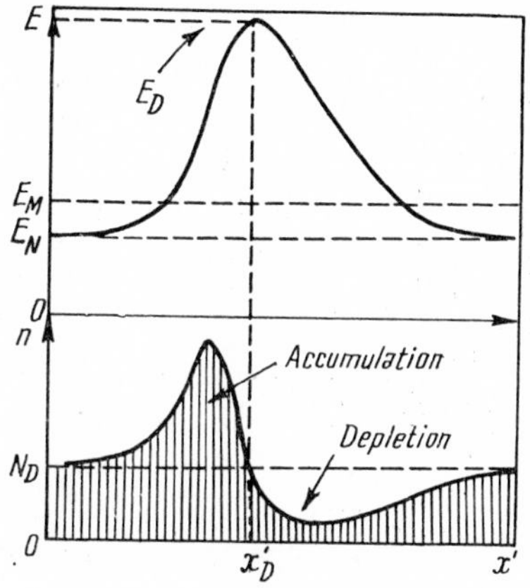

Figure 13.3. Electron and field distribution inside a moving high-field domain.

At this point we consider that this solution describes a dipole domain such that the electric field is the same before and behind the domain and has the value E_N. To be explicit, we shall consider a high-field domain. The qualitative dependence (33.23) is shown in Figure 13.3, where E_D is the maximum field in the domain.

The general solution, equation (13.25) should contain (except the trivial solution $E = E_N$ and $n = N_D$, corresponding to the neutral regions) two particular solutions: one for the accumulation layer and another one describing the depletion layer.

Let us consider $E = E_D$ for the upper limit of the integral in equation (13.25). Because

$$\left.\frac{\mathrm{d}E}{\mathrm{d}x'}\right|_{E=E_D} = 0, \qquad n\Big|_{E=E_D} = N_D \quad (13.26)$$

the left-hand side should be zero and the above integral should be also zero. This is not possible as far as the quantity under the sign of integral depends upon n and this is because the integral should be zero in two cases: (a) the integral is carried out on the leading front of the domain and $n < N_D$; (b) the integral is carried out on the trailing edge of the domain and $n > N_D$. Therefore, the coefficient of N_D/n in equation (13.15) should be zero, or

$$v_D = v_N \quad (D = \text{const.}) \quad (13.27)$$

i.e. the domain velocity is equal to the carrier velocity outside the domain.

Equations (13.25) – (13.27) yield [260], [261]

$$\int_{E_N}^{E_D} (v - v_D)\,\mathrm{d}E = \int_{E_N}^{E_D} [v\,(E) - v_N]\,\mathrm{d}E = 0. \tag{13.28}$$

This is the so-called 'equal-areas rule' and is illustrated in Figure 13.4. As the 'external' velocity v_N varies, the point having the coordinates (E_D, v_D) traces the *dynamic char-*

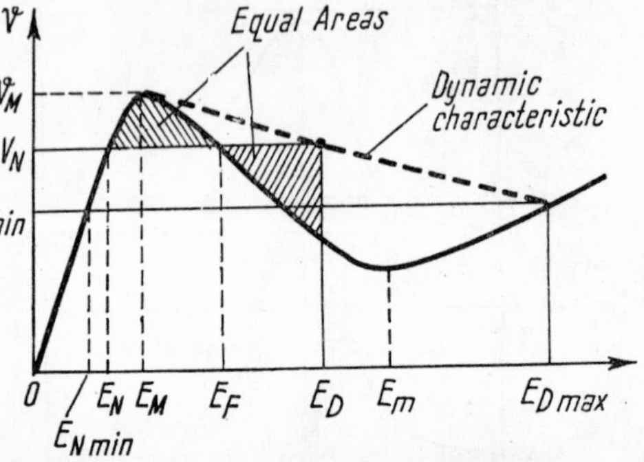

Figure 13.4. The "equal areas rule" and the dynamic $v - E$ characteristic (dashed curve).

acteristic in the $E - v$ plane. Note the fact, as far as $D =$ const. the dynamic characteristic is completely determined by the static characteristic, according to equation (13.28), and does not depend upon N_D or D.

The situation depicted in Figure 13.4 is characterized by an increasing $v - E$ curve in the high-field region (see Figure 13.1b, for InP): therefore the peak field inside the domain cannot exceed the maximum peak field $E_{D\max}$, which is the abscisa of the crossover point of the static and dynamic characteristics. If the carrier velocity continuously decreases or saturates at very high field intensities (such as for GaAs, Figure 13.1a), then $E_{D\max}$ will tend to infinity. Consequently the electric field inside the domain may exceed the critical value required for avalanche multiplication, thus leading to breakdown.

The shape of the domain should be calculated by using

$$\frac{n}{N_D} - \ln\frac{n}{N_D} - 1 = \frac{\varepsilon}{eN_D D}\int_{E_N}^{E} [v(E) - v_D]\,\mathrm{d}E \tag{13.29}$$

$$x' = x'_D + \frac{\varepsilon}{e}\int_{E_D}^{E} \frac{\mathrm{d}E}{n(E) - N_D} \tag{13.30}$$

where x'_D is the position of the interface between the accumulation and the depletion layers of the dipole (maximum electric field) in the coordinate system moving with the dipole. Here, $v(E)$ will be conveniently expressed by using the experimental and/or theoretical data. Sometimes, a piecewise linear approximation is used.

We shall consider below a special case, by assuming

$$D \to 0 \tag{13.31}$$

in equation (13.29) and this will be equivalent to the case when the diffusion current is neglected in equation (13.1)*). Because of $D \to 0$ the right-hand side of equations (13.29) tends to infinity (except for $E = E_N$ and $E = E_D$ and the integral is zero).

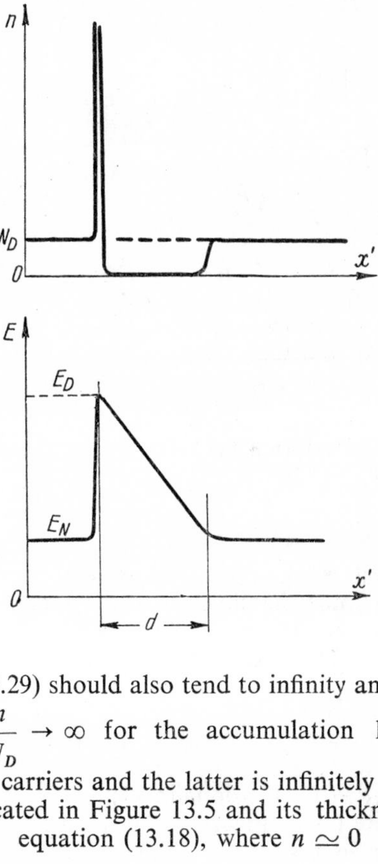

Figure 13.5. Electron and field distribution in a "triangular" high-field domain.

Therefore, the left-hand part of equation (13.29) should also tend to infinity and we have $\frac{n}{N_D} \to 0$ for the depletion layer and $\frac{n}{N_D} \to \infty$ for the accumulation layer: the former is completely depleted of mobile carriers and the latter is infinitely thin. The domain has a triangular shape, as indicated in Figure 13.5 and its thickness d can be easily calculated by using Poisson's equation (13.18), where $n \simeq 0$

$$\mathrm{d} \simeq \frac{\varepsilon}{e N_D} (E_D - E_N) \,. \tag{13.32}$$

*) From a mathematical point of view, the same result will be obtained if N_D is taken very small, i.e. the crystal resistivity is very high.

13.4 Effect of External Bias upon Domain Propagation

The domain configuration can be determined according to equations (13.29) and (13.30), if E_N and E_D are known. If the external field E_N is known, the peak field E_D will be determined by using the equal areas rule, equation (13.28).

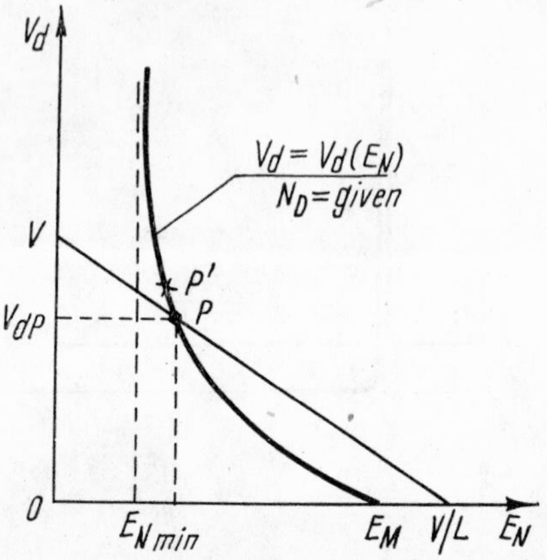

Figure 13.6. Typical $V_d = V_d(E_N)$ dependence (N_D = given) and the "bias line" (13.33) for a device of length L, and a constant applied voltage V. The crossover point P corresponds to the propagation of a stable high-field domain. A new domain forms when the preceding one disappears at the anode [260].

The external field E_N can be found by using the relation

$$V = E_N L + V_d \tag{13.33}$$

where V is the applied voltage and V_d is the domain excess-voltage, equation (13.9). On the other hand, E_N depends upon V_d according to equations (13.9), (13.29) and (13.30) and the $v = v(E)$ dependence. A typical curve $V_d = V_d(E_N)$ is drawn on Figure 13.6*). Note the fact that, according to Figure 13.4, E_N should satisfy the following equation

$$E_{N\min} < E_N < E_M\,. \tag{13.34}$$

For $E_N \to E_M$ the domain is small (V_d small). If the $v = v(E)$ characteristic does not saturate (as in Figure 13.4), the maximum field in the domain is limited to $E_{D\max}$, but the excess voltage V_d may increase arbitrarily, due to the increase of the domain thickness. Therefore V_d tends to infinity for $E_N \to E_{N\min}$ (see Figure 13.6).

Now, assume we apply a voltage bias V to a device of length L. The straight-line, equation (13.33), represented in Figure 13.6 intersects the $V_d = V_d(E_N)$ curve in a single point P. Note that for $V_d \to 0$, the bias field V/L exceeds the threshold field

*) This curve depends not only upon $v = v(E)$, but also upon N_D (for a given E_N, V_d decreases with increasing N_D).

E_M. Thus, we may assume that a domain will form. The point P corresponds to a domain having an excess voltage $V_{d,P}$ (Figure 13.6). It will be shown that this domain is stable indeed. Suppose that V_d has an accidental increase which brings the point P in P' (Figure 13.6). Because E_N decreases, $E_N L$ has a decrease but this is not sufficient to compensate the increase of V_d and satisfy equation (13.33), which requires $V =$ const. Consequently, V_d should decrease and P' should tend back to P, etc.

Note the fact that if V increases, E_N tends to $E_{N\min}$, $v_N \to v_{N\min}$ and $J \to eN_D\, v_{N\min} = J_{\text{sat}}$. The current flowing in the external circuit during the propagation of the domain reaches a (lower) limit value as the bias voltage increases. Such a domain is called *saturated.*

Consequently, the situation depicted in Figure 13.6 describes an *astable* behaviour of the sample. A stable domain forms and propagates through the device. As soon as this domain reaches the anode (collecting electrode), it should disappear. The electric field in the semiconductor increases again above the threshold field and a new domain will form. We assume, for simplicity, that the new domains form only at the cathode and that the time interval required for the domain formation or disappearance is very short as compared to the transit-time of the domain, T_L. The process repeats cyclically with a frequency

$$f_T = \frac{1}{T_L}. \tag{13.35}$$

Every time a domain reaches the anode, a current spike occurs in the external circuit. During the propagation of a stable domain the current remains constant and equal to $eN_D v_N$. The current oscillations for a transferred-electron device biased with a constant voltage are called Gunn oscillations [262]. The efficiency of these oscillations is relatively low*⁾.

Figure 13.7 shows a bias-line, equation (13.33), crossing the $V_d = V_d(E_N)$ characteristic in two points, P_1 and P_2. Note that for $V_d \to 0$ (no domain in the crystal), the electric field is below the threshold field and no domain can occur. Let us suppose that a domain does nevertheless exist. There are two possible states corresponding to a stable dipole-domain. These are given by the crossover points P_1 and P_2. However, only the point P_2 describes a stable dynamic behaviour. The demonstration of this property is identical to that given above for the point P on Figure 13.6. In a similar manner the fact can be demonstrated that a high-field domain corresponding to P_1 in Figure 13.7 is unstable with respect to perturbations. A small perturbation either leads to the disappearance of the (P_1) domain, or determines the transition into the (P_2) state.

The case represented in Figure 13.7 corresponds to the *monostable* mode of operation. A high-field domain can be nucleated by applying a short†⁾ positive voltage pulse ΔV and thus increasing temporarily the bias field above the threshold

*⁾ There are two practical methods to improve this efficiency: (a) to increase the time required for domain nucleation and disappearance (by increasing the resistivity) and thus to broaden the current pulses; (b) to use a negative-mobility semiconductor with a larger peak-to-valley velocity ratio (see Figure 13.1).

†⁾ This pulse should be longer than the time required for domain formation but shorter than the domain transit-time.

field. This domain propagates through the crystal and disappears at the anode giving a negative pulse of current. The pulse width is approximately equal to the domain transit-time. A new domain cannot appear until a new triggering pulse ΔV is applied.

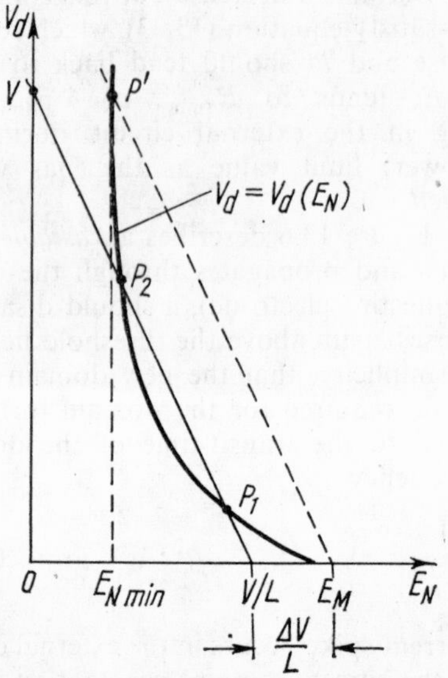

Figure 13.7. Figure illustrating the monostable mode of operation. The point P_2 corresponds to a stable high-field domain launched by a trigger pulse ΔV superimposed on the constant voltage V.

The main application of this type of operation is pulse regeneration. The pulse width cannot be changed in wide limits, because as the d.c. bias V increases the domain velocity $v_D = v_N$ tends to saturate.

The bistable mode of operation is also possible. We should have, in principle, two stable states when no domain moves through the device and the current remains constant for an indefinite time as far as the voltage is kept constant. The steady-state characteristic

$$J = e\,N_D\,v\,(E) \tag{13.36}$$

(current *versus* bias field) is plotted in Figure 13.8 and has exactly the shape of the $v - E$ characteristic*). Consider that the device is biased from a constant-voltage source in series with a resistor. A particular position of the load line is indicated in Figure 13.8. There are two crossover points with the steady-characteristic in the positive conductance ($\mathrm{d}J/\mathrm{d}E > 0$) regions (for a $v-E$ characteristic like that depicted in Figure 13.8). These points may correspond to a high-field (P_2) and a low field (P_1) steady-state, respectively. The intermediate crossover point does not represent a real state, because the negative differential conductivity leads to the formation of a domain.

---

*) This steady-state characteristics coincides here with the neutral characteristic defined in Section 3.7. The necessary condition for this is a large doping-length product.

The existence of the steady-state P_2 is doubtful. Although the bias-field is in the positive-mobility region, the electric field in the vicinity of the cathode may fall below the bias field due to carrier injection (Section 1.4) and thus the cathode region will be in the negative-mobility region and the domain formation is possible*).

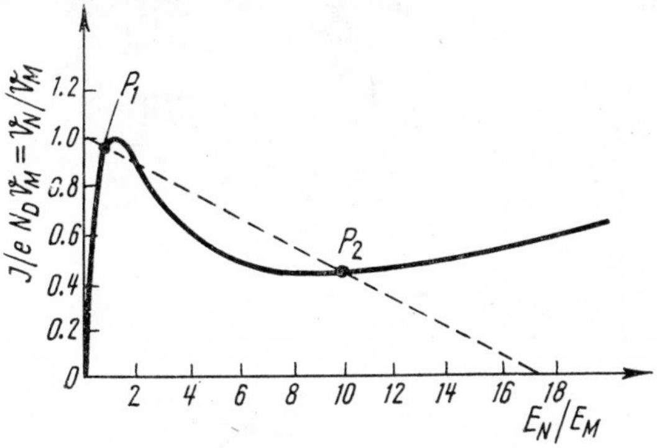

Figure 13.8. Static $v - E$ characteristic and a load line (the device is biased in series with a resistance). The crossover points P_1 and P_2 may correspond to stable states [266].

13.5 Dynamics of High-field Domains

The equations of Section 13.3 apply to the propagation in the semiconductor bulk of a stable completely-formed high-field domain. The domain formation, growing and disappearance should be described by differential equations with partial derivatives (Section 13.2).

A relatively simple mathematical description of domain dynamics can be obtained for the special case when diffusion is neglected. We shall consider below only this case. In the absence of diffusion the *stable* high-field dypole domain is triangular (Section 13.3) or trapezoidal (Problem 13.3), i.e. consists of a depletion layer completely free of mobile carriers, followed by an infinitely-thin accumulation layer.

The starting equations are derived from equations (13.1), (13.2) and (13.31)

$$J = env(E) + \varepsilon \frac{\partial E}{\partial t}, \tag{13.37}$$

$$\frac{\partial E}{\partial x} = \frac{e}{\varepsilon}(n - N_D) \tag{13.38}$$

---

*) Thim discussed recently [263], [264] bistable switching with negative-mobility diodes. The switching properties of these devices were also studied by Engelmann and Heinle [265], [266].

and are well known from other chapters of this book. These equations will be combined and rewritten in a different form (as a set of two coupled ordinary differential equations) which is convenient for describing the dynamics of the domain.

By replacing into equation (13.37) n given by equation (13.38) and by integrating from $x = 0$ to $x = L$ one obtains

$$L\,J(t) = e\int_0^L \left(\frac{\varepsilon}{e}\frac{\partial E}{\partial x} + N_D\right) v\,(E)\,\mathrm{d}x + \varepsilon L\,\frac{\mathrm{d}E_b}{\mathrm{d}t} \tag{13.39}$$

where

$$E_b = V\,(t)/L \tag{13.40}$$

is the bias field, and

$$V\,(t) = \int_0^L E\,\mathrm{d}x \tag{13.41}$$

is the applied voltage. The total current $J(t)$ may be also written*)

$$J\,(t) = \varepsilon\,\frac{\mathrm{d}E_b}{\mathrm{d}t} + \frac{eN_D}{L}\int_0^L v\,(E)\,\mathrm{d}x + \frac{\varepsilon}{L}\int_{E(0)}^{E(L)} v\,(E)\,\mathrm{d}E\,. \tag{13.42}$$

We shall eliminate $J(t)$ between equations (13.37) and (13.42) thus we obtain the equation

$$\frac{\partial E}{\partial t} + \frac{e}{\varepsilon}\left[\frac{\varepsilon}{e}\frac{\partial E}{\partial x} + N_D\right] v\,(E) = \frac{\mathrm{d}E_b}{\mathrm{d}t} + \frac{eN_D}{\varepsilon L}\int_0^L v\,(E)\,\mathrm{d}x + \frac{1}{L}\int_{E(0)}^{E(L)} v\,(E)\,\mathrm{d}E\,. \tag{13.43}$$

The above equation will be used for two particular values of the electric field in the left hand side. First, let us consider $E = E_N$ in the neutral region, outside the domain. We have already demonstrated in Section 13.2 that the 'external' field is uniform and equal on both sides of the domain. By introducing $E = E_N$ in (13.43) we shall also have $\partial E/\partial x = 0$ and $E(L) = E(0)$†). Thus one obtains

$$\frac{\partial E_N}{\partial t} = \frac{\mathrm{d}E_b}{\mathrm{d}t} + \frac{eN_D}{\varepsilon L_0}\int_0^L [v(E) - v_N]\,\mathrm{d}x\,. \tag{13.44}$$

---

*) The integral of equation (13.39) is evaluated at a given moment of time.

†) However, as the domain disappears at the anode, $E(L) \neq E(0)$ and the equation should be corrected accordingly.

In the second case, E in equation (13.43) is put equal to the peak field in the domain E_D. Therefore we have

$$\frac{\partial E_D}{\partial t} - \frac{eN_D}{\varepsilon} v(E_N) = \frac{dE_b}{dt} + \frac{eN_D}{\varepsilon L}\int_0^L [v(E) - v_N]\, dx \tag{13.45}$$

because, according to equation (13.37) where $n = 0$, we have

$$\left.\frac{\partial E}{\partial x}\right|_{E=E_D} = -\frac{eN_D}{\varepsilon}. \tag{13.46}$$

Let us denote by F the following integral

$$F = \frac{eN_D}{\varepsilon}\int_0^L [v(E) - v_N] dx \tag{13.47}$$

and we shall demonstrate that $F = F(E_N, E_D)$.
According to equation (13.10) we also have

$$\frac{dV_d}{dt} = -F, \tag{13.48}$$

where V_d is the excess voltage drop on the dipole domain. It can be now easily checked that equation (13.44) where $\partial E_N/\partial t = dE_N/dt$ is the result of differentiating the bias equation (13.33).

We shall evaluate the integral (13.47) for the special case of triangular domains. One obtains successively

$$\int_0^L v(E)\, dx = v(E_N)(L - d) + \int_{E_D}^{E_N} \frac{v(E)\, dE}{\partial E/\partial x} =$$

$$= v_N L - v_N d + \frac{\varepsilon}{eN_D}\int_{E_N}^{E_D} v(E)\, dE \tag{13.49}$$

where d is the domain thickness, equation (13.22). The final result is

$$F = \frac{eN_D}{\varepsilon}\int_0^L v(E)\, dx - \frac{eN_D L}{\varepsilon} v_N =$$

$$= \int_{E_N}^{E_D} v(E)\, dE - v_N(E_D - E_N) = \int_{E_N}^{E_D} [v(E) - v_N] dE. \tag{13.50}$$

Therefore, the function F (which is *the rate of domain decrease*, as shown by equation 13.48) can be conveniently calculated by using *the static* $v - E$ *characteristic*, and is indeed a function of E_N and E_D.

Because of equation (13.44) and (13.47)

$$\frac{\partial E_N}{\partial t} = \frac{\mathrm{d}E_N}{\mathrm{d}t} = \frac{\mathrm{d}E_b}{\mathrm{d}t} + \frac{F(E_N, E_D)}{L}. \tag{13.51}$$

On the other hand we have

$$\frac{\mathrm{d}E_D}{\mathrm{d}t} = \frac{\partial E_D}{\partial t} + \left.\frac{\partial E}{\partial x}\right|_{E=E_D} \times \frac{\mathrm{d}x_D}{\mathrm{d}t} = \frac{\partial E_D}{\partial t} - \frac{eN_D}{\varepsilon} v_D. \tag{13.52}$$

(x_D being the position of the maximum field plane in the domain and v_D the velocity of this plane) and equations (13.45), (13.47), (13.50) and (13.52) yield

$$\frac{\mathrm{d}E_D}{\mathrm{d}t} = \frac{\mathrm{d}E_b}{\mathrm{d}t} + \frac{F(E_N, E_D)}{L} + \frac{eN_D}{\varepsilon}(v_N - v_D). \tag{13.53}$$

By substracting part by part equations (13.51) and (13.53), one obtains,

$$\frac{\mathrm{d}}{\mathrm{d}t}(E_D - E_N) = \frac{eN_D}{\varepsilon}(v_N - v_D). \tag{13.54}$$

Whereas v_N is equal to $v(E_N)$, the domain velocity v_D differs from the value $v(E_D)$. The difference $v_N - v_D$ can be, however, expressed as a function of E_N and E_D. According to Heinle [267], we differentiate the bias equation (13.33), thus obtaining

$$\frac{\mathrm{d}E_b}{\mathrm{d}t} = \frac{\mathrm{d}E_N}{\mathrm{d}t} + \frac{1}{L}\frac{\mathrm{d}V_d}{\mathrm{d}t}. \tag{13.55}$$

For a triangular domain the excess voltage drop V_d can be easily calculated as

$$V_d = \frac{\varepsilon}{2eN_d}(E_D - E_N)^2. \tag{13.56}$$

By introducing V_d into equation (13.55) and using equation (13.51) and (13.54), one obtains

$$v_N - v_D = -\frac{F(E_N, E_D)}{E_D - E_N} \tag{13.57}$$

and thus equation (13.53) becomes [267]:

$$\frac{dE_D}{dt} = \frac{dE_b}{dt} + \frac{F(E_N, E_D)}{L}\left(1 - \frac{eN_D L/\varepsilon}{E_D - E_N}\right). \tag{13.58}$$

Therefore, we have a set of two ordinary differential equations, namely equations (13.51) and (13.58). The unknown variables are $E_N = E_N(t)$ and $E_D = E_D(t)$. The explicit form of these equations depends upon the particular $v = v(E)$ dependence.

Note the fact that stable high-field domains ($dE_N/dt = 0$, $dE_D/dt = 0$) moving in a device biased with a constant voltage ($dE_b/dt = 0$) are characterized by

$$F(E_N, E_D) = \int_{E_N}^{E_D} [v(E) - v_N]\, dE = 0, \tag{13.59}$$

which is just the 'equal-areas rule' equation (13.28), and also by $v_D = v_N$ which is identical with equations (13.27) (the domain velocity is identical with the electron velocity in the neutral regions).

We shall now discuss the dynamic behaviour ($dE_N/dt \neq 0$, $dE_D/dt \neq 0$), i.e. the domain growth or decay.

We note, following Heinle [267], that equations (13.51), (13.58) are invariant with respect to the following transformation

$$t \to \lambda t, \quad L \to \lambda L, \quad N_D \to N_D/\lambda. \tag{13.60}$$

If the doping-length product $N_D L$ is kept constant, all processes run down λ-times faster if the sample length L is shortened or the doping N_D is increased by the factor λ. For a time-dependent process having the duration t, the rule $tN_D =$ const.holds. Therefore, the domain formation time should be inversely proportional to the doping, for a given $N_D L$ product [267].

Let us consider the domain formation. The magnitude of the electric charge in the accumulation layer is (per unit area)

$$Q_D = \varepsilon(E_D - E_N) \tag{13.61}$$

and because of equation (13.54) and (13.57), Q_D satisfies the equation

$$\frac{dQ_D}{dt} = \frac{eN_D F(E_N, E_D)}{E_D - E_N}, \tag{13.62}$$

Assume for simplicity that: (a) the $v - E$ dependence is approximated by a linear piecewise characteristic, the negative differential mobility being $\mu_d < 0$, (b) the electric field ranges in this negative-mobility region. Thus we have

$$v(E) - v_N = \mu_d(E - E_N), \quad \mu_d < 0 \tag{13.63}$$

and $F(E_N, E_D)$ in equation (13.63) will be calculated according to equation (13.50), thus obtaining

$$\frac{dQ_D}{dt} = -\frac{1}{2\tau_d} Q_D, \ \tau_d = \frac{\varepsilon}{e\mu_d N_D} < 0. \tag{13.64}$$

The domain charge grows exponentially (see also [29])

$$Q_D(t) = \text{const. } \exp\,(t/2|\tau_d|); \tag{13.65}$$

the time constant is twice the differential relaxation time $|\tau_d|$. The proportionality factor should be determined by using an initial condition. For a triangular domain the charge Q_D is proportional to the domain width d (see equations (13.32) and (13.61)). Thus, d will grow exponentially. The initial field non-uniformity may arise from a doping notch, as shown above [260]. The initial domain width should coincide with the notch dimension. However, it can be readily shown that the initial non-uniformity does not generally have the form of a triangular domain and, during an initial time interval, the growing domain should readjust its shape.

The exponential growth predicted by equation (13.65) continues as long as both E_N and E_D find themselves in the negative-mobility region. Assume, for example, that E_N falls below the threshold field. It can be easily found by using equations (13.48) and (13.50) that dV_d/dt decreases and vanishes (asymptotically). By also using equations (13.22) and (13.54) we may state the following rule:

a) If $F(E_N, E_D) > 0$, then $dV_d/dt < 0$, the domain charge will decrease and $v_N < v_D$.

b) If $F(E_N, E_D) < 0$, $dV_d/dt > 0$, the domain will increase and $v_N > v_D$. The increase of the domain can be easily understood, in view of the fact that the electrons entering in the domain move faster than the domain accumulation layer ($v_N > v_D$) and the domain is forced to absorb more charge.

The domain velocity v_D can be found by using a kind of modified equal-areas rule. According to equations (13.50) and (13.57) we have

$$v_D = \frac{\int_{E_N}^{E_D} v(E)\, dE}{E_D - E_N}, \tag{13.66}$$

which may be also written

$$\int_{E_N}^{E_D} [v(E) - v_D]\, dE = 0. \tag{13.67}$$

For the special case $v_N = v_D$ (stable high-field domain), equation (13.67) yields the equal areas rule, equation (13.59). Figure 13.9 illustrates the above considerations.

We shall consider below that the device contains two doping notches and, as the bias field rises above the threshold field, two high-field domains are nucleated.

As long as these field non-uniformities are still small, they retain the shape and dimension of the nucleation centre. The larger domain will absorb a higher voltage. The electric field outside the domains (E_N) is uniform and decreases with increasing t. Both domains are described by a relation of the form of equation (13.48)

$$\frac{\mathrm{d}V_{d_j}}{\mathrm{d}t} = -F(E_N, E_{D_j}) = \frac{eN_D}{\varepsilon} \int_{E_N}^{E_{D_j}} [v_N - v(E)]\,\mathrm{d}E \tag{13.68}$$

where j is the domain index. Assume $V_{d1} > V_{d2}$. Both domains will continue to grow

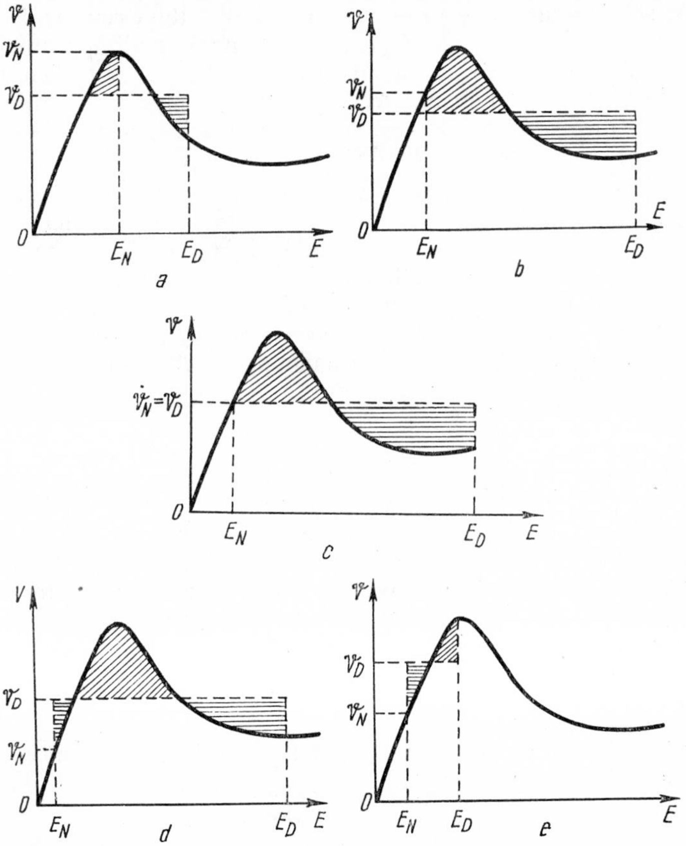

Figure 13.9. Illustrating the domain dynamics. The shaded areas are equal, according to equation (13.67).

a) $v_N > v_D$ and the domain grows, (b) $v_N > v_D$ and the domain charging continues, but with a slower rate, (c) $v_N > v_D$ and the domain is stable (at equilibrium), (d) $v_N \leq v_D$ and the domain discharges, (e) $v_N < v_D$ and the domain nearly disappears (note that the mobility is positive everywhere in the bulk) [29].

as long as F is negative (the integral in equation (13.68) is positive). The domain growing stops when the integral vanishes. This will happen first for the smaller domain (for a certain E_N, $E_{D,1} > E_{D,2}$). This domain will decrease and disappear completely whereas the larger domain continues to grow, absorbing the voltage from the disappearing one.

Let us suppose that the difference between the two nucleation centers is small. The growing domains will have approximately the same dimension. When the smaller domain begins to decrease (the integral (13.68) is zero), the second domain will increase *slowly*, because the corresponding integral (13.68) is small, nearly zero. Therefore, a long time will pass before the larger domain absorbs the voltage drop from the smaller one. During this time interval, the domains continue to travel towards the anode and one of them may disappear at this electrode. Let us assume that this domain is the larger one. Then, at the time of its disappearance, the bulk field E_N rises abruptly and $\mathrm{d}V_d/\mathrm{d}t$ in equation (13.68) changes its sign, marking the growth of the remaining domain. Because of this competition between domains nucleated by centres with comparable sizes, the duration of a cycle of oscillation can be considerably longer than the domain transit-time.

The computer simulation [268] shows that the existence and propagation of multiple domains is undesirable, because the external current changes with every domain disappearing at the anode or in the bulk, the amplitude and period of current oscillations being affected. It is preferable to introduce a pronounced non-uniformity (notch) in the doping profile, in the vicinity of the cathode [76], [268]. This notch will nucleate a large domain which grows rapidly and maintains the outside field E_N below the threshold field, thus annihilating the effect of the non-uniformities of the doping profile.

13.6 An Equivalent Circuit for Domain Dynamics

Robrock [268] suggested the lumped parameter equivalent circuit of Figure 13.10 to describe the domain dynamics. It was assumed that the domain preserves the shape of a stable domain during the time interval it grows or readjusts itself. Robrock [268] shows that the above hypothesis is valid with the following exceptions:

(1) the initial shape of the domain depends upon the nucleation centre,

(2) the domain may show appreciable deviations from the stable-domain shape, when the applied bias experiences large and abrupt changes.

The parallel combination R_N, C_0 corresponds to the neutral semiconductor and supports the voltage drop $E_N L$, whereas the right-hand side characterizes the domain and has on it a voltage drop equal to the excess voltage on the domain V_d. R_N is the low field resistance ($E < E_M$)

$$R_N = \frac{E_N}{eN_D v(E_N)} \frac{L}{A} \tag{13.69}$$

and is almost linear*), $C_0 = \varepsilon\, A/L$ is the geometrical capacitance of the device. The current generator delivers a current equal to

$$I_d = I = eN_D Av\,(E_N). \qquad (13.70)$$

The current I_d depends upon the external field and, therefore, upon the excess voltage drop on the domain V_d, according to a dependence like that shown in Figures 13.6 and 13.7. Therefore we have $I_d = I_d(V_d)$, as shown in Figure 13.10. The domain capacitance C_d is associated to the domain charge Q_D.

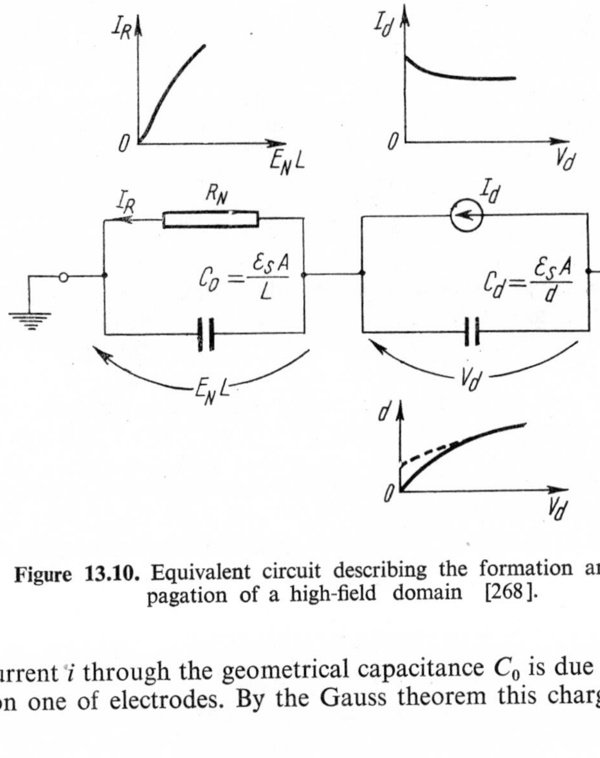

Figure 13.10. Equivalent circuit describing the formation and propagation of a high-field domain [268].

The current i through the geometrical capacitance C_0 is due to the variation of charge upon one of electrodes. By the Gauss theorem this charge is $\varepsilon A E_N$ and we have

$$i = \frac{dQ}{dt} = \frac{\varepsilon A}{L}\frac{d\,(E_N L)}{dt} = C_0\,\frac{dE_N L}{dt} \qquad (13.71)$$

(see Figure 13.10).

*) A is the electrode area.

The current charging the domain capacitance is

$$i_d = \frac{dQ_D}{dt} = \frac{dQ_D}{dV_d}\frac{dV_d}{dt} = C_d \frac{dV_d}{dt} \tag{13.72}$$

where

$$C_d = \frac{dQ_d}{dV_d} \tag{13.73}$$

is the *differential* capacitance of the domain. Let us consider the case of negligible diffusion which is characterized by triangular domains. The domain charge is

$$Q_d = eN_D\, Ad \tag{13.74}$$

(the charge of the depletion layer), where d is the domain width, equation (13.32). By also using equations (13.56) and (13.51), one obtains

$$C_d = \left(\frac{eN_D\,\varepsilon}{2V_d}\right)^{1/2} A = \frac{\varepsilon A}{d} \tag{13.75}$$

(a typical depletion-layer capacitance). The dependence $d = d(V_d)$ is shown in Figure 13.10. For the case of domain formation ($V_d \simeq 0$) the dependence $d = d(V_d)$ should be corrected as shown by the dashed line in Figure 13.10, because the initial dimension of the domain is finite and equal to the width of the nucleation centre.

The equivalent circuit of Figure 13.10 can be utilized for computer calculations. The results are very close to those obtained starting from the basic semiconductor equations. Series combinations of circuits like that shown in Figure 13.10 may be used for computer simulation of device behaviour where more than one domain exists in the bulk [269].

The circuit given in Figure 13.10 can be also used for analytical calculations, as shown below. The total current flowing through the device may be written as [270]

$$I = I_R + C_0\left(\frac{dV}{dt} - \frac{dV_d}{dt}\right) = I_d(V_d) + C_d(V_d)\,\frac{dV_d}{dt} \tag{13.76}$$

and thus one obtains

$$\frac{dV_d}{dt} = \frac{I_R - I_d(V_d) + C_0\,(dV/dt)}{C_0 + C_d(V_d)}\,. \tag{13.77}$$

Consider the case of a constant applied bias ($dV/dt = 0$). Equation (13.77) indicates that the domain grows when the ohmic current I_R exceeds the current corresponding to the stable domain $I_d = I_d(V_d)$. At this moment the external field E_N is above the field E_{Nd} which corresponds to the propagation of stable domain. Because $I_d(V_d) =$

$= eN_D Av(E_{Nd}) \simeq eN_D A\mu E_{Nd}$, where μ is the low-field mobility, and because we suppose $d \ll L$ leading to $C_d \gg C_0$, one obtains

$$\frac{dV_d}{dt} = \frac{d(V_d)}{\tau_r} [E_N - E_{Nd}(V_d)] \tag{13.78}$$

where $\tau_r = \varepsilon/e\mu N_D$ is the dielectric relaxation time. For a triangular domain, we have $d \propto N_D^{-1/2}$ and therefore*) $\frac{dV_d}{dt} \propto N_D^{1/2}$ [270].

13.7 Saturated Domains

Certain interesting properties of transferred-electron (Gunn effect) devices result from the existence of 'saturated' domains (Section 13.4). We have indicated that, if the applied bias increases and becomes sufficiently high, the stable-domain velocity $v_D = v_N$ will decrease and reach a limit value $v_{N\min}$, and so does the current density flowing during the propagation of the stable domain, i.e. $J \to e N_D v_{N\min}$.

The total current flowing through the device may be written [271]

$$I(t) = eN_D(x_D) v_{N\min} A(x_D) \tag{13.79}$$

where x_D is the position of the domain (having negligible dimensions). One assumes that the domain readjusts almost instantaneously to the shape of a stable-domain as it propagates through the crystal and experiences the changes in doping and transversal section A which are indicated by equation (13.79). Because $x_D = x_D(t)$ the current equation (13.79) changes during the domain propagation, according to the variations of the doping and/or of the transversal section. It is possible to obtain the desired shape of the current variation by constructing a device with a suitable doping profile [272] or by using a sample of non-uniform section [271], [273], [274] which is preferable in the present state of technology. For example, a device with a saw-tooth profile would give a saw-tooth output waveform for the current. Note the fact that these waveforms correspond to frequencies in the microwave domain.

Other applications of Gunn devices in digital and logic circuits were envisaged and achieved experimentally [275] – [278]. We do not discuss them here. Instead, we note that the theoretical analysis of these devices (as well as the study of normal Gunn oscillators constructed with coplanar contacts) cannot use a one-dimensional model. The problem of boundary conditions and of domain formation and propagation into a three-dimensional space was recently discussed in a number of papers [271], [269], [280], etc.

---

*) This calculation is not strictly correct and other results can be found in literature. The process of domain cancellation at the anode, requires a separate discussion [269].

Problems

13.1. The existence of a *low-field* dipole domain is also possible (the accumulation layer forms the leading edge of such a domain). Draw the dynamic characteristic $v_D = v_D(E_D)$ in the $v = v(E)$ plane for the low-field dipole domains.

13.2. Discuss the formation of a low-field domain. Examine the rôle of the injecting contact (cathode) by considering: (a) ohmic cathode, (b) imperfect (non-ohmic) cathode (Chapter 3).

13.3. The accumulation and depletion layers of a high-field domain may be separated by a neutral layer. In the limit case of negligible diffusion, this domain has a trapezoidal form. Show that a stable travelling low-field domain can occur only for a $v = v(E)$ dependence like in Figure 13.1b, and not for that indicated in Figure 13.1a. Demonstrate that the stable domain velocity is determined uniquely by the $v = v(E)$ relationship.

13.4. Show that the propagation of a stable high-field domain is impossible under constant bias current conditions (note also the open-circuit a.c. stability of a negative-mobility sample, Chapter 5).

Hint: Use either Figure 13.6, or equation (13.76).

13.5. Discuss the competition between two high-field domains (Section 13.5) by using equation (13.77).

14

Propagation of Accumulation Layers in Negative-mobility Semiconductors

14.1 Accumulation and Depletion Layers

Figure 14.1 shows the field distribution which characterizes the existence of pure accumulation or depletion layers in the bulk. Such a field non-uniformity contains a net space charge and the electric field in the neutral region adjacent to the cathode ($E_{N,1}$) differs from the electric field ($E_{N,2}$) on the opposite side of the layer

$$E_{N,1} \neq E_{N,2} . \tag{14.1}$$

Such a field disturbance cannot occur in the bulk. This was shown by discussing equation (13.4): a field non-uniformity occurring in the neutral bulk must satisfy $E_{N,1} = E_{N,2}$. However, an accumulation layer may be launched by an injecting cathode. If the bias field is below the threshold field and the crystal conductivity is

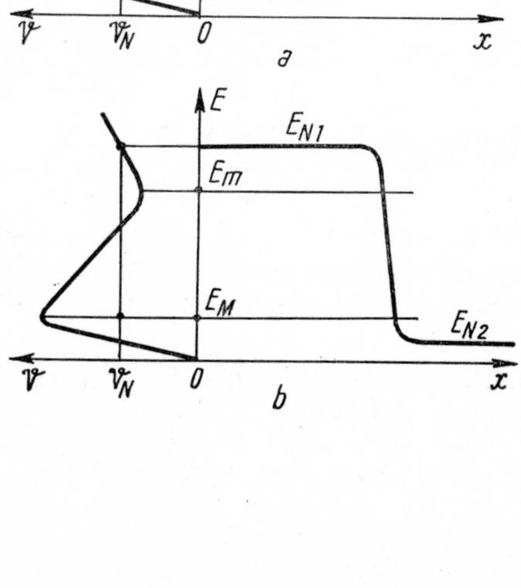

Figure 14.1. Field distribution for: (*a*) an accumulation layer, (*b*) a depletion layer.

sufficiently high, or better, the doping-length product is sufficiently high (Section 1.4), the crystal will exhibit ohmic conduction the semiconductor bulk being neutral, except in the immediate vicinity of the cathode, where an accumulation of mobile electrons will exist.

If the bias field is rised above the threshold field, the primary accumulation layer will grow, being continuously provided with more and more electrons injected from the cathode. Therefore, the electric field on the anode side of the layer will increase above the value of the bias field E_b whereas the electric field on the cathode side will decrease below E_b. One assumes a $v - E$ characteristic of the form indicated in Figure 14.1. Thus $E_{N,2}$ and $E_{N,1}$ may come to a positive mobility region and it is possible to get a 'stable' situation when the accumulation layer detaches itself from the cathode and the charge stored in the accumulation layer

$$Q_a = \varepsilon(E_{N,2} - E_{N,1}) \tag{14.2}$$

remains constant. This layer will propagate towards the anode. By requiring the current continuity in the neutral regions bounding this layer we have

$$J = eN_D\, v(E_{N,1}) = eN_D\, v(E_{N,2}),\ v(E_{N1}) = v(E_{N,2})\,. \tag{14.3}$$

The equality $v(E_{N,1}) = v(E_{N,2})$ characterizing a stable accumulation layer is indicated in Figure 14.1. If the carrier velocity does not increase at high electric fields, the existence of stable accumulation layer is impossible [281]. This is the case of GaAs (Figure 13.1a). Note the fact that the existence of both equations (14.2) and (14.3) is consistent with $E_{N,1}$, $E_{N,2}$= const., independent of time*). It is difficult to justify that a field discontinuity like an accumulation layer can preserve this shape during transit. This is because of the existence of relaxation mechanism and diffusion. We avoid this difficulty by assuming that the thickness of this layer remains very small as compared to the sample length L. Therefore, the applied voltage V is almost equal to the voltage drop on the neutral regions V_N, and we have

$$V \simeq V_N = x_A E_{N,1} + (L - x_A)\, E_{N,2} \tag{14.4}$$

where $x_A = x_A(t)$ is the position of the accumulation layer. Due to equation (14.4) where $E_{N,1}$, $E_{N,2}$ are constant, the existence of stable accumulation (and also depletion, Figure 14.1b) layers under constant bias conditions is impossible.

14.2 Pure Accumulation Mode [282]

Kroemer [282] postulated the existence of the so-called pure accumulation mode. The doping-length product $N_D L$ should have moderate values for this mode of operation; its value should be sufficiently high to allow growing of a considerable

---

*) In fact, by writing equation (14.3) we tacitly neglected the displacement current, i.e. the variation in time of $E_{N,1}$ and $E_{N,2}$.

space charge. However, $N_D L$ should not be too high, so that the bulk dielectric relaxation time will be still comparatively long and only the primary accumulation layer will be important. The space-charge dynamics will be dominated by the propagation and growth of this layer [282].

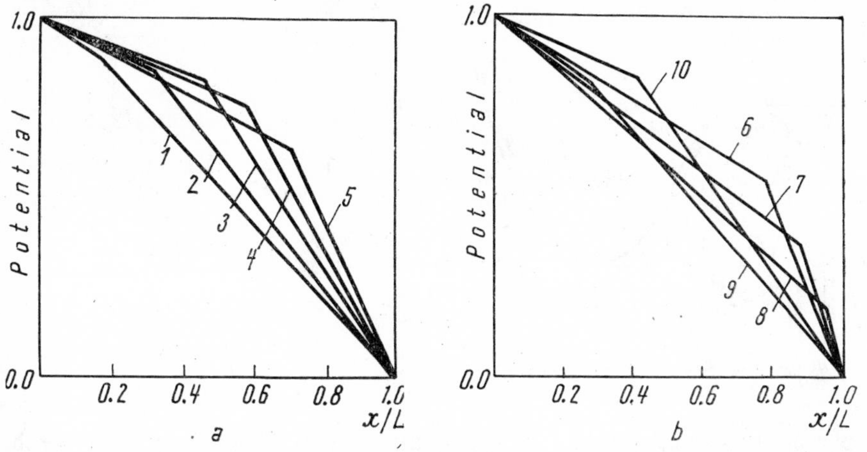

Figure 14.2. Potential distribution at several moments separated by the time interval 0.1 t_0 ($t_0 = L/v_0$, $v_0 = 10^7$ cm s^{-1}), indicating the propagation of an accumulation layer (Kroemer [282]).

The theory developed by Kroemer [282] is based upon the following assumptions:

(a) uniform (homogeneous) semiconductor bulk;

(b) ideal ohmic (injecting) contacts;

(c) one-dimensional geometry.

The device is biased with a constant voltage. The values of the 'external' field ($E_{N,1}$, $E_{N,2}$) should change in time, to keep $V =$ const. during the transit of accumulation layer, as shown at the end of the previous section. However, one assumes [282] for simplicity, that the displacement current is negligible in the neutral regions. This should be valid for large impurity concentrations N_D only. Thus, the condition (14.3) should be again satisfied all the time $E_{N,1}$ and $E_{N,2}$ change during the layer propagation.

Assume that the bias voltage is rised slightly above the threshold value $E_M L$. The cathode accumulation layer grows and moves towards the anode. This is shown by the variation of the potential distribution, reproduced in Figure 14.2 after Kroemer [282]. The details of the numerical computations do not concern us here. The time interval t_0 is taken arbitrarily as being the transit-time of an electron with the velocity of 10^7 cm s^{-1}.

Kroemer [282] distinguishes two phases of the domain propagation. During a first time interval, $v(E_{N,1})$ decreases following the low-field region of the $v = v(E)$

characteristic, whereas $E_{N,2}$ increases towards E_m (Figure 14.3) and the region between the layer and the anode is biased in the negative-mobility region ($v(E_{N,1}) = v(E_{N,2})$). The current J decreases reaching the minimum value $J = e\, N_D v_m$ at the end of phase 1. Let x_1 be the position of the accumulation layer at the end of

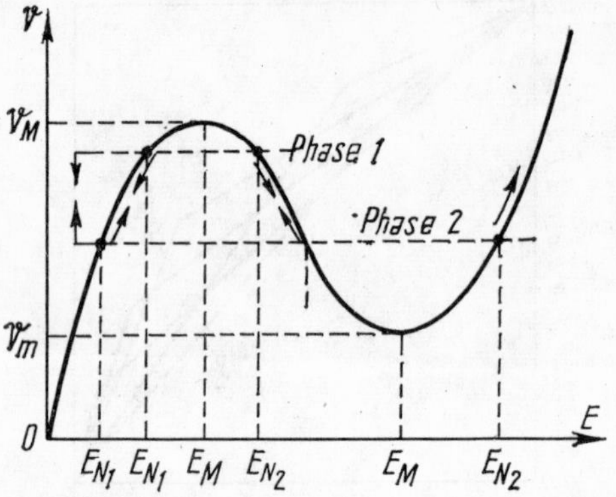

Figure 14.3. Diagram illustrating the dynamics of a pure accumulation layer [282].

this phase. By assuming that the accumulation layer is infinitely thin we have (Figure 14.3, E' is the minimum cathode field)

$$V = x_1 E' + (L - x_1)\, E_m \tag{14.5}$$

or

$$\frac{x_1}{L} = \left(E_m - \frac{V}{L}\right)(E_m - E')^{-1}\,. \tag{14.6}$$

Suppose now that V/L is larger than but nearly equal to the threshold field. The ratio x_1/L will depend upon the $v = v(E)$ characteristic only. x_1/L is comparable to unity and this means that a large part of the interelectrode distance is covered by the layer during the phase 1. This is an unfavourable situation because real crystals contain non-uniformities which may lead to growing field non-uniformity (domains) in the region between the layer and the anode, where the mobility is negative. Kroemer [282] simulated on a computer the behaviour of a non-uniformly doped sample and found that numerous domains occur in the anode region. The competition between domains (Section 13.5) determines fluctuations of the external current*).

We note that the electron velocity in the accumulation layer during phase 1 is larger than the electron velocity in the neutral regions. The 'electronic gas' will be compressed at the leading edge and rarefied at the trailing edge of the accumula-

*) If a large domain is nucleated in the bulk immediately after the voltage bias rises (slowly) above the threshold value $E_M L$, then this domain will absorb voltage and the cathode field may be always below E_M and no accumulation layer will be formed.

tion layer. Therefore the accumulation layer propagates faster than the electrons it comprises. At the beginning of phase 1, the electron velocity is close to v_M and the layer velocity exceeds v_M, as shown by Figure 14.4, reproduced after Kroemer [282]. Figure 14.4 also shows that the electron density in the accumulation layer becomes

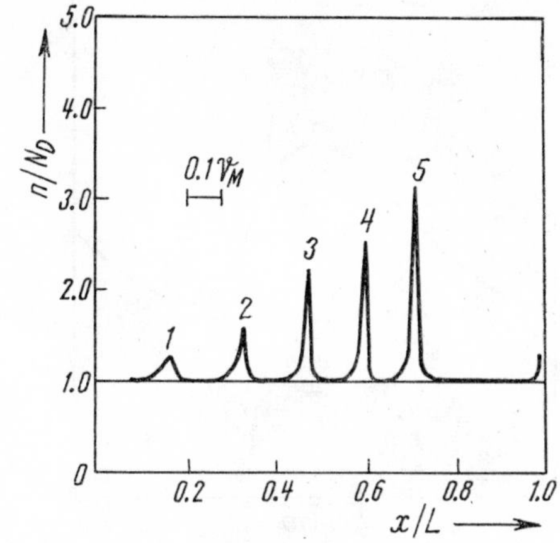

Figure 14.4. Electron density versus distance, indicating the propagation of the accumulation layer [282].

much larger than the background ion density N_D, and also that the total charge contained in the accumulation layer increases in time*).

The phase 2 is characterized by the following processes. $E_{N,2}$ rises above E_m and $E_{N,1}$ should also increase to preserve the equality $v(E_{N,1})=v(E_{N,2})$. After Kroemer [282] this increase continues until $E_{N,1}$ reaches E_M and a new layer forms at the cathode. The old and the new layer may coexist for a time (see curve 9 in Figure 14.2). Let x_2 be the position of the old layer when $E_{N,1}$ reaches E_M. A simple calculation, which is similar to that used for deriving equation (14.6), yields (E'' is the maximum anale field)

$$\frac{x_2}{L} = \left(E'' - \frac{V}{L}\right)(E'' - E_M)^{-1} < 1 \qquad (14.7)$$

and this formula proves that x_2 may be close but is always smaller than L. If the drift velocity saturates at large fields (Figure 13.1a), then E'' in equation (14.7) should formally tend to infinity and $x_2 \to L$.

It is interesting to note at this point that the fact that an accumulation layer reaching the anode should indeed disappear under constant bias conditions is not evident. Thim [264] indicated that if the drift velocity saturates at large fields and if the sample conductivity is sufficiently high†), the accumulation layer which reaches the anode will readjust its shape and will not disappear, being sustained by the flow of electrons from the bulk.

*) This is in contradiction to the initial hypothesis that the drift currents entering and leaving the layer are equal. We suppose that the difference between these currents is small.

†) Thim [264] calculated $N_D > 5 \cdot 10^{14}\,\mathrm{cm}^{-2}$ for GaAs. A doping non-uniformity below 10% is also required.

Figure 14.5 shows the variation of the external current computed by Kroemer [282] for the 'pure accumulation mode' of operation. The current increase occurs during phase 2 and its decrease during phase 1. The oscillation amplitude is relatively large, comparable with the steady-state component. The higher the peak-to-valley

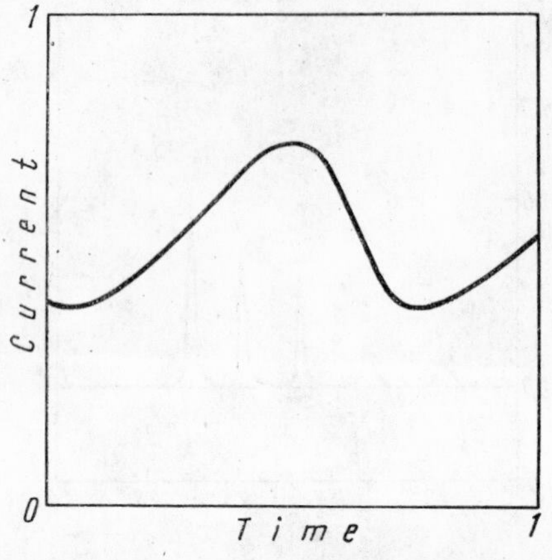

Figure 14.5. Variation of the external current during the formation, propagation and disappearance of an accumulation layer [282].

velocity ratio v_M/v_m, the higher the ratio of oscillation amplitude to the steady-state current. Whereas the Gunn oscillations discussed in Section 13.4 consist of spikes of current, the oscillations of pure accumulation mode are relatively close to a sinusoid and the efficiency is relatively high.

14.3 Limited-space-charge Accumulation Mode

The limited space-charge accumulation (LSA) mode is obtained by biasing the negative-mobility diode (with supercritical doping-length product) with a high-frequency oscillating signal superimposed on a steady-state voltage. The proper bias is indicated in Figure 14.6: note that for a short time during each period the bias field falls below the threshold field. The signal frequency is properly chosen such that during one period an appreciable growth of field non-uniformities (bulk space-charge) is not possible. As the field is below the threshold field, the small space-charge accumulated in the rest of period disappears by dielectric relaxation mechanism. The electric field is practically uniform. During most of the period, the crystal exhibits a negative differential conductivity. This negative conductance generates an a.c. power in the microwave region. The oscillator circuit consists in the negative-mobility device in series with a d.c. bias and a parallel LC circuit tuned at a proper frequency. This oscillation mode is typical large-signal and the oscillations are not self-starting.

Let us first suppose that the space charge is completely absent. The electric field in the semiconductor bulk is uniform and equal to the bias field

$$E(x, t) = E_b(t) = E_{b,0} + E_{b,1} \sin 2\pi f_0 t \,. \tag{14.8}$$

An electron moving at a velocity v in an electric field E absorbs the instantaneous power eEv. The average power absorbed from the d.c. source is [283]

$$P_{d.c.} = \frac{eE_{b,0}}{T_0} \int_0^{T_0} v \mathrm{d}t, \quad T_0 = \frac{1}{f_0} \tag{14.9}$$

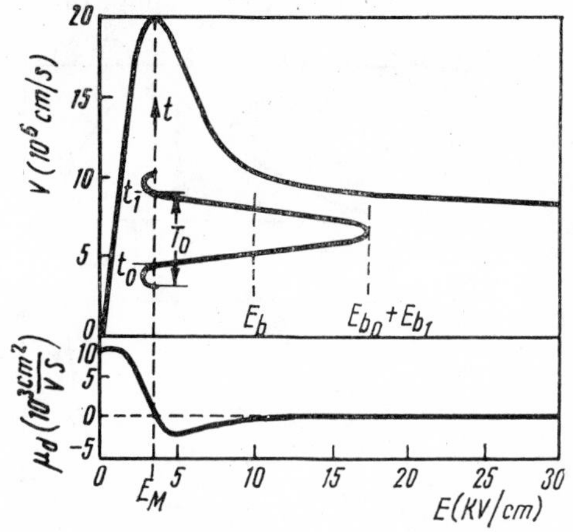

Figure 14.6. Field-dependence of electron velocity and electron mobility in GaAs [283], indicating the biasing for the LSA mode.

and the radio frequency power absorbed by the same electron is

$$P_{a.c.} = \frac{eE_{b,1}}{T_0} \int_0^{T_0} v \sin (2\pi f_0 t) \, \mathrm{d}t \tag{14.10}$$

where $v = v(E) = v(t)$ according to Figure 14.6. By properly choosing $E_{b,0}$ and $E_{b,1}$ it is possible to obtain $P_{a.c.} < 0$, i.e. the electron delivers energy to the radio frequency field, or the d.c. energy is converted into a.c. energy. The conversion efficiency is

$$\eta_c = -(P_{a.c.}/P_{d.c.}) \, . \tag{14.11}$$

Figure 14.7 shows the dependence $\eta_c = \eta_c(E_{b,1})$ for a d.c. bias field chosen as indicated in Figure 14.6 ($E_{b,0} = 10$ kV cm^{-1}). The same figure represents the ratio $R_{a.c.}/R_0$ as a function of $E_{b,1}$. R_0 is the ohmic resistance

$$R_0 = \frac{L}{eN_D \mu_0 A} \tag{14.12}$$

(μ_0 is the low-field mobility) and $R_{a.c.}$ is computed as

$$R_{a.c.} = \frac{(E_1 L)^2/2}{N_D L A P_{a.c.}} \tag{14.13}$$

where P_{ac} is the power corresponding to one electron and $N_D LA$ the total number of electrons in the sample [283].

In the case of sustained (sinusoidal) oscillations the load resistance should be equal to the absolute value of the device (negative) resistance. Figure 14.7 shows that the load should be matched to the diode to get the maximum conversion efficiency. The maximum value obtainable for a given semiconductor also depends upon $E_{b,0}$ and upon the exact shape of the $v - E$ characteristic.

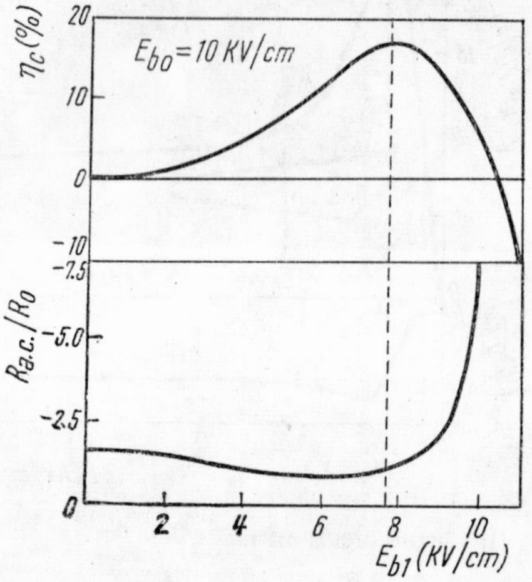

Figure 14.7. Dependence of the conversion efficiency and the a.c. resistance upon the amplitude of the applied electric field (GaAs, average electric-field equal to 10 kV cm^{-1}).

We shall discuss below, following Heinle [284], the conditions required for the limitation of the space charge growth. The integral equation (13.43) obtained in the previous Section will be used. The electric field is non-uniform and satisfies the following equation

$$\int_0^L E(x, t)\,\mathrm{d}x = V(t) = LE_b(t). \tag{14.14}$$

One assumes that the deviations from the bias field are relatively small, i.e.

$$E(x, t) = E_b(t) + \Delta E(x, t) \tag{14.15}$$

where ΔE is a first-order quantity. By introducing equation (14.15) in (13.43) and neglecting the second-order quantities one obtains

$$\frac{\partial \Delta E}{\partial t} + v(E_b)\frac{\partial \Delta E}{\partial x} + \frac{e\mu_d N_D}{\varepsilon}\Delta E =$$

$$= \frac{e\mu_d N_D}{\varepsilon L}\int_0^L \Delta E \mathrm{d}x + \frac{v(E_b)}{L}[E(L) - E(0)] + \frac{\mu_d}{L}\int_{E(0)}^{E(L)} \Delta E\,\mathrm{d}E. \tag{14.16}$$

The order of magnitude of the quantities in the right-hand-side (RHS) of the above equality should be carefully evaluated. By introducing $E(x, t)$ of equation (14.15) into equation (14.16) one obtains

$$\frac{1}{L}\int_0^L \Delta E \, dx = 0 . \tag{14.17}$$

The second term in the RHS of equation (14.16) is first order, because $E(L)$ and $E(0)$ are very close to $E_b(t)$ and their difference is small. The third term should be negligible, because, at given t

$$\int_{E(0)}^{E(L)} \Delta E dE = \int_{E(0)}^{E(L)} \Delta E(x, t) \frac{\partial E}{\partial x} dx = \int_{E(0)}^{E(L)} \frac{e}{\varepsilon} (n - N_D) \Delta E \, dx \tag{14.18}$$

where $n - N_D$ is very small because the deviations from neutrality are first-order. Therefore equation (14.16) becomes

$$\frac{\partial}{\partial t} \Delta E + v(E_b) \frac{\partial \Delta E}{\partial x} + \frac{eN_D\mu_d(E_b)}{\varepsilon} \Delta E = \frac{v(E_b)}{L} [E(L, t) - E(0, t)] . \tag{14.19}$$

Let us now suppose that the electric field non-uniformity has the form of an infinitely thin accumulation layer (negligible diffusion), i.e. a field discontinuity. $E(x, t)$ is equal to $E_1(t)$ between the cathode and the accumulation layer and to $E_2(t)$ between this layer and the anode. One introduces

$$\Delta E_j(t) = E_j(t) - E_b(t), \quad j = 1,2 \tag{14.20}$$

in equation (14.19), and equation (14.21) follows ($j = 1,2$)

$$\frac{d\Delta E_j}{dt} = - \frac{eN_D \mu_d (E_b)}{\varepsilon} \Delta E_j + \frac{v(E_b)}{L} (\Delta E_1 - \Delta E_2). \tag{14.21}$$

The two ordinary differential equations contained in equation (14.21) form a system of equations with ΔE_1 and ΔE_2 as dependent variables. Substituting

$$\gamma(t) = \Delta E_2 - \Delta E_1, \text{ and } \bar{\gamma}(t) = \Delta E_2 + \Delta E_1 \tag{14.22}$$

one obtains

$$\gamma(t) = \gamma_0 \exp \{ - \frac{eN_D}{\varepsilon} \int_{t_0}^t \mu_d [E_b(\xi)] d\xi \} \tag{14.23}$$

$$\bar{\gamma}(t) = \gamma(t) \left\{ \frac{\bar{\gamma}}{\gamma_0} - \frac{2}{L} \int_{t_0}^t v[E_b(\xi)] d\xi \right\} \tag{14.24}$$

where t_0 is the time when E_b exceeds the threshold field E_M. Note the fact that $\varepsilon\,\gamma(t) = \varepsilon\,(E_2 - E_1)$ is just the excess mobile charge (per unit area) contained in the accumulation layer.

We shall first impose the condition

$$\Gamma_0 = \frac{\gamma(t_0+T_0)}{\gamma_0} < 1 \tag{14.25}$$

which states that the growth of the space charge does not cumulate from one period to another. According to equation (14.23), equation (14.25) is equivalent to

$$\frac{1}{T}\int_{t_0}^{t_0+T_0} \mu_d\,[E_b(t)]\,\mathrm{d}t > 0 \tag{14.26}$$

a relation which was derived by Copeland [283] by using a different method. For a given $v = v\,(E)$, the above condition is satisfied if both $E_{b,0}$ and $E_{b,1}$ are properly chosen.

The second condition may be formulated as follows: the growth of the space charge during the time interval (t_0, t_1) when the field is in the range of negative mobi-i ty should be restricted to a certain upper value, i.e.

$$\Gamma_1 = \frac{\gamma\,(t_1)}{\gamma_0} \leqslant C_1 > 1. \tag{14.27}$$

By using equations (14.23) and (14.27), one obtains an upper limit for the ratio of doping to frequency

$$\frac{N_D}{f_0} \leqslant \frac{\varepsilon}{e}\left[\frac{1}{-\dfrac{1}{T_0}\displaystyle\int_{t_0}^{t_1}\mu_d\,\mathrm{d}t}\right]\ln C_1 \tag{14.28}$$

(the right-hand side does not depend upon the frequency f_0).

The LSA mode of oscillation has the advantage that the operating frequency is not limited by transit-time effects. Therefore, the sample length L can be made large, and the device will admit a higher voltage and higher power dissipation. This mode of operation is preferable if appreciable power is necessary at very high frequencies (several tens of GHz) [285].

However, the device fabrication and circuit adjustment are much more difficult that for other modes of operation. The device should be constructed by using an active layer with very low non-uniformities. Any appreciable non-uniformity may lead to domain formation (Chapter 13). These domains produce high local fields and avalanche multiplication breakdown is possible. The oscillation frequency and amplitude should be carefully chosen and maintained in order to provide a reasonable efficiency [286] – [288].

Problems

14.1. Show that the propagation velocity of an infinitely thin accumulation layer (field discontinuity) is given by

$$\frac{\mathrm{d}x_a}{\mathrm{d}t} = \frac{\int_{E_1}^{E_2} v(E)\mathrm{d}E}{E_2 - E_1}. \tag{P.14.1}$$

14.2. Derive condition (14.26) directly, by examining the effect of the dielectric relaxation mechanism upon small space-charge perturbations.

14.3. Find the condition required for the accumulation layer to travel a relatively short distance before being quenched during the minimum-field portion of the cycle (LSA operation).

Appendices

Appendix to Section 3.8.

Derivation of Incremental Resistance

By differentiating equations (3.43) and (3.44) with respect to E_c one obtains, respectively

$$\frac{eN}{\varepsilon}\,\frac{dV}{dE_c} = \int_{E_c}^{E_a} \frac{-\varphi(E)\,E\,\dfrac{dJ}{dE_c}\,dE}{[J\varphi(E)-1]^2} + \frac{dE_a}{dE_c}\left[\frac{E_a}{J\varphi(E_a)-1}\right] - \frac{E_c}{J\varphi(E_c)-1} \tag{3.8.1}$$

$$0 = \int_{E_c}^{E_a} \frac{-\varphi(E)\,\dfrac{dJ}{dE_c}\,dE}{[J\varphi(E)-1]^2} + \frac{dE_a}{dE_c}\,\frac{1}{J\varphi(E_a)-1} - \frac{1}{J\varphi(E_c)-1} \tag{3.8.2}$$

where

$$\varphi(E) = \frac{1}{j_n(E)}\,. \tag{3.8.3}$$

By eliminating dE_a/dE_c between equations (3.8.1) and (3.8.2) one obtains

$$\frac{eN}{\varepsilon}\,\frac{dV}{dE_c} = \frac{dJ}{dE_c}\int_{E_c}^{E_a} \frac{(E_a-E)\,\varphi(E)\,dE}{[J\varphi(E)-1]^2} + \frac{E_a-E_c}{J\varphi(E_c)-1}\,. \tag{3.8.4}$$

The equation (3.56) from the text can be readily obtained by using equations (3.45), (3.8.3) and (3.8.4).

Appendix to Section 3.8.

Negative Incremental Resistance due to Trapping at the Cathode Interface

We shall show that the sufficient criterion for a static negative resistance can be satisfied by injecting carriers in a negative mobility material in the presence of trapping at the cathode interface.

The control characteristic $j_c(E)$ is given by

$$j_c(E) = ep(0)\, v(E) = \frac{eN_v v(E)}{\beta\left(\dfrac{e\Sigma}{\varepsilon E} - 1\right)} \tag{3.8.5}$$

where equation (3.18) was used. The crossover field E_x is determined by $j_c'(E_x) = j_n(E_x)$, which yields

$$N_v = N_A\beta\left(\frac{e\Sigma}{\varepsilon E_x} - 1\right) \tag{3.8.6}$$

or

$$E_x = \frac{e\Sigma}{\varepsilon\left(\dfrac{N_v}{N_A\beta} + 1\right)} \tag{3.8.7}$$

and, finally,

$$\frac{E_x}{E_{cr}} = \frac{1}{1 + N_v/N_A\beta} < 1, \quad E_{cr} = \frac{e\Sigma}{\varepsilon} \tag{3.8.8}$$

where E_{cr} is a critical field (for $E_c \to E_{cr}$ in equation (3.18), one obtains $p(0) \to \infty$).

The sufficient criterion requires a negative slope of the control characteristic at the crossover point, i.e.

$$\left.\frac{dj_c}{dE}\right|_{E=E_x} < 0\,. \tag{3.8.9}$$

By using equation (3.8.5), one obtains ($\mu_d = dv/dE$)

$$\frac{\beta}{e\varepsilon N_v} \frac{dj'_c}{dE} = \frac{\mu_d E(e\Sigma - \varepsilon E) + e\Sigma v(E)}{(e\Sigma - \varepsilon E)^2} \tag{3.8.10}$$

and the sign of dj_c/dE is the sign of the denumerator in the right-hand side of equation (3.8.10). One can find that the condition (3.8.9) is equivalent to

$$\mu_d (E_x) E_x \left(1 - \frac{E_x}{E_{cr}}\right) + v(E_x) < 0. \tag{3.8.11}$$

This inequality can be satisfied only if $\mu_d < 0$. Therefore, we must have $\mu_d < 0$ and

$$\left|\mu_d(E_x)\right| > \frac{v(E_x)}{E_x\left(1 - \frac{E_x}{E_{cr}}\right)}. \tag{3.8.12}$$

The most favourable situation occurs for $E^x \ll E_{cr}$ and from inequality (3.8.12), the magnitude of the differential mobility should exceed the d.c. mobility v/E (both being calculated at the crossover field).

Consider a $v - E$ characteristic having a region of negative-mobility. Assume that this region can be approximated by a straight line and let E_m be the maximum field corresponding to this region (the 'valley' field) and E_f the crossover of the extrapolated negative-mobility region with the E axis. It can be easily shown that for equation (3.8.12) to be satisfied, E_m should be at least equal to $E_f/2$. This condition seems to be satisfied for InP but not for GaAs (at room temperature, see for example Figure 13.1).

Appendix to Sections 3.9 and 3.10.

Parametric Equations of the Steady-state Characteristic

The normalized current, equation (3.77), and the normalized voltage, equation (3.78), can be obtained as shown below. We have

$$L = \int_0^L \mathrm{d}x = \int_0^{T_L} v\, \mathrm{d}T_x = \mu \int_0^{T_L} E(T_x)\, \mathrm{d}T_x \tag{3.9.1}$$

where $E(T_x)$ must be replaced by equation (3.74). The result of integration is

$$\bar{J} = \frac{J}{e\, NL\omega_r} = \left[\theta_r + \left(\frac{\Delta - 1}{\Delta}\right)\left[\exp(-\theta_r) - 1\right)\right]^{-1} \tag{3.9.2}$$

where ω_r and Δ are given by equation (3.66) and (3.72), respectively. Note that for $\Delta = 1$ and $\theta_r = 1$, one obtains $\bar{J} = 1$.

The voltage drop V is

$$V = \int_0^L E\, \mathrm{d}x = \int_0^{T_L} vE\, \mathrm{d}T_x = \mu \int_0^{T_L} [E(T_x)]^2\, \mathrm{d}T_x \cdot \tag{3.9.3}$$

From equations (3.74) and (3.93) we obtain

$$\bar{V} = \frac{V}{eNL^2/\varepsilon} =$$

$$= \frac{\left[\theta_r + 2\left(\frac{\Delta - 1}{\Delta}\right)(\exp(-\theta_r) - 1) - \frac{1}{2}\left(\frac{\Delta - 1}{\Delta}\right)^2\left(\exp(-2\theta_r) - 1\right)\right]}{\left[\theta_r + \frac{\Delta - 1}{\Delta}\left(\exp(-\theta_r) - 1\right)\right]^2} \tag{3.9.4}$$

and $\bar{V} = 1$ for $\Delta = 1$ and $\theta_r = 1$.

In Section (3.10), the mobility is no longer constant, but $v = v(E)$ is approximated by equation (3.87). We shall use again $L = \int_0^L \mathrm{d}x = \int_0^{T_L} v\, \mathrm{d}T_x$, but v will be replaced by equation (3.92). Equations (3.95) and (3.98) will be also used and the final result will be $\bar{J} = \bar{J}(\theta_d, \Delta)$, which is identical with $\bar{J}$ given by equation (3.92). However, note the fact that the normalization current in equation (3.104) is different from that used in equations (3.77) and (3.9.2).

The normalized voltage $\overline{V'}$ defined by equation (3.105) may be obtained as follows:

$$V = \int_0^L E\,\mathrm{d}x = \int_0^{T_L} v\,E\mathrm{d}T_x = \int_0^{T_L} v\frac{v - v_d}{\mu_d}\,\mathrm{d}T_x\,. \qquad (3.9.5)$$

In the above equation $v = v(T_x)$ is given by equations (3.92). The final result is

$$\overline{V'}\,(\theta_d, \Delta) = \overline{V}(\theta_d\,, \Delta) \qquad (3.9.6)$$

where $\overline{V}$ is given by equation (3.9.4).

Appendix to Section 5.6.

Effect of Diffusion on Small-signal Behaviour of SCLC Diodes

In this Appendix we shall suggest a set of transport coefficients which can be used for a semiconductor region where the charge carriers (holes) are injected with zero drift velocities, i.e. $v(0) = 0$. We assume that the charge carriers overcome the potential maximum ($x = 0$, $E = 0$, $v = 0$) by diffusion. Therefore, the direct current density can be finite, even if the carrier density at the injection plane $p_0(0)$ is finite (condition 5.52).

We shall calculate the a.c. transport coefficients defined in Section 4.6, by assuming that the steady-state carrier distribution is known, or, more precisely

$$p_0 = p_0(T_x) = \text{given}, \; p_0(0) = \text{finite}. \tag{5.6.1}$$

According to equations (4.65) — (4.70) we have successively

$$K_{22} = \frac{p_0(T_L) - N}{p_0(0) - N} \exp(-i\theta) \tag{5.6.2}$$

$$\begin{aligned} K_{21} &= \frac{K_{22}}{\varepsilon} \int_0^{T_L} \frac{p_0(0) - N}{p_0(\eta) - N} \exp(i\omega\eta)\, d\eta = \\ &= \left[\frac{p_0(T_L) - N}{\varepsilon}\right] \exp(-i\theta) \int_0^{T_L} \frac{\exp\; i\omega\eta}{p_0(\eta) - N}\, d\eta \end{aligned} \tag{5.6.3}$$

$$\begin{aligned} K_{12} &= \int_0^{T_L} \frac{v_0(T_x)\,[p_0(\eta) - N]}{p_0(0) - N} \exp(-i\omega\eta)\, d\eta = \\ &\frac{1}{p_0(0) - N} \int_0^{T_L} \left[\frac{J_0}{\varepsilon} - N v_0(\eta)\right] \exp(-i\omega\eta)\, d\eta = \\ &\frac{J_0}{e[p_0(0) - N]} \left[\frac{1}{i\omega}(1 - \exp - i\theta) - N \int_0^{T_L} \frac{\exp - i\omega\eta}{p_0(\eta)}\, d\eta\right]. \end{aligned} \tag{5.6.4}$$

K_{11} can be calculated in an analogous manner. The important result of these calculations is that K_{22} and K_{12} are finite, although the d.c. injection velocity is zero (note that by simply replacing $v_0(0) = 0$ in equations (4.56) and (4.58), one obtains zero).

Consequently, the a.c. impedance of an SCLC diode can be calculated by considering a finite a.c. conductivity of the emitting plane (see equation 5.5 from the text). The device impedance will be given by (4.18) where $K_{12} \neq 0$ and $\sigma_1(0) =$ finite.

Because $K_{12} \neq 0$ and $K_{22} \neq 0$, the semiconductor region $0 < x < L$ with $v_0(0) = 0$, will no longer be 'separated', or 'isolated', with respect to the a.c. régime, from the processes occurring at $x < 0$. As will be shown later (see Appendix to Section 9.2) the injection noise is no longer totally suppressed if K_{12} becomes finite, $\neq 0$.

Such calculations are, of course, approximate because the diffusion current is taken into account in the steady-state régime but neglected in a.c. calculations. Note, however, the fact that the diffusion should be important only in the close vicinity of the injecting plane, $x = 0$ and, therefore, we may assume that neglecting the a.c. diffusion current has little effect upon the transport coefficients which are obtained, as shown, by 'integrating' the effect of carrier transit-time from $x = 0$ to $x = L$.

Appendix to Section 6.6.

Numerical Example of Calculation for an ECL Negative-mobility Amplifier

Let us consider the following values (which roughly correspond to GaAs at room temperature). In equation (3.86) we introduce $v_d = 3.10^7$ cm s^{-1} and $\mu_d = -2500$ cm^2 V s^{-1}, such that $E_d = v_d/\mu_d = -12$ kV cm^{-1}. Then, we have $v_M = 2.25 \times 10^7$ cm s^{-1} and $v_m = 1.125 \times 10^7$ cm s^{-1}. The relative permittivity is taken as 12.5 such that $\varepsilon = 1.106 \times 10^{-12}$ F cm^{-1}.

We shall consider a very low doping-length product, namely $NL = 10^{10}$ cm^{-2}. We calculate $L/\tau_d = (e\mu_d/\varepsilon)NL = 3.62.10^6$ cm s^{-1}. By using the above values for v_M and v_m, we obtain, according to equations (3.99), (3.100) and (3.101)

$$3.11 < |\varphi_0|, |\varphi_L| < 6.22. \tag{6.6.1}$$

This double inequality will be used to define the domain in the $\Delta - \theta_d$ plane of Figure 3.25, where the approximation (3.86) is valid (i.e. any point from the above domain corresponds to a physical situation characterized by d.c. field in the negative mobility region). Assume accumulation of the drift region, i.e. $\Delta > 1$. In this case we must have $v_0(0) < v_M$ and $v_0(L) > v_m$ or $|\varphi_0| < 6.22$ and $|\varphi_L| > 3.11$. The permissible domain can be approximately determined in Figure 3.25, and this domain corresponds to the doping-length product specified above.

Let us consider the length of the drift region $L = 10\ \mu$m. Because $NL = 10^{10}$ cm^{-2}, we obtain $N = N_{\text{donors}} = 10^{13}$ cm^{-3}. The steady-state current J_0 will be normalized to $|eNL/\tau_d| = 58$ mA mm^{-2} (see equation (3.104)). The quantity eNL^2/ε, used in equation (3.105) for voltage normalization will be equal to 1.45 V. Note that E_dL in equation (3.105) is -12 V.

We shall choose from Figure 3.25 $|\bar{J}_0| = 5$ and $|\bar{V}'| = 4$, which correspond, approximately, to $\Delta = 1.2$ and $|\theta_d| = 0.25$. We get a bias current $J_0 = 290$ mA mm^{-2} and the voltage drop $V = 6.2$ volts.

Figure 6.9 yields, approximately, $\bar{G}_{p\infty} = 1.2$. and $\bar{Q} = -3.5$. The v.h.f. parallel conductance $G_{p\infty}$ is normalized to $1/R_d$ (see equations (6.52) and (6.54)), and one obtains $|1/R_d| = |eN\mu_d/L| = 4$ mho cm$^{-2} = 40$ mmho mm^{-2}. The v.h.f. device conductance will be of the order of $-1.2\times 40 = -48$ mmho mm^{-2}. The v.h.f. device capacitance is the geometrical capacitance and has the value of 11.06 pF mm^{-2}.

The transit-time frequency is

$$f_T = \frac{1}{T_L} = \frac{1}{\theta_d \tau_d} = \frac{1}{0.25\ \tau_d} \simeq 14.5\ \text{GHz}. \tag{6.6.2}$$

At the transit-time frequency the parallel conductance is of the order of the above calculated v.h.f. negative conductance, and the quality factor is roughly equal to $Q = -\theta \times 3.5 = -7\pi = \simeq -22$. For a transit-angle $\theta = 4\pi$, the equivalent circuit is almost identical with the v.h.f. circuit, and the quality factor is $Q = -44$.

Appendix to Section 7.3.

The Effect of Diffusion upon the Space-charge Wave

Figure 7.3 indicates that the amplitude of $\underline{\mathbf{p}}_1$ at the cathode is infinite. This gives a finite value for $\underline{\mathbf{J}}_{1p} = e\, v_0 \underline{\mathbf{p}}$ where $v_0(0)$ should be zero (SCLC injection and $E_0(0) = 0$). On the other hand we have $\underline{\mathbf{v}}_1(0) = 0$ and $p_0(0) \to \infty$ and thus $\underline{\mathbf{J}}_{1p} = e\, v_0 \underline{\mathbf{p}}_1$ is also finite.

If we assume (see equation (5.52)) that $p_0(0)$ is finite at the plane of potential maximum ($x = 0$) and the finite current at this plane is carried by diffusion, then the a.c. particle current at $x = 0$ is given by equation (5.54) and is due to velocity modulation only

$$\underline{\mathbf{j}}_1 (0) = e\, \mu p_0(0) \underline{\mathbf{E}}_1 (0) = \sigma_1(0) \underline{\mathbf{E}}_1(0). \tag{7.3.1}$$

Therefore, the amplitude of the space-charge wave at $x = 0$ is equal to zero, and in fact, we cannot speak about an *injected* space-charge wave.

It is interesting to note at this point that Kroemer's analysis [83] which takes diffusion current into account (both the steady-state and the a.c. component) leads to mathematical conditions which are equivalent to zero density modulation at the cathode and indicates a reduction of the oscillating character of the conductance-frequency dependence above the transit-time frequency. This observation is consistent with the idea that the high-frequency oscillations of the conductance are due to the existence of an injected space-charge wave.

Appendix to Section 8.2

Solution of Equation (8.28)

By using substitution (8.31), equation (8.28) becomes

$$\frac{d^2\Psi_1}{dz^2} + \frac{1}{z}\frac{d\Psi_1}{dz} - \left[\frac{1}{4z^2} + \frac{i\omega T_0}{a^2}z^{-1/2}\right]\Psi_1 = 0. \tag{8.2.1}$$

It may be shown (see for example [170]) that the general solution of this equation is of the form

$$\Psi_1 = \mathbf{Z}_{2/3}\left(\frac{4}{3a}\sqrt{-i\omega T_0}\, z^{3/4}\right) \tag{8.2.2}$$

where $\mathbf{Z}_\nu$, is the general solution

$$u = \mathbf{Z}_\nu(y) = A\,\mathbf{J}_\nu(y) + B\mathbf{Y}_\nu(y) \tag{8.2.3}$$

of the Bessel equation

$$\frac{d^2u}{dy^2} + \frac{1}{y}\frac{du}{dy} + \left(1 - \frac{\nu^2}{y^2}\right)u = 0. \tag{8.2.4}$$

In equation (8.2.3), $\mathbf{J}_\nu(y)$ and $\mathbf{Y}_\nu(y)$ are the Bessel functions of the first and second kind, respectively, and A, B are constants. Because ν is not an integer, u may be also written

$$u = A_0\,\mathbf{J}_\nu(y) + B_0\,\mathbf{J}_{-\nu}(y). \tag{8.2.5}$$

Thus, equation (8.2.2) becomes

$$\Psi_1 = A_0\,\mathbf{J}_{2/3}\left(\frac{4i}{3a}\sqrt{i\omega\, T_0}\, z^{3/4}\right) + B_0\,\mathbf{J}_{-2/3}\left(\frac{4i}{3a}\sqrt{i\omega\, T_0}\, z^{3/4}\right) \tag{8.2.6}$$

and then the solution may be expressed in terms of modified Bessel functions, as shown by equation (8.30) from the text. By definition, the modified Bessel function of the order ν is given by

$$\mathbf{I}_\nu(x) = \sum_{k=0}^{\infty} \frac{(x/2)^{\nu+2k}}{\Gamma(k+1)\,\Gamma(k+\nu+1)} \tag{8.2.7}$$

where Γ is Euler's function satisfying

$$\Gamma(\alpha+1) = \alpha\,\Gamma(\alpha)\,. \tag{8.2.8}$$

Expression (8.37) of a.c. Particle Current

Equation (8.17) may be split into two equations: one corresponding to the d.c. quantities and another to the a.c. quantities. The latter may be written

$$\mathbf{I}_{c,1}(x) = -\frac{\mu\varepsilon_i W}{h}\frac{\mathrm{d}}{\mathrm{d}x}[\underline{\Psi}_0\underline{\Psi}_1) \tag{8.2.9}$$

and because of equation (8.31)

$$\mathbf{I}_{c,1}(z) = \frac{\varepsilon_i\mu W}{hL}\,a\,\frac{\mathrm{d}}{\mathrm{d}z}[\underline{\Psi}_0(z)\underline{\Psi}_1(z)]\,. \tag{8.2.10}$$

According to equation (8.26) we have $\underline{\Psi}_0(z) = (V_{G0} - V_T)\,z^{1/2}$ and equation (8.2.10) becomes

$$\mathbf{I}_{c,1} = a g_{ms}\frac{\mathrm{d}}{\mathrm{d}z}\{z^{1/2}\underline{\Psi}_1(z)\} \tag{8.2.11}$$

where g_{ms} is the transconductance in saturation, equation (8.38). Here $\underline{\Psi}_1(z)$ should be replaced by its expression (8.30). By using the relationship

$$\frac{\mathrm{d}}{\mathrm{d}x}x^\nu\,\mathbf{I}_\nu(x) = x^\nu\,\mathbf{I}_{\nu-1}(x) \tag{8.2.12}$$

satisfied by the modified Bessel functions, one obtains equation (8.37) from the text.

Appendix to Section 9.2.

Injection Noise in SCLC Diodes

The open-circuit noise factor F_∞ defined by equation (9.16), where K_{12} is given by equation (9.18) should vanish for SCLC diodes, where $E_0(0)$ and $v_0(0) = 0$ at $x = 0$ (the cathode plane). However, if the finite conductivity of the cathode plane is considered, as in Section 5.6 and the corresponding Appendix, the injection noise is not totally suppressed. This occurs because $p_0(0) =$ finite (see equation (5.52)) and the continuity of the steady-state current at $x = 0$ is provided by diffusion. According to Appendix to Section 5.6, K_{12} will be finite (equation (5.6.4)) and so will do F_∞. Let us calculate the low-frequency noise factor $F_\infty(0)$. From equation (5.6.4) we obtain for $\omega = 0$

$$K_{12}(0) = \frac{J_0 T_L/e - NL}{p_0(0) - N} = \frac{J_0 T_L - e\,NL}{e p_0(0) - eN}\,. \tag{9.2.1}$$

Note the fact that the denumerator of the above fraction is just the difference between the total mobile charge and the total fixed charge inside the semiconductor region $0 < x < L$; both are calculated for an electrode of unit area. By using equation (9.16) where $\sigma_1(0)$ is given by equation (5.55), one obtains

$$F_\infty(0) = \frac{J_0 T_L - eNL}{e p_0(0)\,\mu[e\,p_0(0) - N]}\,. \tag{9.2.2}$$

By assuming that the injected mobile charge is everywhere much larger than the fixed space charge, F_∞ (0) may be approximated by

$$F_\infty(0) \simeq \frac{J_0 T_L}{\mu e^2 p_0^2(0)}\,. \tag{9.2.3}$$

If the injection noise is full-shot noise, then, according to equation (9.14), the open-circuit mean-square noise voltage is (for a device of unit area, at low frequencies)

$$\overline{V^2_{1(\text{c.c.})}} = F_\infty(0) \times 2e\,J_0(\Delta f) = \frac{2J_0^2(\Delta f)\,T_L}{e\mu p_0^2(0)} \propto \left[\frac{J_0 T_L^{1/2}}{p_0(0)}\right]^2. \tag{9.2.4}$$

For a saturated velocity diode $\overline{V^2_{1(0.c.)}}$ should be proportional to the square of the ratio between the hole density in the bulk and the hole density at the $x = 0$ plane. At higher frequencies, $K_{12} = K_{12}(\omega)$ and the noise measured in the external circuit will be modified by transit-time effects. The frequency dependence of K_{12} is given by equation (9.19) and discussed in the text. The frequency dependence of the normalised factor $\overline{F_\infty}$ is indicated by equation (9.20).

The suppresion (or reduction) of current fluctuations at the potential maximum (or minimum) which characterizes the SCLC régime, is known from vacuum tubes. The mechanism of noise reduction in an SCLC vacuum diode can be understood from the following quotation from reference [74], p. 200. 'Bursts of emitted current in excess of the average depress the potential minimum, causing more electrons to be returned to the cathode, thereby reducing the amount of current passing the minimum. A small decrease in emitted current has just the opposite effect. Every fluctuation in the emitted current has associated with it an almost-compensating fluctuation produced by the potential minimum. As a consequence, the noise current in the stream and that induced in the external circuit is only the small difference between the two fluctuations. The ratio of this residual noise to full-shot noise is called the shot noise reduction factor. When the transit-time between cathode and potential minimum is large enough to introduce a time lag between the emission fluctuation and the compensating action of the potential minimum, then the compensation breaks down......' [74].

Llewellyn-Peterson equations were used to evaluate the external noise due to velocity fluctuations at the virtual cathode of an SCLC plane-parallel vacuum diode (see for example, section 6.06 of reference [74]). A similar approach is used in this Appendix to evaluate the effect of random fluctuations of the particle current*) by using the transport coefficients.

*) Due to the frequent collisions with the lattice, the fluctuations of the velocity of carriers leaving the virtual cathode are unimportant for SCLC solid-state diodes.

Appendix to Section 10.4.

Evaluation of the Voltage Drop

Let us introduce

$$f(t, \tau) = \int_{\tau}^{t} \delta(\xi)\, d\xi \qquad (10.4.1)$$

(where δ is given by equation (10.60)) and also*)

$$t - \tau = T. \qquad (10.4.2)$$

According to equations (10.27), (10.28, (10.59) and (10.4.1) we have

$$E_{t,\tau} = \frac{J_0}{\varepsilon} \int_{\tau}^{t} [1 + \delta(\xi)]\, d\xi = \frac{J_0}{\varepsilon} [(t - \tau) + f(t, \tau)] = \frac{J_0}{\varepsilon} [T + f(t, t - T)]. \qquad (10.4.3)$$

$$x_{t,\tau} = \frac{\mu J_0}{\varepsilon} \int_{\tau}^{t} (t - \xi) [1 + \delta(\xi)]\, d\xi = \frac{\mu J_0}{\varepsilon} (U_1 + U_2) \qquad (10.4.4)$$

$$U_1 = \int_{\tau}^{t} (t - \xi)\, d\xi, \qquad U_2 = \int_{\tau}^{t} (t - \xi)\, \delta(\xi)\, d\xi. \qquad (10.4.5)$$

After a change of variable $t - \xi = \eta$, U_1 and U_2 become, respectively

$$U_1 = \int_{\tau}^{T} \eta\, d\eta = U_1(T) = \frac{T^2}{2} \qquad (10.4.6)$$

$$U_2 = \int_{\tau}^{T} \eta\, \delta(t - \eta)\, d\eta = U_2(T, t). \qquad (10.4.6')$$

---

*) T should not be confused with the period of the applied signal, as defined before.

$E_{t,\tau}$ may be written

$$E_{t,\tau} = \frac{J_0}{\varepsilon} \mathcal{E} \tag{10.4.7}$$

where

$$\mathcal{E} = T + f(t, t - T) \tag{10.4.8}$$

and $V(t)$ becomes

$$V(t) = \frac{\mu J_0^2}{\varepsilon^2} \mathcal{F} \tag{10.4.9}$$

where

$$\begin{aligned} \mathcal{F} &= \int_0^{T_L} \mathcal{E} \frac{\partial (U_1 + U_2)}{\partial T} \, \mathrm{d}T = \int_0^{T_L} \mathcal{E} \frac{\mathrm{d}U_1}{\mathrm{d}T} \, \mathrm{d}T + \int_0^{T_L} \mathcal{E} \frac{\partial U_2}{\partial T} \, \mathrm{d}T = \\ &= \int_0^{T_L} \mathcal{E} \, T \, \mathrm{d}T + \mathcal{E} (U_2)_{T=T_L} - \int_0^{T_L} U_2 (T, t) \frac{\partial \mathcal{E}}{\partial T} \, \mathrm{d}T. \end{aligned} \tag{10.4.10}$$

Here T_L is the total transit-time of the hole emitted at the moment τ

$$T_L = t - \tau (L, t). \tag{10.4.11}$$

T_L may be also written

$$T_L = T_L^0 + \Delta T \tag{10.4.12}$$

where ΔT is the variation (or difference) with respect to the d.c. ($\nu = 0$) transit-time T_L^0, which is the same for all injected holes. ΔT should be small, first-order in ν.

The integrals appearing in equation (10.4.10) can be evaluated as follows. Let us consider first the integral of an arbitrary function $\psi(T)$.

$$\int_0^{T_L} \psi (T) \, \mathrm{d}T = \int_0^{T_L^0} \psi (T) \, \mathrm{d}T + \int_{T_L - \Delta T}^{T_L} \psi (S) \, \mathrm{d}T. \tag{10.4.13}$$

Because ΔT is small, the second integral in the right-handside may be approximated by

$$\int_{T_L - \Delta T}^{T_L} \psi(T) \, \mathrm{d}T = \psi(T_L) \, \Delta T - \frac{1}{2} \left[\psi'(T_L) \, \Delta T \right] \Delta T \tag{10.4.14}$$

(only the first- and second-order quantities are taken into account). Therefore

$$\int_0^{T_L} [\mathcal{E}\, T]\, \mathrm{d}T = \int_0^{T_L^0} \mathcal{E}\, T\, \mathrm{d}T + [\mathcal{E}T]_{T=T_L^0} \times \Delta T - \tag{10.4.15}$$

$$-\frac{(\Delta T)^2}{2} \left\{ \frac{\partial}{\partial T} [\mathcal{E}\, T] \right\}_{T=T_L} + \mathcal{O}\,(\nu^3)$$

$$\int_0^{T_L} [\mathcal{E}\, T]\, \mathrm{d}T = \int_0^{T_L^0} \mathcal{E}\, T\, \mathrm{d}T + [E\, T]_{T=T_L^0} \times \Delta T - \frac{1}{2}\, \mathcal{E}\,|_{T=T_L^0} \times \tag{10.4.15'}$$

$$\times (\Delta T)^2 - \frac{T_L^0}{2} \frac{\partial \mathcal{E}}{\partial T}\bigg|_{T=T_L^0} \times (\Delta T)^2 + \mathcal{O}\,(\nu^3)$$

(the terms of third- and higher-order of magnitude are incorporated in $\mathcal{O}\,(\nu^3)$)

$$\int_0^{T_L} U_2(T, t) \frac{\partial \mathcal{E}}{\partial T}\, \mathrm{d}T = \int_0^{T_L^0} U_2\,(T,\; t) \frac{\partial \mathcal{E}}{\partial T}\, \mathrm{d}T + \tag{10.4.16}$$

$$+ U_2(T_L, t) \left(\frac{\partial \mathcal{E}}{\partial T} \right)\bigg|_{T=T_L^0} \times \Delta T + \mathcal{O}\,(\nu^3)\,.$$

In evaluating the last integral, use was made of the fact that $\partial U_2/\partial T$ and $\partial^2 \mathcal{E}/\partial T^2$ are second and third-order quantities and occur multiplied by $(\Delta T)^2$.

By introducing the above results into equation (10.4.10) one obtains

$$\mathcal{F} = \int_0^{T_L^0} \mathcal{E}\, T \mathrm{d}T - \int_0^{T_L^0} U_2\,(T, t) \frac{\partial \mathcal{E}}{\partial T}\, \mathrm{d}T + \mathcal{E}\,|_{T=T_L} \times \Big[\, U_2(T_L, t) + T_L \Delta T - \tag{10.4.17}$$

$$- \frac{1}{2}\, (\Delta T)^2 \,\Big] - \frac{\partial \mathcal{E}}{\partial T}\bigg|_{T=T_L^0} \times \left[\frac{T_L^0}{2} \Delta T + U_2\,(T_L, t) \right] \Delta T + \mathcal{O}\,(\nu^3).$$

Then, equation (10.4.4) yields

$$L = \frac{\mu J_0}{\varepsilon} [U_1\,(T_L) + U_2\,(T_L, t)] = \frac{\mu}{\varepsilon} \bar{J}_0\, U_1(T_L^0) \tag{10.4.18}$$

and by using equation (10.64) one obtains

$$U_1\,(T_L) + U_2\,(T_L, t) = \frac{\bar{J}_0}{J}\, U_1\,(T_L^0) = (1 - \nu^2 \rho)\, U_1\,(T_L^0) + \mathcal{O}\,(\nu^4) \tag{10.4.19}$$

and also

$$U_1(T_L) - U_1(T_L - \Delta T) + U_2(T_L, t) = T_L \Delta t - \frac{1}{2}(\Delta T)^2 +$$
$$+ U_2(T_L, t) + \mathcal{O}(\nu^3) = -\nu^2\rho\, U_1(T_L^0) + \mathcal{O}\,\nu^4. \tag{10.4.20}$$

We shall calculate successively the terms appearing in the expression (10.4.17). One of these is

$$\mathcal{E}|_{T=T_L} \times \left[U_2(T_L, t) + T_L \Delta T - \frac{1}{2}(\Delta T)^2 \right] = -\nu^2 \times \mathcal{E}|_{T_L = T_L^0} \times$$
$$\times \frac{(T_L^0)^2}{2} + \mathcal{O}(\nu^3) = -\nu^2\rho \frac{(T_L^0)^3}{2} + \mathcal{O}(\nu^3). \tag{10.4.21}$$

The last term in the right-hand-side of equation (10.4.17) will be evaluated now. We first calculate

$$\frac{\partial \mathcal{E}}{\partial T} = 1 + \frac{\partial}{\partial T} \qquad\qquad + \frac{\partial}{\partial T}\int_{t-T}^{t} \delta(\xi)\,\mathrm{d}\xi$$
$$= 1 + \frac{\partial}{\partial T}\int_0^T \delta(t - T)\,\mathrm{d}T = 1 + \delta(t - T). \tag{10.4.22}$$

$\delta(t - T)$ is a small quantity, of the first order of magnitude. However, because in equation (10.4.17) $\partial\mathcal{E}/\partial T$ multiplies a second order quantity (see below), we shall approximate

$$\frac{\partial \mathcal{E}}{\partial T} \simeq 1\,. \tag{10.4.23}$$

We evaluate now the following quantity appearing in equation (10.4.17)

$$\mathcal{H} = \left[\frac{T_L^0}{2} \Delta T + U_2(T_L, t) \right] \Delta T \tag{10.4.24}$$

$$U_2(T_L, t) \simeq \int_\tau^t (t - \xi)\,\mathrm{d}\xi = U_1(T_L) \simeq U_1(T_L^0 - \Delta T) -$$
$$- U_1(T_L^0) = -\left.\frac{\mathrm{d}U_1}{\mathrm{d}T}\right|_{T = T_L^0} \times \Delta T = -T_L^0 \Delta T \tag{10.4.25}$$

$$\mathcal{H} = -\frac{T_L^0}{2}(\Delta T)^2. \tag{10.4.26}$$

Equation (4.17) becomes

$$\mathcal{F} = \int_0^{T_L^0} \mathcal{E} T \, \mathrm{d}T - \int_0^{T_L^0} U_2 (T, t) \frac{\partial \mathcal{E}}{\partial T} \mathrm{d}T - \\ - \nu^2 \rho \frac{(T_L^0)^3}{2} + \frac{T_L^0}{2} (\Delta T)^2 + \mathcal{O} (\nu^3). \tag{10.4.27}$$

We shall now find the integrals appearing in the above expression. We have

$$U_2 (T,t) \frac{\partial \mathcal{E}}{\partial T} = \int_0^T \eta \delta (t - \eta) \, \mathrm{d}\eta \left[1 + \delta (t - T) \right] \tag{10.4.28}$$

$$\mathcal{E} T = T^2 + T \int_0^T \delta (t - \eta) \, \mathrm{d}\eta \tag{10.4.29}$$

and therefore $\mathcal{F}$ is

$$\mathcal{F} = \int_0^{T_L^0} T^2 \, \mathrm{d}T + \int_0^{T_L^0} T \, \mathrm{d}T \int_0^T \delta (t - \eta) \, \mathrm{d}\eta - \int_0^{T_L^0} \mathrm{d}T \int_0^T \eta \delta(t - \eta) \, \mathrm{d}\eta - \\ \int_0^{T_L^0} \delta(t - T) \, \mathrm{d}T \int_0^T \eta \delta(t - \eta) \, \mathrm{d}\eta - \nu^2 \rho \frac{(T_L^0)^3}{2} + \frac{T_L^0}{2} (\Delta T)^2 + \mathcal{O} (\nu^3) \tag{10.4.30}$$

and may be also written

$$\mathcal{F} = \frac{(T_L^0)^3}{3} + \int_0^{T_L^0} \mathrm{d}T \int_0^T \mathrm{d}\xi \int_0^{\xi} \delta (t - \eta) \, \mathrm{d}\eta - \\ - \int_0^{T_L^0} \delta(t - T) \, \mathrm{d}T \int_0^T \eta \delta(t - \eta) \mathrm{d}\eta - \nu^2 \rho \frac{(T_L^0)^3}{2} + \frac{T_L^0}{2} (\Delta T)^2 + \mathcal{O} (\nu^3). \tag{10.4.31}$$

We have to calculate

$$P_1 = \int_0^{T_L^0} \mathrm{d}T \int_0^T \mathrm{d}\xi \int_0^{\xi} \delta (t - \eta) \, \mathrm{d}\eta \tag{10.4.32}$$

$$P_2 = \int_0^{T_L^0} \delta(t - T) \, \mathrm{d}T \int_0^T \eta \delta(t - \eta) \, \mathrm{d}\eta \tag{10.4.33}$$

where δ is given by equation (10.60), in the text. One finds

$$P_1 = \frac{(T_L^0)^3}{3} \sum_{n=1}^{\infty} \nu_n [S_2 (n\omega T_L^0) \sin \psi_n (t) + S_1 (n\omega T_L^0) \cos \psi_n (t)]. \quad (10.4.34)$$

In the expression of P_2 it is sufficient to evaluate δ to first-order. One obtains

$$\delta(\xi) \simeq \nu_1 \sin \psi_1 (\xi), \ \nu_1 = \alpha' \nu, \ \psi_1 = \omega\xi + \alpha_1 \quad (10.4.35)$$

and P_2 becomes

$$P_2 = \frac{\nu_1^2 (T_L^0)^3}{4} [M_1 + M_2 \cos^2 \psi_1(t) + M_3 \sin^2 \psi_1 (t)] \quad (10.4.36)$$

where M_1, M_2, M_3 are defined by

$$M_1 = \frac{2}{\theta} [A (\theta) \sin \theta - B (\theta) \cos \theta] \quad (10.4.37)$$

$$M_2 = \frac{2}{\theta} [- A (\theta) \sin \theta - B(\theta) \cos \theta + B (2\theta)] \quad (10.4.38)$$

$$M_3 = \frac{2}{\theta} [A (\theta) \cos \theta - B (\theta) \sin \theta - A (2\theta)] \quad (10.4.39)$$

and θ is the transit-angle, equation (10.54), rewritten here for convenience

$$\theta = \omega T_L^0. \quad (10.4.40)$$

We note that $A(\theta)$, $B(\theta)$ were introduced in equation (10.33) and their explicit form is

$$A (\theta) = \frac{\cos \theta}{\theta^2} + \frac{\sin\theta}{\theta} - \frac{1}{\theta^2} \quad (10.4.41)$$

$$B (\theta) = \frac{\sin\theta}{\theta^2} - \frac{\cos \theta}{\theta}. \quad (10.4.42)$$

Therefore, one obtains

$$M_1 = M_1(\theta) = \frac{2}{\theta^3}(\theta - \sin\theta) \qquad (10.4.43)$$

$$M_2 = M_2(\theta) = \frac{2\sin\theta}{\theta^3} - \frac{3}{2}\frac{\sin 2\theta}{\theta^3} + \frac{\cos 2\theta}{\theta^2} \qquad (10.4.44)$$

$$M_3 = M_3(\theta) = -\frac{2\cos\theta}{\theta^3} + \frac{\sin 2\theta}{\theta^2} + \frac{3}{2}\frac{\cos 2\theta}{\theta^3}. \qquad (10.4.45)$$

Before introducing F given by equation (10.4.31) into the expression of $V(t)$ we have to evaluate $(\Delta T)^2$ with the required accuracy (to second-order) T_L is a function of ν and may be developed in series

$$T_L(\nu) = T_L^0 + \nu\left(\frac{\partial T_L}{\partial \nu}\right)_{\nu=0} + \frac{\nu^2}{2}\left(\frac{\partial^2 T_L}{\partial \nu^2}\right)_{\nu=0} + \ldots\ldots \qquad (10.4.46)$$

$$(\Delta T)^2 = (T_L - T_L^0)^2 \simeq \nu^2\left(\frac{\partial T_L}{\partial \nu}\right)^2_{\nu=0}. \qquad (10.4.47)$$

$\left(\frac{\partial T_L}{\partial \nu}\right)$ may be calculated as follows. By taking into account equations (10.31) and (10.64), one obtains*)

$$L = \frac{\mu}{2\varepsilon}\bar{J}_0(1 + \nu^2\rho_0)T_L^2\left\{1 + 2\sum_{n=1}^{\infty}\nu_n\,[A(n\omega T_L)\sin\psi_n(t) - \right.$$

$$\left. - B(n\,\omega\,T_L)\cos\psi_n(t)\,]\right\} \qquad (10.4.48)$$

or

$$\frac{2L\varepsilon}{\mu\bar{J}_0} = (1 + \nu^2\rho_0)\,T_L^2\,[1 + 2\zeta(\omega T_L)] \qquad (10.4.49)$$

where ζ is defined by equation (10.44). For $\nu = 0$ we obtain

$$(T_L^0)^2 = \frac{2\varepsilon L}{\mu\bar{J}_0} \qquad (10.4.50)$$

which is implicitly contained in equation (10.34).

*) In equation (10.31), α was replaced by $\omega[t - \tau(L, t)] = \omega T_L$.

As shown by equation (10.61), ν_1 should be of the form

$$\nu_1 = \alpha_0^{(1)} \nu + \alpha_1^{(1)} \nu^3 + \ldots\ldots \tag{10.4.51}$$

and equation (10.4.49) yields

$$T_L \simeq T_L^0 (1 + \nu^2 \rho_0)^{-1/2} [1 - \zeta (\omega T_L)] \tag{10.4.52}$$

because ζ is small. By using equation (10.44) we have

$$T_L \simeq \frac{T_L^0}{\sqrt{1+\nu^2 \rho_0}} \{1 - \alpha' \nu [A(\theta) \sin \psi_1 (t) - B(\theta) \cos \psi_1 (t)]\} \tag{10.4.53}$$

and

$$\left(\frac{\partial T_L}{\partial \nu}\right)_{\nu=0} = - \alpha' T_L^0 [A(\theta) \sin \psi_1 - B(\theta) \cos \psi_1]. \tag{10.4.54}$$

By using successively equations (10.4.47) and (10.4.54) one obtains

$$\frac{T_L^0 (\Delta T)^2}{2} = \frac{T_L^0}{2} \nu^2 \left(\frac{\partial T_L}{\partial \nu}\right)^2_{\nu=0} = \frac{\nu^2 (\alpha')^2 (T_L^0)^3}{4} \{A^2(\theta) + B^2(\theta) + $$

$$+ [B^2(\theta) - A^2(\theta)] \cos^2 \psi_1(t) - 2A(\theta) B(\theta) \sin 2\psi_1(t)\}. \tag{10.4.55}$$

We are now in position to evaluate $V(t)$. By taking into account equations (10.4.9), (10.4.31), etc. we may write

$$V(t) = \frac{\mu J_0^2}{\varepsilon^2} \frac{(T_L^0)^3}{3} \{1 + \sum_{n=1}^{\infty} \nu_n [S_2 (n\omega T_L^0) \sin \psi_n (t) + S_1 (n\omega T_L^0) \cos \psi_n (t)] +$$

$$+ \nu^2 (\alpha')^2 \frac{3}{4} [A^2 + B^2 - M_1 + (B^2 - A^2 - M_2) \cos 2\psi_1 (t) - \tag{10.4.56}$$

$$- (2AB + M_3) \sin 2\psi_1 (t)] - \frac{3}{2} \nu^2 \rho_0\}.$$

We introduce the following notations ($\omega T_L^0 = \theta$)

$$A(n\theta) = A_n,\ B(n\theta) = B_n,\ S_1(n\theta) = S_{1n},\ \ S_2(n\theta) = S_{2n} \tag{10.4.57}$$

$$\alpha_2 - 2\alpha_1 = \gamma. \tag{10.4.58}$$

Equation (10.64) also yields

$$J_0^2 = \bar{J}_0^2 (1 + 2\nu^2 \rho_0). \qquad (10.4.59)$$

It is permissible to write

$$\psi_2 = 2\psi_1 + \gamma. \qquad (10.4.60)$$

Then, we calculate

$$\nu_2 (S_{22} \sin \psi_2 + S_{12} \cos \psi_2) = \nu_2 \varphi_1 \cos 2\psi_1 + \nu_2 \varphi_2 \sin 2\psi_2 \qquad (10.4.61)$$

where

$$\varphi_1 = S_{22} \sin \gamma + S_{12} \cos \gamma = (S_{12} \cos 2\alpha_1 - S_{22} \sin 2\alpha_1) \times$$

$$\times \cos \alpha_2 + (S_{12} \sin 2\alpha_1 + S_{22} \cos 2\alpha_1) \sin \alpha_2 \qquad (10.4.62)$$

$$\varphi_2 = S_{22} \cos \gamma - S_{12} \sin \gamma = (S_{12} \sin 2\alpha_1 + S_{22} \cos 2\alpha_1) \times$$

$$\times \cos \alpha_2 - (S_{12} \cos 2\alpha_1 - S_{22} \sin 2\alpha_1) \sin \alpha_2. \qquad (10.4.63)$$

By using equations (10.4.50) and (10.4.59), and then replacing $\bar{J}_0$ as a function of $\bar{V}_0$, which is defined by equation (10.37)

$$\bar{V}_0 = \frac{2}{3} \sqrt{\frac{2L^3}{\varepsilon\mu}}\, \bar{J}_0 = \frac{\mu \bar{J}_0^2}{\varepsilon^2} \frac{1}{3} \left(\frac{2L\varepsilon}{\mu \bar{J}_0}\right)^{3/2}, \qquad (10.4.64)$$

one obtains

$$\frac{\mu J_0}{\varepsilon^2} \frac{(T_L^0)^3}{3} = \frac{\mu \bar{J}_0^2}{3\varepsilon^2} \left(\frac{2L\varepsilon}{\mu \bar{J}_0}\right)^{3/2} (1 + 2\nu^2 \rho_0) = \bar{V}_0 (1 + 2\nu^2 \rho_0). \qquad (10.4.65)$$

Finally, by neglecting all third- and higher-order quantities in equation (10.4.56) one obtains equation (10.65) given in the text.

Appendix to Section 10.7.

Second-harmonic Calculation

The second harmonic will be calculated for $\theta_r = 0$, starting from the results derived in Section 10.5 and Appendix to Section 10.4. With reference to equations (10.72), (10.73), (10.4.62) and (10.4.63) one obtains [228]

$$\varphi_1^2 + \varphi_2^2 = S_{12}^2 + S_{22}^2 \tag{10.7.1}$$

$$\nu_2 = \beta\nu^2 = \frac{3}{4}\,\nu^2\,\frac{\sqrt{(M_3 + 2A_1B_1)^2 + (M_2 + A_1^2 - B_1^2)}}{(S_{11}^2 + S_{21}^2)\sqrt{S_{12}^2 + S_{22}^2}}. \tag{10.7.2}$$

The amplitude and phase of the second harmonic of the current is determined if $\nu_2 \cos \alpha_2$ and $\nu_2 \sin \alpha_2$ are known. $\cos \alpha_2$ and $\sin \alpha_2$ may be found after some algebraic manipulation [229] involving equations (10.70), (10.72), (10.73), (10.76)|, (10.4.62) and (10.4.63).

References

1. N. F. Mott and R. W. Gurney, 'Electronic processes in ionic crystals' Oxford University Press, Oxford, 1940.
2. A. Rose, 'Space-charge-limited currents in solids', *Phys. Rev.*, **97**, 1538–1544 (1955).
3. M. A. Lampert, 'Simplified theory of space-charge-limited currents in insulators with traps', *Phys. Rev.*, **103**, 1648–1656 (1956).
4. M. A. Lampert, 'Simplified theory of one-carrier currents with field-dependent mobilities', *J. Appl. Phys.*, **29**, 1082–1090 (1958).
5. M. A. Lampert, 'Injection currents in insulators', *Proc. I.R.E.*, **50**, 1781–1796 (1962).
6. G. T. Wright, 'Space-charge-limited currents in insulating materials', *Nature*, 102, 1296–1297 (1958).
7. A. M. Conning, A. A. Kayali and G. T. Wright, 'Space-charge-limited dielectric diodes', *Journ. I.E.E.*, **5**, 595 (1959).
8. G. T. Wright, 'A proposed space-charge-limited dielectric triode', *J. Brit. I.R.E.*, **20**, 337–350 (1960).
9. G. T. Wright, 'Mechanisms of space-limited-current in solids' *Solid-St. Electron.*, **2**, 165 – 189, (1961).
10. J. Shao and G. T. Wright, 'Characteristic of the space-charge-limited dielectric diode at very high frequencies', *Solid-St. Electron.*, **3**, 291–303 (1961).
11. W. Shockley, 'Transistor electronics: imperfections, unipolar and analog transistors', *Proc. I.R.E.*, **40**, 1289–1313 (1952).
12. W. Shockley, 'A unipolar field-effect transistor', *Proc. I.R.E.*, **40**, 1365–1370, (1962).
13. W. Shockley and R. C. Prim, 'Space-charge limited emission in semiconductors, *Phys. Rev.*, **90**, 753–758 (1953).
14. W. Shockley, 'Negative resistance arising from transit-time in semiconductor diodes', *Bell Syst. Techn. J.*, **33**, 799–826 (1954).
15. D. Dascalu, 'Space-charge-limited minority hole flow in very high resistivity silicon', *Rev. Roum. Phys.*, **17**, 675–708 (1972).
16. W. Shockley, 'High-frequency negative resistance device', *U.S. Pat.* 2 794 917, June 4, 1957.
17. H. Yoshimura, 'Space-charge-limited and emitter-current-limited injections in space charge region of semiconductors' *I.E.E.E. Trans. Electron. Dev.*, **ED–11**, 412–422 (1964).
18. C. A. Lee and G. C. Dalman, 'Local oscillator noise in a silicon Pt-n-p+ microwave diode source', *Electron. Lett.*, **7**, 565–566 (1971); 'A low-noise Ku-band silicon Pt-n-p+ diode oscillator', *Paper presented at the Internat. Electron Dev. Meeting*, Oct. 11–13, 1971, Washington.

19. C. P. SNAPP and P. WEISSGLAS, 'Experimental comparison of silicon $p^+ n p^+$ and Cr $n p^+$ transit-time oscillators', *Electron. Lett.*, 7, 753–744 (1971).

20. J. FREYER, M. CLAASSEN and W. HARTH, 'Fabrication of an epitaxial-silicon Pd-*n*-Pd microwave generator', *AEÜ*, **26**, 150–151 (1972).

21. S. M. SZE, D. C. COLEMAN JR. and A. LOYA, 'Current transport in metal-semiconductor-metal (MSM) structures', *Solid-St. Electron.*, **14**, 1209–1218 (1972).

22. D. J. COLEMAN, JR. and S. M. SZE, 'A low-noise metal-semiconductor-metal (MSM) microwave oscillator', *Bell Syst. Techn. J.*, **50**, 1695–1699 (1971).

23. H. EHRENREICH, 'Band structure and electron transport of GaAs', *Phys. Rev.*, **120**, 1951–1963 (1960).

24. G. GIBBONS and H. T. MINDEN, 'Avalanche and Gunn-effect microwave oscillators', *Solid State Technol.*, **13**, 37–48, Febr. 1970.

25. V. SZEKELY and K. TARNEY, 'Intervalley scattering model of the Gunn domain', *Electron. Lett.*, **4**, 592–594 (1968).

26. W. R. CURTICE and J. J. PURCELL, 'Analysis of the LSA mode including effects of space charge and intervalley transfer time', *I.E.E.E. Trans. Electron. Dev.*, **ED–17**, 1048–1060 (1970).

27. J. R. HAUSER and M. A. LITTLEJOHN, 'Approximations for accumulation and inversion space-charge layers in semiconductors', *Solid-St. Electron.*, **11**, 667–674 (1968).

28. H. KROEMER, 'Negative conductance in semiconductors', *I.E.E.E. Spectrum*, **5**, No. 1, 47–56 (1968).

29. J. E. CARROLL, 'Hot electron microwave generators', Edward Arnold (Publishers) Ltd. London, 1970.

30. W. SHOCKLEY, 'The theory of *pn* junctions in semiconductors and *pn* junction transistors', *Bell Syst. Techn. J.*, **28**, 435 (1949).

31. J. E. PARROTT, 'Reformulation of basic semiconductor transport equations', *Solid-St. Electron.* **14**, 885–899 (1971).

32. C. GOLDBERG, 'Electric current in a semiconductor space-charge region', *J. Appl. Phys.*, **40**, 4612–4614 (1969).

33. K. BLÖTEKJAER, 'High-frequency conductivity, carrier waves, and acoustic amplification in drifted semiconductor plasmas', *Ericsson Technics*, **22**, 125–183 (1966).

34. R. STRATTON, 'Carrier heating or cooling in a strong built-in field', *J. Appl. Phys.*, **40**, 4582–4583 (1969).

35. K. BLOTEKJAER, 'Waves in semiconductors with nonconstant mobility' *Electron. Lett.*, **4**, 357–358 (1968).

36. K. BLOTEKJAER, 'Transport equations for electrons in two-valley semiconductors', *I.E.E.E. Trans. Electron. Dev.*, **ED–17**, 38–47, (1970).

37. CH. KITTEL, 'Introduction to solid state physics', John Wiley and Sons, Inc., New York, 1971; Ch. 8.

38. J. C. MCGRODDY and P. GUÉRET, 'Dynamic bulk negative differential conductivity in semiconductors', *Solid-St. Electron.*, **14**, 1219–1224 (1971).

39. J. M. ANDREWS and M. P. LEPSELTER, 'Reverse current-voltage characteristics of metal-silicide Schottky contacts', *Solid-St. Electron.*, **13**, 1011–1023 (1970).

40. M-A. NICOLET, V. RODRIGUEZ and D. STOLFA, 'Unipolar interface-charge-limited current', *Surf. Sci.*, **10**, 146–164 (1968).

41. C. A. Mead, E. H. Snow and B. E. Deal, 'Barrier lowering and field penetration at metal-dielectric interfaces', *Appl. Phys. Lett.*, **10**, 53–55 (1966).

42. S. S. Perlman, 'Barrier height diminution in Schottky diodes due to electrostatic screening', *I.E.E.E. Trans. Electron Dev.*, **ED–16**, 450–454 (1969).

43. G. W. Eimers and E. H. Stevens, 'A composite model for Schottky diode barrier height', *I.E.E.E. Trans Electron. Dev.*, **ED–18**, 1185–1186 (1971).

44. R. F. Broom, 'Doping dependence of the barrier height of palladium-silicide Schottky-diodes', *Solid-St. Electron.*, **14**, 1087–1092 (1971).

45. A.Y.C.Yu, 'Electron tunnelling and contact resistance of metal-silicon contact barriers', *Solid-St. Electron.*, **13**, 239–247, (1970).

46. C. Y. Chang, 'Carrier transport across metal-semiconductor barriers', *Solid-St. Electron.*, **13**, 727–740 (1970).

47. V. L. Rideout and C. R. Crowell, 'Effects of image force and tunnelling on current transport in metal-semiconductor (Schottky barrier) contacts', *Solid-St. Electron.*, **13**, 993–1009 (1970).

48. M. Claassen and W. Harth, 'Field-emission-controlled transit-time negative resistance', *Electron. Lett.*, **6**, 512–513 (1970).

49. A. Semichon and J. Michel, 'Microwave oscillation of a tunnel transit-time diode', *Presented at the 1970 MOGA Conf., Amsterdam, the Netherlands*, Sept. 7–11, 1970.

50. F. A. Padovani and R. Stratton, 'Field and thermionic-field emission in Schottky barriers', *Solid-St. Electron.*, **9**, 158–163 (1966).

51. E. Rossiter, P. Mark and M. A. Lampert, 'Unusual field phenomena associated with majority-carrier injection from a point contact', *Solid-St. Electron.*, **13**, 491–503 (1970).

52. H. Kroemer, 'The Gunn effect under imperfect cathode boundary conditions', *I.E.E.E Trans. Electron. Dev.*, **ED–15**, 819–837 (1958).

53. A. Van Der Ziel, 'Normalized characteristics of *n*-ν-*n* devices', *Solid-St. Electron.*, **10**, 267–268 (1967).

54. D. Dascalu, M. Badila and N. Marin, 'Exact small signal theory of space-charge-limited majority carrier flow in semiconductors', *Rev. Roum. Phys.*, **15**, 1197–1199 (1970).

55. O. J. Marsh and C. R. Viswanathan, 'Space-charge-limited current of holes in silicon and techniques for distinguishing double and single injection', *J. Appl. Phys.*, **38**, 3155–3144 (1967).

56. S. Okazaki and M. Hiramatsu, 'Observations of space-charge-limited currents in *p*-type silicon', *Solid-St. Electron.*, **10**, 273–279 (1967).

57. N. Marin, *Graduation Thesis, Electronics Dept., Polytechnical Institute of Bucharest*, 1972.

58. E. I. Adirovichi, 'Electric field and currents in dielectrics' (in Russian), *Solid-State Physics* **2**, 1410 (1960); E. I. Adirovichi and L. A. Dubrovski, 'Dielectric electronics and the parabolic law of space-charge-limited currents' (in Russian), *U.S.S.R. Sci. Acad. Reports*, **164**, 771 (1964).

59. N. Sinharay and B. Meltzer, 'Characteristics of insulator diodes determined by space-charge and diffusion', *Solid-St. Electron.*, **7**, 125–136 (1964).

60. D. J. Page, 'Some computed and measured characteristics of CdS space-charge-limited diodes', *Solid-St. Electron.*, **9**, 255–264 (1966).

61. M. A. Lampert and F. Edelman, 'Theory of one-carrier, space-charge-limited currents including diffusion and trapping', *J. Appl. Phys.*, **35**, 2971–2982 (1964).
62. H. Kroemer, 'Generalized proof of Shockley's positive conductance theorem', *Proc. I.E.E.E.*, **58**, 1844–1845 (1970).
63. P. S. Hauge, 'Static negative resistance in Gunn effect materials with field-dependent carrier diffusion', *I.E.E.E. Trans. Electron. Dev.*, **ED–18**, 390–391 (1971).
64. G. Döhler, 'Shockley's positive conductance theorem for Gunn materials with field-dependent diffusion', *I.E.E.E. Trans. Electron Dev.*, **ED–18**, 1190–1192 (1971).
65. G. T. Wright, 'Depletion layer formation, space-charge injection and current-voltage characteristics for the silicon $p-n-p$ ($n-p-n$) structure', *Solid-St. Electron.*, **15**, 381–386 (1972).
66. B. I. Timan and V. M. Fesenko, 'Space-charge-limited currents in the presence of donors', *Soviet Physics-Semiconductors*, **2**, 228–230 (1968).
67. M. A. Lampert, 'Volume-controlled current injection in insulators', *Rep. Progr. Phys.*, **27**, 329–367 (1964).
68. M. A. Lampert and P. Mark, 'Current injection in solids', Academic Press, New York, 1970.
69. R. S. Muller, 'Theoretical admittance variation with frequency in insulators having traps subjected to charge injection', in *Physics of Semiconductors, Proc. 7th Int. Conf.* (Paris, 1964), M. Hulin, editor., Academic Press, New York, 1964, pp. 631–638.
70. D. Dascalu, 'Trapping and transit-time effects in high-frequency operation of space-charge-limited dielectric diodes. I. Diode impedance', *Rev. Roum. Phys.*, **13**, 513–529 (1968).
71. W. E. Benham, 'Theory of internal action of thermionic systems at moderately high frequencies', Part I, *Phil. Mag.*, **5**, 641–662 (1928); 'Frequency variations of valve oscillators' *Experimental Wireless and Wireless Engineer*, **7**, 42 (1930); Part II of 1928 paper, *Phil. Mag.*, **11**, 457–517, *February Suppl.* 1931.
72. I. Müller, 'Elektronenschwingungen in Hochvakuum', *Hochfrequenztechnik, u. Elekt. Akustik*, **14**, 156–167 (1933).
73. F. B. Lewellyn, 'Vacuum tube electronics at ultra-high frequencies', *Proc. I.R.E.*, **21**, 1532–1573 (1933); 'Operation of ultrahigh frequency vacuum tubes', *Bell Syst. Techn. J.*, **14**, 632–665 (1935).
74. Ch. K. Birdsall and W. B. Bridges, 'Electron dynamics of diode regions', Academic Press, New York, 1966.
75. R. P. Wadhwa and M. L. Sisodia, 'Transit-time effects in the presence of traps in space-charge-limited solid-state devices', *Int. J. Electronics*, **23**, 83–97 (1967).
76. D. E. McCumber and A. G. Chynoweth, 'Theory of negative-conductance amplification and of Gunn instabilities in "two-valley" semiconductors', *I.E.E.E. Trans. Electron Dev.*, **ED–13**, 4–21 (1966).
77. M. Draganescu, 'Small-signal high-frequency behaviour of the ideal dielectric diode' (in Russian), *Bull. Polytech. Inst. Bucharest*, **27**, No. 5, 131–138 (1965).
78. A. Van Der Ziel and S. T. Hsu, 'High-frequency admittance of space-charge-limited solid-state diodes', *Proc. I.E.E.E.* (*Lett.*), **54**, 1194 (1966).
79. R. Holstrom and H. Derfler, 'Space-charge waves and stability of electron diodes', *I.E.E.E. Trans. Electron Dev.*, **ED–13**, 539–544 (1966).
80. D. Dascalu, 'Simple method for the calculation of vacuum diode high-frequency characteristics', *Bull. Polytech. Inst. Bucharest*, **29**, No. 2, 127–131 (1967).

81. A. Shumka and M.-A. Nicolet, 'Impedance of space-charge-limited-currents with field-dependent mobility', *Solid-St. Electron.*, **7**, 106–107 (1964).

82. D. Dascalu, 'Small-signal theory of space-charge-limited diodes', *Int. J. Electron.*, **21**, 183–200 (1966).

83. H. Kroemer, 'Detailed theory of the negative conductance of bulk negative mobility amplifiers, in the limit of zero ion density', *I.E.E.E. Trans. Electron. Dev.*, **ED–14**, 476–492 (1967).

84. F. B. Llewellyn and L. C. Peterson, 'Vacuum-tube networks', *Proc. I.R.E.*, **32**, 144–166 (1944).

85. D. Dascalu, 'The space-charge factor and the equivalent Llewellyn-Peterson equations for semiconductor regions', *Rev. Roum. Phys.*, **17**, 643–647 (1972).

86. G. T. Wright, 'Transistor transit-time oscillator', *Electron. Lett.*, **3**, 234–235 (1967).

87. G. T. Wright, 'Efficiency of the transistor transit-time oscillator', *Electron. Lett.*, **4**, 217–219 (1968).

88. G. T. Wright, 'Punch-through transit-time oscillator', *Electron. Lett.*, **4**, 543–544 (1968).

89. G. T. Wright, '2-port impedance parameters of the junction transistor', *Electron. Lett.*, **5**, 375–376 (1969).

90. G. T. Wright, 'Transit-time oscillator with velocity-limited injection', *Electron. Lett.*, **7**. 449–451 (1971).

91. D. Dascalu, 'Effects of traps distributed in energy on the a.c. response of thin film transistor, and dielectric diode', *Solid-St. Electron.*, **9**, 1020–1021 (1966).

92. D. Dascalu, 'Trapping and transit-time effects in high-frequency operation of space-charge limited dielectric diodes. II. Frequency characteristics', *Solid-St. Electron.*, **11**, 391–400 (1968); also paper presented at 'Semiconductor Device Research Conference', *Bad Nauheim* (*Germany*), 17–21 April, 1967.

93. D. Dascalu, 'Small-signal theory of unipolar injection currents in solids', *I.E.E.E. Trans. Electron Dev.*, **ED–19**, 1239–1251, (1972).

94. G. C. Dacey, 'Space-charge limited hole current in germanium', *Phys. Rev.*, **90**, 759–763 (1953).

95. B. L. Gregory and A. G. Jordan, 'Experimental investigations of single injection in compensated silicon at low temperatures', *Phys. Rev.*, **134**, A 1378–A 1376 (1964).

96. B. L. Gregory and A. G. Jordan, 'Single-carrier injection in silicon at 76 and 300 K', *J. Appl. Phys.*, **35**, 3046–3047 (1964).

97. J. M. Brown and A. G. Jordan, 'Injection and transport of added carriers in silicon at liquid-helium temperatures', *J. Appl. Phys.*, **37**, 337–346 (1966).

98. S. Denda and M-A. Nicolet, 'Pure space-charge-limited electron current in silicon', *J. Appl. Phys.*, **37**, 2412–2424 (1966).

99. U. Büget and G. T. Wright, 'Space-charge-limited current in silicon', *Solid-St. Electron.*, **10**, 199–207 (1967).

100. A. K. Hagenlocher, 'Space-charge-limited currents in high-resistivity *p*-type silicon', *Appl. Phys. Lett.*, **10**, 119–121 (1967).

101. A. Yamashita, 'Space-charge-limited currents and the velocity-field characteristic in *n*-type GaAs', *Japanese J. Appl. Phys.*, **7**, 1084–1091 (1968).

102. V. Rodriguez and M-A. Nicolet, 'Drift velocity of electrons in silicon at high electric fields from 4.2 to 300 K', *J.Appl.Phys.* 496–498 (1969).

103. D. Dascalu, 'Unipolar injection in semiconductor electron devices' (in Romanian), Publishing House of the Romanian Academy, Bucharest, 1972.
104. D. Dascalu, 'Space-charge effects upon unipolar conduction in semiconductor regions', *J. Appl. Phys.*, **44**, 3609–3616 (1973).
105. D. Dascalu, 'Small-signal impedance of space-charge-limited semiconductor diodes', *Electron Lett.*, **5**, 230–231 (1969).
106. D. Dascalu, 'Space-charge-limited currents in semiconductors', *Ph. D. dissertation*, Polytechnical Inst. Bucharest, 1970.
107. D. Dascalu, 'High-frequency impedance of silicon space-charge-limited diodes', *Solid-St. Electron.*, **12**, 444–446 (1969).
108. S. H. Chisholm and C. S. Yeh, 'Evidence of transit-time effects in silicon *n-v-n* space-charge-limited current solid-state devices', *Electron. Lett.*, **4**, 498–499 (1968).
109. D. Dascalu, 'Experimental evidence of transit-time effects in silicon punch-through diodes', *Electron. Lett.*, **5**, 196–197, (1969).
110. R. Baron, M.-A. Nicolet and V. Rodriguez, 'Differential step response of unipolar space-charge-limited current in solids', *J. Appl. Phys.*, **37**, 4156–4158 (1966).
111. M. Draganescu and D. Dascalu, 'Detailed equivalent scheme for the dielectric diode at small signals', *Bull. Polytech. Inst. Bucharest*, **27**, No. 5, 145–151 (1965).
112. M. Draganescu, *Ph. D. dissertation*, Polytech. Inst. Bucharest, (1956).
113. M. Draganescu, 'A new equivalent circuit for diodes at frequencies corresponding to $\theta < 0.1\ \pi$', *Revue de Physique* (*Romanian Academy*), **4**, 61–68 (1959).
114. M. Draganescu, 'Diffusion capacitance of a *p-n* junction', *St. cercet. fiz.* (*Romania*), **13**, 49–57 (1962).
115. C. D. Bulucea, 'Diffusion capacitance of *p-n* junctions and transistors', *Electron. Lett.*, **4**, 559–561 (1968).
116. E. M. Cherry, 'Active-device capacitances', *I.E.E.E. Trans. Electron Dev.*, **ED–18**, 1166–1168 (1971).
117. H. Kroemer, 'External negative conductance of a semiconductor with negative differential mobility', *Proc. I.E.E.E.* (*Lett.*), **53**, 1246–1247 (1965).
118. J. Nigrin, 'Exact small-signal study on space-charge varactor', *Solid-St. Electron.*, **13**, 1267–1281 (1970).
119. H. A. Hoyen, Jr. J. A. Strozier, Jr. and Che-Yu Li, 'Space charge limited a.c. conduction with nonbloking electrodes', *Surf. Sci.*, **20**, 258–268 (1970).
120. Y. Suematsu and Y. Nishimura, 'Wave theory of the negative resistance element due to Gunn effect', *Proc. I.E.E.E.* (*Lett..*), **56**, 322–324 (1966).
121. M. M. Atalla and J. L. Moll, 'Emitter-controlled negative resistance in GaAs', *Solid-St. Electron.*, **21**, 619–629 (1969).
122. D. Dascalu, 'The wave theory of semiconductor with traps', *Rev. Roum. Phys.*, **16**, 651–654 (1971).
123. S. H. Chisholm and C. S. Yeh, 'High-frequency admittance of *n-ν-n* space-charge-limited current solid-state devices'', *Proc. I.E.E.E.* (*Lett.*), **56**, 2178–2180 (1968).
124. T. Misawa, 'Negative resistance due to transit-time in space-charge-limited current structure', *presented at the Solid State Device Res. Conf., Santa Barbara, Calif.*, June 19, 1967.
125. D. Dascalu, 'Space-charge-waves and high-frequency negative resistance of space-charge-limited diodes', *Int. J. Electron.*, **25**, 301–330 (1968).

126. R. STRATTON and E. L. JONES, 'Effect of carrier heating on the diffusion currents in space-charge-limited current flow', *J. Appl. Phys.*, **38**, 4596–4608 (1967).

127. N. B. SULTAN and G. T. WRIGHT, 'Punch-through oscillator-a new microwave solid state source', *Electron.Lett.*, **8**, 24–26 (1972).

128. C. P. SNAPP and P. WEISSGLAS, 'On the microwave activity of punch-through injection transit-time structures', *Report, Microwave Inst. Foundation, Stockholm*, Febr. 1972.

129. D. J. COLEMAN, 'Transit-time oscillations in barrit diodes', *J. Appl. Phys.*, **43**, 1812–1818 (1972).

130. D. R. HAMILTON, J. K. KNIPP and J. B. H. KUPER, 'Klystrons and microwave triodes', Dover Public. Inc., New York, 1966; Ch. 5.

131. F. STERZER, 'Transferred electron (Gunn) amplifiers and oscillators for microwave applications', *Proc. I.E.E.E.*, **59**, 1155–1163 (1971).

132. Y. SUEMATSU and Y. NISHIMURA, 'Small-signal admittance of bulk semiconductor devices at higher frequencies', *Proc. I.E.E.E.* (*Lett.*), **56**, 242–243 (1968).

133. S. MAHROUS, P. N. ROBSON and H. L. HARTNAGEL, 'The stability and reflection gain of sub-critically doped Gunn diodes', *Solid-St. Electron.*, **11**, 965–977 (1968).

134. R. HOLSTROM, 'Small-signal behaviour of Gunn diodes', *I.E.E.E. Trans. Electron. Dev.* **ED–14**, 464–469 (1967).

135. S. MAHROUS and H. L. HARTNAGEL, 'Gunn-effect domain formation controlled by complex load', *Brit. J. Appl. Phys.*, *Ser. 2*, **2**, 1–5 (1969).

136. R. CHARLTON, K. R. FREEMAN and G. S. HOBSON, 'Stabilisation mechanism for supercritical transferred electron amplifiers', *Electron. Lett.*, **7**, 575–577 (1971).

137. H. W. THIM, 'Noise reduction in bulk negative-resistance amplifiers', *Electron. Lett.*, **7**, 106–108 (1971).

138. M. SHOJI, 'Small-signal impedance of bulk semiconductor amplifier having a non-uniform doping profile', *I.E.E.E. Trans. Electron. Dev.*, **ED–14**, 323–329 (1967).

139. M. A. LAMPERT, A. MANY and P. MARK, 'Space-charge-limited currents injected from a point contact', *Phys. Rev.*, **135**, A1444–1453 (1964).

140. M-A. NICOLET, 'Unipolar space-charge-limited current in solids with non-uniform spacial distribution of shallow traps', *J. Appl. Phys.*, **37**, 4224–4235 (1966).

141. I. LUNDSTRÖM and R. L. WIERICH, 'Negative resistance from velocity-saturated non-uniform semiconductor samples', *Electron. Lett.*, **7**, 209–211 (1971).

142. D. DASCALU, 'The operation mechanism of space-charge-limited current diodes' (in Romanian), *Bull. Polyteh. Inst. Bucharest*, **28**, No. 5, 123–138 (1966).

143. D. DASCALU, 'A high-frequency negative resistance in dielectric diodes with a high density of shallow traps', *Brit. J. Appl. Phys.*, **18**, 875–886 (1967).

144. D. DASCALU, 'Effects of traps on small-signal operation of the dielectric diode' (in Romanian), *St. Cercet. Fiz.*, (*Romania*), **17**, 875–886 (1965).

145. J. H. INGOLD, 'Electron flow in gas diodes. I. Transition from inertia-limited flow to mobility limited flow', *J. Appl. Phys.*, **40**, 55–62 (1969); 'Electron flow in gas diodes. II. Mobility-limited flow for collision frequency proportional to electron speed', *J. Appl. Phys.*, **40**, 62–66 (1969).

146. M. M. ATALLA, and R. W. SOSHEA, 'Hot-carrier triodes with thin film metal base', *Solid-St. Electron.*, **6**, 245–250 (1963).

147. G. T. WRIGHT, 'The space-charge-limited dielectric triode', *Solid-St. Electron.*, **5**, 117–126 (1962).

148. S. BROJDO, T. J. RILEY and G. T. WRIGHT, 'The heterojunction transistor and the space-charge-limited triode', *Brit. J. Appl. Phys.*, **16**, 133–136 (1965).

149. S. BROJDO, 'Characteristics of the dielectric diode and triode at very high frequencies', *Solid-St. Electron.*, **6**, 611–629 (1963).

150. D. DASCALU, 'High-frequency behaviour of unipolar space-charge-limited transistors' (in Romanin), *Bull. Polyteh. Inst. Bucharest*, **29**, No. 3, 149–159 (1967).

151. M. T. BRIAN, personal communication (The University of Birmingham, England, 1969).

152. G. O. LADD, Jr. and D. L. FEUCHT, 'Performance potential of high frequency heterojunction transistors', *I.E.E.E. Trans. Electron Dev.*, **ED–17**, 413–420 (1970).

153. T. HARIU, S. ONO and Y. SHIBATA, 'Wide-band performance of the injection-limited Gunn diode', *Electron. Lett.*, **6**, 666–667 (1970).

154. S. P. YU, W. TANTRAPORN and J. D. YOUNG, 'Transit-time negative conductance in GaAs bulk-effect diodes', *I.E.E.E. Trans. Electron Dev.*, **ED–18**, 88–93 (1971).

155. K. P. WELLER, 'Small-signal theory of a transit-time negative resistance utilizing injection from a Schottky barrier', *R.C.A. Review*, **32**, 372–383 (1971).

156. W. HARTH and M. CLASSEN, 'Microwave Baritt diodes', paper communicated by the authors (*Institut für Hochfrequenztechnik*, *TU Braunschweig*, W.-Germany) 1972.

157. J. HELMCKE, H. HERBST, M. CLAASSEN and W. HARTH, 'F.M.-noise and bias-current fluctuations of a silicon Pd-n-p^+ microwave oscillator', *Electron. Lett.*, **8**, 158–159 (1972).

158. H. HERBST and W. HARTH, 'Frequency modulation sensitivity and frequency pushing factor of a Pd-n-p^+ punch-through microwave diode', *Electron. Lett.*, **8**, 358–359 (1972).

159. S. G. LIU and J. J. RISKO, 'Low noise punch through p-n-v-p, p-n-p and p-n-metal microwave diodes', *R.C.A. Rev.*, **32**, 636–644 (1971).

160. H. A. HAUS, H. STATZ and R. A. PUCEL, 'Noise measure of metal-semiconductor-metal, Schottky-barrier microwave diodes', *Electron. Lett.*, **7**, 667–669 (1971).

161. D. DASCALU, 'Emitter-current-limited injection in negative-mobility semiconductors in the limit of zero doping', *Electron. Lett.*, **8**, 185–186 (1972).

162. D. DASCALU, 'The effect of space-charge upon the emitter-controlled negative-mobility semiconductors', to be published.

163. F. B. LLEWELLYN, 'Electron inertia effects', Cambridge Univ. Press, London, New York, 1941.

164. D. DASCALU, 'Transit-time effects in bulk negative-mobility amplifiers', *Electron. Lett.*, **4**, 581–583 (1968).

165. J. E. SCHROEDER and R. S. MÜLLER, 'IGFET analysis through numerical solution of Poisson's equation', *I.E.E.E. Trans. Electron Dev.*, **ED–15**, 954–961 (1968).

166. D. FROHMAN-BENTCHKOWSKY and A. S. GROVE, 'Conductance of MOS transistors in saturation', *I.E.E.E. Trans. Electron Dev.*, **ED–16**, 108–113 (1969).

167. M. REISER, 'Two-dimensional analysis of substrate effects in junction F.E.T s', *Electron. Lett.*, **6**, 353–355 (1970).

168. D. VANDORPE, J. BOREL, G. MERCKEL and P. SAINTOT, 'An accurate two-dimensional numerical analysis of the MOS transistor', *Solid-St. Electron.*, **15**, 547–557 (1972).

169. A. POPA, 'An injection level dependent theory of the MOS transistor in saturation', *I.E.E.E. Trans. Electron Dev.*, **ED–19**, 774–781 (1972).

170. See for example: A. ANGOT, 'Compléments de Mathématiques', Éditions de la Revue D'Optique, Paris, 1961; par. 7.5.36.

171. J. R. Burns, 'High-frequency characteristics of the insulated-gate field-effect transistor', *R.C.A. Rev.*, **28**, 385–418 (1967).

172. M. B. Das, 'High-frequency network properties of MOS transistors including substrate resistivity effects', *I.E.E.E. Electron Dev.*, **ED–16**, 1049–1069 (1969).

173. J. R. Burns, 'Large-signal transit-time effects in the MOS transistor', *R.C.A. Rev.*, **30**, 15–35 (1969).

174. R. F. Pierret, 'Frequency variation of the MOS-transistor $V_D = 0$ admittance', *Solid-St. Electron.*, **11**, 253–260 (1968).

175. J. A. Geurst and H.J.C.A. Nunnink, 'Numerical data on the high-frequency characteristics of thin film transistors', *Solid-St. Electron.*, **8**, 769–771 (1965).

176. I. Richer, 'The equivalent circuit of an arbitrarily doped field-effect transistor', *Solid-St. Electron.*, **8**, 381–393 (1965).

177. J. A. Geurst, 'Calculation of high-frequency characteristics of thin-film transistors', *Solid-St. Electron.*, **8**, 88–90 (1965).

178. H. C. de Graaff, 'High-frequency measurements of thin-film transistors', *Solid-St. Electron.*, **10**, 51–56 (1967).

179. E. M. Cherry, 'Small-signal high-frequency response of the insulated-gate field-effect transistor', *I.E.E.E. Trans. Electron Dev.*, **ED–17**, 569–577 (1970).

180. A. van der Ziel and J. W. Ero, 'Small-signal high-frequency theory of field-effect transistors', *I.E.E.E. Trans. Electron Dev.*, **ED–11**, 128–135 (1964).

181. J. R. Hauser, 'Small-signal properties of field-effect devices', *I.E.E.E. Trans. Electron Dev.*, **ED–12**, 605–618 (1965).

182. D. B. Candler and A. G. Jordan, 'A small-signal high-frequency analysis of the insulated-gate field-effect transistor', *Int. J. Electron.*, **19**, 181–196 (1965).

183. B. Reddy and F. N. Trofimenkoff, 'FET high-frequency analysis', *Proc. I.E.E.*, **113**, 1755–1762 (1966).

184. J. A. van Nielen, 'A simple and accurate approximation to the high-frequency characteristics of insulated-gate field-effect transistors', *Solid-St. Electron.*, **12**, 826–829 (1969).

185. J. S. T. Huang, 'Charge control approach to the small signal theory of field-effect devices', *I.E.E.E. Trans. Electron Dev.*, **ED–16**, 775–781 (1969).

186. D. H. Treleaven and F. N. Tromfimenkoff, 'MOS FET equivalent circuit at pinch-off', *Proc. I.E.E.E. (Corresp.)*, **54**, 1223–1224 (1966).

187. J. W. Haslett and F. N. Tromfimenkoff, 'Small-signal high-frequency equivalent circuit for the metal-oxide-semiconductor field-effect transistor', *Proc. I.E.E.*, **116**, 699–702 (1969).

188. M. Shoji, 'Analysis of high-frequency thermal noise of enhancement mode MOS field-effect-transistors', *I.E.E.E. Trans. Electron Dev.*, **ED–13**, 520–524 (1966).

189. J. A. Geurst, 'Calculation of high-frequency characteristics of field-effect transistors', *Solid-St. Electron.*, **8**, 563–566 (1965).

190. M. B. Das, 'Generalised high-frequency network theory of FET's', *Proc. I.E.E.*, **114**, 50–59 (1967).

191. W. R. Bennett, 'Electrical noise', McGraw-Hill, New York, 1960.

192. A. van der Ziel, 'Noise in solid-state devices and lasers', *Proc. I.E.E.E.*, **58**, 1178–1206 (1970).

193. A. van der Ziel, 'Noise: sources, characterization, measurement', Prentice-Hall, Englewood Cliffs, N.J., 1970.

194. F.N.H. Robinson, 'Thermal noise, shot noise and statistical mechanics', *Int. J. Electron.*, **26**, 227–235 (1969).

195. A. Shumka, 'Thermal noise in space-charge-limited solid-state diodes', *Solid-St. Electron.*, **13**, 751–754 (1970).

196. A. van der Ziel, 'Low frequency noise suppression in space charge limited solid state diodes', *Solid-St. Electron.*, **9**, 123–127 (1966).

197. A. van der Ziel, 'H.F. thermal noise in space-charge-limited solid-state diodes', *Solid-St. Electron.*, **9**, 1139–1140 (1966).

198. V. Sergiescu, 'Statistical correlations of charge-carriers and transport noise in solids', *Ph.D. dissertation*, Institute of Physics of the Romanian Academy, 1968.

199. P. W. Webb and G. T. Wright, 'The dielectric triode: a low-noise solid-state amplifier', *J. Brit. Instn. Radio Engrs.*, **23**, 111–112 (1962).

200 A. van der Ziel, 'Thermal noise in space-charge-limited diodes', *Solid-St. Electron.*, **9**, 899–900 (1966).

201. A. van der Ziel and K. M. van Vliet, 'H.F. thermal noise in space-charge limited solid state diodes-II', *Solid-St. Electron.*, **11**, 508–509 (1968).

202. S. T. Hsu, A. van der Ziel and E. R. Chenette, 'Noise in space-charge-limited solid-state devices', *Solid-St. Electron.*, **10**, 129–135 (1967).

203. S. T. Liu, 'Thermal noise in space-charge-limited solid-state diodes', *Solid-St. Electron.*, **10**, 253–254 (1967).

204. S. Yamamoto, S. T. Liu and A. van der Ziel, 'Thermal noise in Ge *p-v-p* SCL diodes', *Appl. Phys. Lett.*, **11**, 140–141 (1967).

205. S. Yamamoto and A. van der Ziel, 'Noise parameters of a silicon space-charge-limited triode', *Solid-State Electron.*, **11**, 572—574 (1968).

206. S. Yamamoto, S. T. Liu and A. van der Ziel, 'Noise in *p-v-p* diodes at room temperature', *Solid-St. Electron.*, **11**, 707–710 (1968).

207. M-A. Nicolet, 'Thermal noise in single and double injection devices', *Solid-St. Electron.*, **14**, 377–380 (1971).

208. A. Shumka, 'Thermal noise in space-charge-limited solid-state diodes with field-dependent mobility and hot carriers', *Solid-St. Electron.*, **14**, 367–369 (1971).

209. D. N. Bougalis and A. van der Ziel, 'Hot electron effects in single-injection silicon SCL diodes', *Solid-St. Electron.*, **14**, 265–272 (1971).

210. W. Shockley, J. A. Copeland and R. P. James, in 'Quantum theory of atoms, molecules and the solid state' (P.-O. Loewdin, editor), Academic Press, New York (1966); p. 537.

211. V. Sergiescu and A. Friedman, 'Position correlations and space-charge noise suppression in solids at low frequencies', *Brit. J. Appl. Phys.*, **17**, 1409–1423 (1966).

212. A. Friedman and A. B. Fazakas, 'Space charge noise suppression in solids at low injection level', *Rev. Roum. Phys.*, **12**, 457–467 (1967).

213. V. Sergiescu, 'The decomposition of the transport noise under space charge conditions in solids', *Physica*, **40**, 110–124 (1968).

214. V. Sergiescu, 'SCLC-noise as a statistical carrier correlation effect', *Solid-St. Electron.*, **14**, 357–359 (1971).

215. A. Friedmann, 'Noise of hot carriers', *Solid.St. Electron.*, **14**, 361–363 (1971).

216. A. G. Jordan and N. A. Jordan, 'Theory of noise in metal oxide semiconductor devices', *I.E.E.E. Trans. Electron Dev.*, **ED–12**, 148–156 (1965).

217. H. E. Halladay and A. van der Ziel, 'On the high frequency excess noise and equivalent circuit representation of the MOS-FET with *n*-type channel', *Solid-St. Electron.*, **12**, 161–176 (1969).

218. C. T. Sah, S. Y. Wu and F. H. Hielscher, 'The effects of fixed bulk charge on the thermal noise in metal-oxide-semiconductor transistors', *I.E.E.E. Trans. Electron Dev.*, **ED–13**, 410–414 (1966).

219. L. D. Yau and C. T. Sah, 'On the excess white noise in MOS transistors', *Solid-St. Electron.*, **12**, 927–936 (1969).

220. F. M. Klaassen, 'High-frequency noise of the junction field-effect transistor', *I.E.E.E. Trans., Electron Dev.*, **ED–14**, 368–373 (1967).

221. P. S. Rao, 'The effect of the substrate upon the gate and drain noise of MOS FET's', *Solid-St. Electron.*, **12**, 549–555 (1969).

222. P. S. Rao and A. van der Ziel, 'Noise and y-parameters in MOS FET's' *Solid-St. Electron.* **14**, 939–944 (1971).

223. J. W. Haslett and F. N. Trofimenkoff, 'Gate noise in MOS FET's at moderately high frequencies', *Solid-St. Electron.*, **14**, 239–245 (1971).

224. F. M. Klaassen, 'On the influence of hot carrier effects on the thermal noise of field-effect transistors', *I.E.E.E. Trans. Electron. Dev.*, **ED–17**, 858–862 (1970).

225. W. Baechtold, 'Noise behavior of Schottky barrier gate field-effect transistors at microwave frequencies', *I.E.E.E. Trans. Electron Dev.*, **ED–18**, 97–104 (1971).

226. G. A. Grinberg, 'Selected problems from mathematical theory of electric and magnetic phenomena' (in Russian), Moscow, 1948.

227. D. Dascalu, 'Basic equations for non-linear theory of space-charge-limited currents in semiconductors', *Rev. Roum. Phys.*, **15**, 1197–1199 (1970).

228. D. Dascalu, 'Transit-time effects in space-charge-limited solid-state diode operation', *Rev. Roum. Phys.*, **12**, 701–720 (1967).

229. Gh. Brezeanu, *Graduation Thesis*, Electronics Dept., Polytechn. Institute of Bucharest, 1972.

230. D. Dascalu and Gh. Brezeanu, 'High-frequency detection properties of silicon n^+-ν-n^+ devices', *Rev. Roum. Phys.* **18**, 677–680 (1973).

231. D. Dascalu and Gh. Brezeanu, 'Theory of v.h.f detection and frequency multiplication with SCLC silicon diodes', to be published.

232. D. Dascalu, 'Detection characteristics at very high frequencies of the space-charge-limited solid-state diode', *Solid-St. Electron.*, **9**, 1143–1145 (1966).

233. A. M. Cowley and H. O. Sorensen, 'Quantitative comparison of solid-state microwave detectors', *I.E.E.E. Trans. Microwave Theory Techn.*, **MTT–14**, 588–600 (1966).

234. D. Dascalu, 'Square-law silicon punch-through diodes and their application as high-level parabolic detectors', *Rev. Roum. Phys.*, **17**, 401–404 (1972).

235. G. T. Wright, 'Transit-time effects in the space-charge-limited silicon microwave diode', *Solid-St. Electron.*, **9**, 1–6 (1966).

236. D. Dascalu, 'Detection properties of space-charge-limited dielectric diodes in the presence of trapping', *Solid-St. Electron.*, **10**, 729–731 (1967).

237. D. Dascalu, 'The behaviour of the dielectric diode as a low-level detector' (in Romanian), *St. Cercet. Fiz.*, **20**, 7–23 (1968).

238. A. H. Taub and N. Wax, 'Theory of the parallel plane diode', *J. Appl. Phys.*, **21**, 974–980 (1950).

239. D. Dascalu, 'Theory of saturated-velocity semiconductor diode', *Rev. Roum. Phys.*, **17**, 1203–1206 (1972).

240. H. W. Rüegg, 'A proposed punch-through microwave negative-resistance diode', *I.E.E.E. Trans. Electron Dev.*, **ED–15**, 577–585 (1968).

241. U. B. Sheorey, I. Lundström and E. A. Ash, 'Analysis of punch-through injection for a transit-time negative resistance diode', *Int. J. Electron.*, **30**, 19–32 (1971).

242. A. Semichon and E. Constant, 'Sur un mécanisme général d'oscillation et son application aux sémi-conducteurs', *Compt. Rend. Acad. Sci. Paris*, **270**, 665–668 (1970).

243. A. Many and G. Rakavy, 'Theory of transient space-charge-limited currents in solids in the presence of trapping', *Phys. Rev.*, **126**, 1980–1988 (1962).

244. R. B. Schilling and H. Schachter, 'Oscillatory modes associated with one carrier transient space-charge-limited currents', *J. Appl. Phys.*, **38**, 1643–1646 (1967).

245. R. B. Schilling and H. Schachter, 'Neglecting diffusion in space-charge-limited currents', *J. Appl. Phys.*, **38**, 841–844 (1967).

246. M. A. Lampert and R. B. Schilling, 'Space-charge-limited current transient including diffusion', *Phys. Rev. Lett.*, **18**, 493–495 (1967).

247. P. Mark and W. Helfrich, 'Space-charge-limited currents in organic crystals', *J. Appl. Phys.*, **33**, 205–215 (1962).

248. A. Many, S. Z. Weisz and M. Simhony, 'Space-charge-limited currents in iodine single crystals', *Phys. Rev.*, **126**, 1989–1995 (1962).

249. H. Lemke and G. O. Müller, 'Transient behaviour of space-charge-limited currents in *p*-type silicon', *Phys. Stat. Sol.*, **24**, 127–133 (1967).

250. S. Z. Weisz, A. Cobas, S. Trester and A. Many, 'Electrode-limited and space-charge-limited transient currents in insulators', *J. Appl. Phys.*, **39**, 2296–2302 (1968).

251. P. N. Keating and A. C. Papadakis, 'The drift of carriers from a narrow region of instantaneous ionisation in an insulator', *Proc. Internat. Conf. Semicond., Paris*, 1964, 519–525.

252. A. C. Papadakis, 'Theory of transient space-charge-perturbed currents in insulators', *J. Phys. Solids*, **28**, 641–647 (1967).

253. D. N. Poenaru, 'Electric pulses in semiconductor nuclear detectors' (in Romanian), Publish. House of Romanian Academy, 1968.

254. T. J. O'Reilly, 'The transient response of insulated-gate field-effect transistors', *Solid-St. Electron.*, **8**, 947–956 (1965).

255. M. B. Das, 'Switching characteristics of metal-oxide-semiconductor and junction-gate field-effect transistors', *Proc. I.E.E.*, **114**, 1223–1230 (1967).

256. R. H. Crawford, 'MOS FET in circuit design', McGraw-Hill, New York, 1967; Ch. 4.

257. A. G. Zhdan, T. U. Musabekov, V. B. Sandomirsky, M. I. Elinson and M. E. Chugunova, 'Investigations of transient space-charge-limited currents at saw-tooth voltage pulses', *Solid-St. Electron.*, **11**, 577–582 (1968).

258. D. Colliver and B. Prew, 'Indium phosphide: is it practical for solid state microwave sources?', *Electronics*, **45**, No. 8, 110–113 (1972).

259. P. P. Bohn, 'A nonsaturating velocity-field approximation for improved invariant domain analysis', *Proc. I.E.E.E.*, **58**, 1397–1398 (1970).

260. K. Kurokawa, 'The dynamics of high-field propagating domains in bulk semiconductors', *Bell Syst. Techn. J.*, **46**, 2235–2259 (1967).

261. P. N. Butcher, W. Fawcett and C. Hilsum, 'A simple analysis of stable domain propagation in the Gunn effect', *Brit. J. Appl. Phys.*, **17**, 841–850 (1966).

262. J. A. Copeland, 'Bulk negative-resistance semiconductor devices', *I.E.E.E. Spectrum*, **4**, No. 5, 71–77 (1967).

263. H. Thim, 'Experimental verification of bistable switching with Gunn diodes', *Electron. Lett.*, **7**, 246–247 (1971).

264. H. Thim, 'Stability and switching in overcritically doped Gunn diodes', *Proc. I.E.E.E.*, **59** (1971).

265. R. W. H. Engelmann and W. Heinle, 'Proposed Gunn-effect switch', *Electron. Lett.*, **4**, 190–192 (1968).

266. R. Engelmann and W. Heinle, 'Pulse discrimination by Gunn-effect switching', *Solid-St. Electron.*, **14**, 1–16 (1971).

267. W. Heinle, 'Principles of a phenomelogical theory of Gunn-effect domain dynamics', *Solid-St. Electron.*, **11**, 583–598 (1968).

268. R. B. Robrock, II, 'A lumped model for characterizing single and multiple domain propagation in bulk GaAs', *I.E.E.E. Trans. Electron. Dev.*, **ED–17**, 93–103 (1970).

269. R. P. Robrock, II, 'Extension of the lumped bulk device model to incorporate the process of domain dissolution', *I.E.E.E. Trans. Electron Dev.*, **ED–17**, 103–107 (1970).

270. R. P. Robrock, II, 'Effects of device parameters on domain dynamics in bulk GaAs', *I.E.E.E. Proc. (Letters)*, **59**, 804–805 (1970).

271. M. Shoji, 'Functional bulk semiconductor oscillators', *I.E.E.E. Trans. Electron Dev.*, **ED–14**, 535–546 (1967).

272. R. B. Robrock, II, 'Analysis and simulation of domain propagation in nonuniformly doped bulk GaAs', *I.E.E.E. Trans. Electron Dev.*, **ED–16**, 647–653 (1969).

273. C. P. Sandbank, 'Synthesis of complex electronic functions by solid state bulk effects', *Solid-St. Electron.*, **10**, 369–380 (1967).

274. R. S. Engelbrecht, 'Solid-state bulk phenomena and their application to integrated electronics', *I.E.E.E.J. Solid-St. Circ.*, **SC–3**, 210–212 (1968).

275. J. A. Copeland, T. Hayashi and M. Uenohara, 'Logic and memory elements using two-valley semiconductors', *Proc. I.E.E.E. (Letters)*, **55**, 584–585 (1967).

276. H. L. Hartnagel and S. H. Izadpanah, 'High-speed computer logic with Gunn-effect devices', *The Radio Electron. Eng.*, **36**, 247–255 (1968).

277. S. H. Izadpanah and H. L. Hartnagel, 'Gunn-effect pulse and logic devices', *The Radio Electron. Eng.*, **39**, 329–339 (1970).

278. W. F. Fallmann, H. Hartnagel and G. P. Srivastava, 'A Gunn-effect shift register', *Arch. Elektr. Übertrag.*, **24**, 473–474 (1970).

279. M. Shoji, 'Two-dimensional Gunn-domain dynamics', *I.E.E.E. Trans. Electron Dev.*, **ED–16**, 748–758 (1969); 'Theory of transverse extension of Gunn domains', *J. Appl. Phys.*, **41**, 774–778 (1970).

280. P. L. Shah and T. A. Rabson, 'Effect of various boundary conditions on the instabilities in transffered electron bulk oscillators', *J. Appl. Phys.*, **42**, 783–798 (1971); 'Combined doping and geometry effects on transferred-electron bulk instabilitites', *I.E.E.E. Trans. Electron Dev.*, **ED–18**, 170–174 (1971).

281. J. A. Copeland, 'Stable space-charge layers in two-valley semiconductors', *J. Appl. Phys.*, **37**, 3602–3609 (1966).

282. H. KROEMER, 'Non-linear space-charge domain dynamics in a semiconductor with negative differential mobility', *I.E.E.E. Trans. Electron Dev.*, **ED−13**, 27−40 (1966).

283. J. A. COPELAND, 'LSA oscillator-diode theory', *J. Appl. Phys.*, **38**, 3096−3101 (1967).

284. W. HEINLE, 'Simple theory for l.s.a. operation of Gunn-effect semiconductors', *Electron. Lett.*, **3**, 429−430 (1967).

285. J. S. HEEKS, G. KING and C. P. SANDBANK, 'Transferred-electron bulk effects in gallium arsenide', *Electrical Commun.*, **43**, 334−345 (1968).

286. J. A. COPELAND, 'Doping uniformity and geometry of LSA oscillator diodes', *I.E.E.E. Trans. Electron Dev.*, **ED−14**, 497−500 (1967).

287. W. O. CAMP, JR., 'High-efficiency GaAs transferred electron device operation and circuit design', *I.E.E.E. Trans. Electron. Dev.*, **ED−17**, 1175−1184 (1971).

288. F. STERZER, 'Transferred electron (Gunn) amplifiers and oscillators for microwave applications', *Proc. I.E.E.E.*, **59**, 1155−1163 (1971).

Supplementary References

289. A. SJÖLUND, "Small-signal analysis of punch-through injection microwave devices", *Solid-St. Electron.*, **16**, 559−569 (1973).

290. K. KAWARADA and Y. MIZUSHIMA, "Large-signal analysis of negative-resistance diode due to punch-through injection and transit-time effect", *Japanese J. Appl. Phys.*, **12**, 423−433 (1973).

291. J. L. CHU and S. M. SZE, "Microwave oscillations in PNP reach-through BARITT diodes", *Solid-St. Electron.*, **16**, 85−91 (1973).

292. P. ANTOGNETTI, A. CHIABRERA and G. R. BISIO, "Small-signal theory of transit-time diodes", *Solid-St. Electron.*, **16**, 345−350 (1973).

293. N. B. SULTAN and G. T. WRIGHT, "Electronic tuning of the punch-through injection transit-time (PITT) microwave oscillator", *I.E.E.E. Trans. Microwave Theory Tech.*, **MTT−20**, 773−775 (1972).

294. G. T. WRIGHT and N. B. SULTAN, "Small-signal design theory and experiment for the punch-through injection transit-time oscillator", *Solid-St. Electron.*, **16**, 535−544 (1973).

295. C. P. SNAPP and P. WEISSGLAS, "On the microwave activity of punch-through injection transit-time structures", *I.E.E.E. Trans. Electron. Dev.*, **ED−19**, 1109−1118 (1972).

296. J. A. C. STEWART and J. WAKEFIELD, "Simulation of Pt-*n*-*p* silicon punch-through device", *Electron. Lett.*, **8**, 378−379 (1972).

297. A. SJÖLUND, "Small-signal noise analysis of p^+-n-p^+ BARITT diodes", *Electron. Lett.*, **9**, 2−4 (1973).

298. H. STATZ, R. A. PUCEL and H. A. HAUS, "Velocity fluctuation noise in metal-semiconductor-metal diodes", *Proc. I.E.E.E.* (Corresp.), **60**, 644−645 (1971).

299. B. KÄLLBÄCK, "Noise performance of gallium-arsenide and indium-phosphide injection-limited diodes", *Electron. Lett.*, **9**, 11−12 (1973).

300. B. KÄLLBÄCK, "Noise properties of the injection-limited Gunn diode", *Electron. Lett.*, **8**, 476−477 (1972).

301. H. JOHNSON, "A unified small-signal theory of uniform-carrier-velocity semiconductor transit-time diodes", *I.E.E.E. Trans. Electron Dev.*, **ED−19**, 1156−1166 (1972).

302. M. MATSUMURA, "Small-signal admittance of BARITT diodes", *I.E.E.E. Trans. Electron. Dev.*, **ED−19**, 1131−1133 (1972).

303. V. L. Dalal and M. A. Lampert, "Transient space-charge phenomena in semiconductors at high electric fields", *Solid-St. Electron.*, **16**, 689–699 (1973).

304. L. N. Dworsky and R. I. Harrison, "Trapped plasma oscillations in unipolar semiconductor structures", *I.E.E.E. Trans. Electron Dev.*, **ED–19**, 836–838 (1972).

305. F. Sterzer, "Static negative differential resistance in bulk semiconductors", *R.C.A. Rev.*, **32**, 497–502 (1971).

306. A. Gisolf and R. J. J. Zijlstra, "Lattice interaction noise of hot carriers in single injection solid state diodes", *Solid-St. Electron.*, **16**, 571–580 (1973).

307. J. Golder, M.-A. Nicolet, "Thermal noise measurements on space-charge-limited hole current in silicon", *Solid-St. Electron.*, **16**, 581–585 (1973).

308. M. Reiser, "A two-dimensional numerical FET model for DC, AC and large-signal analysis", *I.E.E.E. Trans. Electron Dev.*, **ED–20**, 35–43 (1973).

309. P. L. Hower and N. G. Bechtel, "Current saturation and small-signal characteristics of GaAs field-effect transistors", *I.E.E.E. Trans. Electron Dev.*, **ED–20**, 213–220 (1973).

310. H. J. Sigg, G. D. Vendelin, T. P. Cauge and J. Kocsis, "D-MOS transistor for microwave applications", *I.E.E.E. Trans. Electron Dev.*, **ED–19**, 45–53 (1972).

311. M. B. Das, "FET noise sources and their effects on amplifier performance at low frequencies", *I.E.E.E. Trans. Electron Dev.*, **ED–19**, 338–348 (1972).

312. R. D. Kässer, "Noise factor contours for field-effect transistors at moderately high frequencies", *I.E.E.E. Trans. Electron Dev.*, **ED–19**, 164–171 (1972).

313. J. W. Haslett and E. J. M. Kendall, "Gate flicker noise in MOSFET's", *Proc. I.E.E.E.* (Corresp.), **60**, 1111–1112 (1972).

314. C. Hu and R. S. Muller, "A resistive-gated IGFET tetrode", *I.E.E.E. Trans. Electron Dev.*, **ED–19**, 418–425 (1971).

315. S. Tezner, "Gridistor development for the microwave power region", *I.E.E.E. Trans. Electron Dev.*, **ED–19**, 355–364 (1972).

316. S. Tezner, "Sur les limites en fréquence et en puissance des triodes à effet de champ", *Annales Télécommun.*, **26**, 303–314 (1971).

317. J. E. Meyer, "MOS models and circuit simulation", *R.C.A. Rev.*, **32**, 42–63 (1971).

318. G. A. Armstrong and J. A. Magowan, "Derivation of simple expressions for interelectrode capacitances of I.G.F.E.T.'s as a function of bias condition", *Electron Lett.*, **7**, 281–283 (1971).

319. J. G. Ruch, "Electron dynamics in short channel field-effect transistors", *I.E.E.E. Trans. Electron Dev.*, **ED–19**, 652–654 (1972).

320. R. J. Strain and N. L. Schryer, "A nonlinear diffusion analysis of charge-coupled-device transfer", *Bell Syst. Tech. J.*, **50**, 1721–1740 (1971).

321. H.-S. Lee and L. G. Heller, "Charge-control method of charge-coupled device transfer analysis", *I.E.E.E. Trans. Electron Dev.*, **ED–19**, 1270–1279 (1972).

322. R. J. Strain, "Properties of an idealized traveling-wave charge-coupled device", *I.E.E.E. Trans. Electron Dev.*, **ED–19**, 1119–1130 (1972).

323. L. G. Heller and H.-S. Lee, "Digital signal transfer in charge-transfer devices", *I.E.E.E. Journ. Solid-St. Circuits*, **SC–8**, 116–125 (1973).

324. H. Tateno and S. Kataoka, "Static negative resistance due to geometrical effect of a GaAs bulk element", *Proc. I.E.E.E.* (Corresp.), **60**, 919–920 (1972).

325. K. Tomizawa, H. Tateno and S. Kataoka "Computer analysis on the static negative resistance due to the geometrical effect of a GaAs bulk element", *I.E.E.E. Trans. Electron Dev.*, **ED–19**, 1299–1300 (1972).

326. K. W. Böer and P. L. Quinn, "Inhomogeneous field distribution in homogeneous semiconductors having an N-shaped differential conductivity", *Phys. Stat. Sol.*, **17**, 307–316 (1966).

327. M. P. Shaw, P. R. Solomon and H. L. Grubin, "The influence of boundary conditions on current instabilities in GaAs", *I.B.M. Journ. Res. Dev.*, **13**, 587–590 (1969).

328. K. W. Böer and G. Dohler, "Influence of boundary conditions on high-field domains in Gunn devices", *Phys. Rev.*, **186**, 793–800 (1969).

329. E. M. Conwell, "Boundary conditions and high-field domains in GaAs", *I.E.E.E. Trans. Electron Dev.*, **ED–17**, 262–270 (1970).

330. H. L. Grubin, "Exact solutions of the linearized equation for current flow in negative differential mobility elements", *I.E.E.E. Trans. Electron Dev.*, **ED–19**, 1294–1296 (1972).

331. P. R. Solomon, M. P. Shaw and H. L. Grubin, "Analysis of bulk negative differential mobility element in a circuit containing reactive elements", *J. Appl. Phys.*, **43**, 159–172 (1972).

332. H. L. Grubin, M. P. Shaw and P. R. Solomon, "On the form and stability of electric-field profiles within a negative differential mobility semiconductor", *I.E.E.E. Trans. Electron Dev.*, **ED–20**, 63–78 (1973).

333. T. Ohmi and S. Hasuo, "Unified treatment of small-signal space-charge dynamics in bulk effect devices", *I.E.E.E. Trans. Electron. Dev.*, **ED–20**, 303–316 (1973).

334. M. A. Lampert, "Stable space-charge layers associated with bulk, negative differential conductivity: further analytic results", *J. Appl. Phys.*, **40**, 335–340 (1969).

335. Y. S. Ryabinkin, "An attempt to find the Gunn effect under space-charge conditions", *Soviet Phys. Semicond.*, **2**, 977–978 (1969).

336. P. S. Hauge, "Domain velocity in semiconductors with field-dependent carrier diffusion", *I.E.E.E. Trans. Electron Dev.*, **ED–17**, 386–387 (1970).

337. E. J. Aas, "Topological study of domain and layer propagation in a Gunn diode", *J. Appl. Phys.*, **40**, 4673–4675 (1969).

338. E. J. Aas, "Comment on A topological theory of domain velocity in semiconductors", *I.B.M. Journ. Res. Dev.*, **14**, 689–690 (1970).

339. H. D. Rees, "Time response of the high-field electron distribution function in GaAs", *I.B.M. Journ. Res. Dev.*, **13**, 537–542 (1969).

340. D. Jones and H. D. Rees, "Electron relaxation effects in transferred-electron devices revealed by new simulation method", *Electron. Lett.*, **8**, 263–264 (1972).

341. D. Jones and H. D. Rees, "Accumulation transit mode in transferred-electron oscillators", *Electron. Lett.*, **8**, 567–568 (1972).

342. P. K. Yang and T. Sugano, "Experimental observation of accumulation mode operation of bulk Ge oscillations", *I.E.E.E. Trans. Electron Dev.*, **ED–18**, 383–384 (1971).

343. N. Suzuki, H. Yanai and T. Ikoma, "Simple analysis and computer simulation on lateral spreading of space charge in bulk GaAs", *I.E.E.E. Trans. Electron Dev.*, **ED−19**, 364−375 (1972).

344. B. L. Gelmont and M. S. Shur, "Analytical theory of stable domains in high-doped Gunn diodes", *Electron. Lett.*, **6**, 385−387 (1970).

345. M. E. Levinstein and M. S. Shur, "Calculation of the parameters of the microwave Gunn generator", *Electron. Lett.*, **4**, 233−235 (1968).

346. A. E. Mircea, "Computer optimized design of pulsed Gunn oscillators", *I.E.E.E. Trans. Electron Dev.*, **ED−19**, 21−26 (1972).

347. W. O. Camp Jr., "High efficiency GaAs transferred electron device operation and circuit design", *I.E.E.E. Trans. Electron Dev.*, **ED−18**, 1175−1184 (1971).

348. A. C. Kak, R. L. Gunshor and C. P. Jethwa, "Equivalent-circuit representation for stably propagating domains in bulk GaAs", *Electron. Lett.*, **6**, 711−712 (1970).

349. R. L. Gunshor and A. C. Kak, "Lumped-circuit representation of Gunn diodes in domain mode", *I.E.E.E. Trans. Electron Dev.*, **ED−19**, 765−770 (1972).

350. A. C. Kak and R. L. Gunshor, "The transient behaviour of high-field dipole domains in transferred electron device", *I.E.E.E. Trans. Electron Dev.*, **ED−20**, 1−5 (1973).

351. A. Mircea, "A simple criterion for LSA oscillation", *I.E.E.E. Trans. Electro Dev.*, **ED−18** 449 (1971).

352. B. C. Taylor and W. Fawcett, "Detailed computer analysis of LSA operation in CW transferred electron devices", *I.E.E.E. Trans. Electron Dev.*, **ED−17**, 907−915 (1970).

353. P. Jeppesen and B. I. Jeppsson, "LSA relaxation oscillator principles", *I.E.E.E. Trans. Electron Dev.*, **ED−18**, 439−449 (1971).

354. P. Jeppesen and B. Jeppsson, "The influence of diffusion on the stability on the supercritical transferred electron amplifier", *Proc. I.E.E.E.* (Corresp.), **60**, 452−454 (1972).

355. K. Kumabe and H. Kanbe, "Mechanism of coupling between space-charge waves and microwaves in a bulk GaAs travelling-wave amplifier", *Rev. Electr. Communic. Lab.*, **18**, 913−920 (1970).

356. R. H. Dean, "Reflection amplification in thin layers of *n*-GaAs", *I.E.E.E. Trans. Electron Dev.*, **ED−19**, 1148−1156 (1972).

357. M. Meyer and T. van Duzer, "Travelling-wave amplification and power flow in conducting solids", *I.E.E.E. Trans. Electron Dev.*, **ED−17**, 193−199 (1970).

358. M. E. Hines, "Theory of space-harmonic travelling-wave interactions in semiconductors", *I.E.E.E. Trans. Electron Dev.*, **ED−16**, 88−97 (1969).

359. P. Guéret, "Limits of validity of the 1−dimensional approach in space-charge-wave and Gunn-effect theories", *Electron. Lett.*, **6**, 197−198 (1970).

360. J. L. Tezner, "Etude bi-dimensionnele des instabilités dans une lame mince d'arseniure de gallium en contact avec un autre semi-conducteur à resistance positive", *Solid-St. Electron.*, **13**, 1471−1481 (1970).

361. R. W. H. Engelmann, "Plane-wave approximation of carrier waves in semiconductor plates with nonisotropic mobility", *I.E.E.E. Trans. Electron Dev.*, **ED−18**, 587−591 (1971).

362. W. HEINLE and R. W. ENGELMANN, "Stability criterion for semiconductor plates with negative AC mobility", *Proc. I.E.E.E.* (Corresp.), **60**, 914–915 (1972).

363. P. GUÉRET, "Convective and absolute instabilities in semiconductors exhibiting negative differential mobility", *Phys. Rev. Lett.*, **27**, 256–258 (1971).

364. T. A. RAMSTAD, "Theory of space-charge wave excitation in $n^+ - n - n^+$ semiconductor structures", Technical report No. AE–179 from Electronics Res. Lab., Trondheim Univ.

365. J. FALNES, "Analysis of $n^+ - n -$semiconductor structures", Technical report No. AE–167 (1971) from Electronics Res. Lab., Trondheim Univ.; "Excitation of space charge waves in a one-carrier semiconductor with variable doping", *Int. J. Electron.*, **33**, 481–505 (1972).

Author Index

Numbers refer to pages on which the author's name is quoted. See also References (p. 373).

Subject Index